Advances in
MICROBIAL ECOLOGY

Volume 9

ADVANCES IN MICROBIAL ECOLOGY

A Continuation Order Plan is available for this series. A continuation order will bring delivery of
each new volume immediately upon publication. Volumes are billed only upon actual shipment.
For further information please contact the publisher.

Advances in
MICROBIAL ECOLOGY

Volume 9

Edited by

K. C. Marshall

University of New South Wales
Kensington, New South Wales, Australia

PLENUM PRESS · NEW YORK AND LONDON

The Library of Congress cataloged the first volume of this title as follows:

Advances in microbial ecology. v. 1–
  New York, Plenum Press c1977–
  v. ill. 24 cm.
  Key title: Advances in microbial ecology, ISSN 0147-4863
  1. Microbial ecology—Collected works.
QR100.A36   576′.15   77-649698

ISBN 0-306-42184-4

© 1986 Plenum Press, New York
A Division of Plenum Publishing Corporation
233 Spring Street, New York, N.Y. 10013

Printed in the United States of America

Contributors

John H. Andrews, Department of Plant Pathology, University of Wisconsin, Madison, Wisconsin 53706

Tom Fenchel, Department of Ecology and Genetics, University of Aarhus, DK-8000 Aarhus C, Denmark

Robin F. Harris, Departments of Soil Science and Bacteriology, University of Wisconsin, Madison, Wisconsin 53706

J. Gwynfryn Jones, Freshwater Biological Association, Ambleside, Cumbria LA22 OLP, England

Bo Barker Jørgensen, Institute of Ecology and Genetics, University of Aarhus, DK-8000 Aarhus C, Denmark

Staffan Kjelleberg, Department of Marine Microbiology, University of Göteborg, S-413 19 Göteborg, Sweden

David J. Lane, Department of Biology and Institute for Molecular and Cellular Biology, Indiana University, Bloomington, Indiana 47405

J. M. Lynch, Plant Pathology and Microbiology Department, Glasshouse Crops Research Institute, Littlehampton, West Sussex BN17 6LP, England

D. J. W. Moriarty, CSIRO Marine Laboratories, Cleveland, Queensland 4163, Australia

Gary J. Olsen, Department of Biology and Institute for Molecular and Cellular Biology, Indiana University, Bloomington, Indiana 47405

Norman R. Pace, Department of Biology and Institute for Molecular and Cellular Biology, Indiana University, Bloomington, Indiana 47405

Niels Peter Revsbech, Institute of Ecology and Genetics, University of Aarhus, DK-8000 Aarhus C, Denmark

Mel Rosenberg, School of Dental Medicine and Department of Human Microbiology, Sackler Faculty of Medicine, Tel-Aviv University, Ramat-Aviv 69978, Israel

David A. Stahl, Department of Biology and Institute for Molecular and Cellular Biology, Indiana University, Bloomington, Indiana 47405;

v

present address: Department of Veterinary Pathobiology, University of Illinois, Urbana, Illinois 61801

J. M. Whipps, Agricultural Research Council, Letcome Laboratory, Wantage, Oxon OX12 9JT, England; *present address:* Plant Pathology and Microbiology Department, Glasshouse Crops Research Institute, Littlehampton, West Sussex BN17 6LP, England

Preface

Advances in Microbial Ecology was established by the International Committee on Microbial Ecology (ICOME) to provide a means for in-depth, critical, and even provocative reviews to emphasize current trends in the rapidly expanding area of microbial ecology. *Advances in Microbial Ecology* is now recognized as a major source of information and inspiration both for practicing and for prospective microbial ecologists. The majority of reviews published in *Advances* have been prepared by leaders in particular areas following invitations provided by the Editorial Board. Although the Board intends to continue its policy of soliciting reviews, individual microbial ecologists are encouraged to submit outlines of unsolicited contributions to any member of the Editorial Board for consideration for inclusion in *Advances.*

Volume 9 of *Advances in Microbial Ecology* covers a particularly broad range of topics related to microbial ecology. The potential for applying ribosomal RNA sequence analysis for the definition of natural microbial populations is considered by N. R. Pace, D. A. Stahl, D. J. Lane, and G. J. Olsen. Other reviews on techniques include the application of microelectrode technology to microbial ecosystems by N. P. Revsbech and B. B. Jørgensen and the use of rates of nucleic acid synthesis to determine bacterial growth rates in natural aquatic habitats by D. J. W. Moriarty. The contribution by T. Fenchel discusses the ecology of heterotrophic microflagellates. J. H. Andrews and R. F. Harris present the concept of *r*- and *K*-selection and its relevance to microbial ecology. Recent studies on the rhizosphere are discussed by J. M. Whipps and J. M. Lynch in the context of the influence of the rhizosphere on crop production. M. Rosenberg and S. Kjelleberg consider the importance of hydrophobic interactions in the behavior of bacteria at various interfaces. Finally, J. G. Jones reviews the role of bacteria in the cycling of iron in freshwater ecosystems.

K. C. Marshall, Editor
R. Atlas
B. B. Jørgensen
J. H. Slater

Contents

Chapter 1

The Analysis of Natural Microbial Populations by Ribosomal RNA Sequences

Norman R. Pace, David A. Stahl, David J. Lane, and Gary J. Olsen

Chapter 2

The Ecology of Heterotrophic Microflagellates

Tom Fenchel

Chapter 3

r- and *K*-Selection and Microbial Ecology

John H. Andrews and Robin F. Harris

Chapter 4

Iron Transformations by Freshwater Bacteria

J. Gwynfryn Jones

Chapter 5

The Influence of the Rhizosphere on Crop Productivity

J. M. Whipps and J. M. Lynch

Chapter 6

Measurement of Bacterial Growth Rates in Aquatic Systems from Rates of Nucleic Acid Synthesis

D. J. W. Moriarty

Chapter 7

Microelectrodes: Their Use in Microbial Ecology

Niels Peter Revsbech and Bo Barker Jørgensen

Chapter 8

Hydrophobic Interactions: Role in Bacterial Adhesion

Mel Rosenberg and Staffan Kjelleberg

The Analysis of Natural Microbial Populations by Ribosomal RNA Sequences

NORMAN R. PACE, DAVID A. STAHL,
DAVID J. LANE, and GARY J. OLSEN

1. Introduction

Recombinant DNA methodology and rapid nucleotide sequence determinations have changed the face of cell biology in the past few years. This technology offers powerful new tools to the microbial ecologist as well. In this chapter we describe technical strategies we are developing which use these methods to analyze phylogenetic and quantitative aspects of mixed, naturally occurring microbial populations.

The procedures we are developing use ribosomal RNA (rRNA) sequences to define and enumerate the components of mixed, natural populations. In one approach, suitable for mixed populations of limited complexity (less than about ten different organisms), we isolate 5S rRNA, sorting out the various species-specific molecules by high-resolution gel electrophoresis. Individual 5S rRNA types then are sequenced and, with reference to existing files of 5S rRNA sequences, the phylogenetic affinities of organisms contributing the analyzed 5S rRNAs are defined.

In a second approach toward analyzing mixed populations, an approach that seems to have no upper limit as to the complexity of the

NORMAN R. PACE, DAVID A. STAHL, DAVID J. LANE, and GARY J. OLSEN • Department of Biology and Institute for Molecular and Cellular Biology, Indiana University, Bloomington, Indiana 47405; *present address for D. A. S.*: Department of Veterinary Pathobiology, University of Illinois, Urbana, Illinois 61801 populations, 16S rRNA genes are "shotgun-cloned" using DNA purified from natural samples. It does not matter that the original DNA was from a mixed population of organisms; the rRNA gene clones are selected on an individual basis, as isolated recombinant bacteriophage. The different types of cloned rRNA genes then are sorted out in the laboratory and submitted to limited sequence analysis using a technique which provides immediate access to regions of the 16S rRNA gene particularly useful for phylogenetic evaluations. Again, by comparison of sequences with existing reference collections of complete and partial rRNA sequences, the phylogenetic affinities of the organisms in the original population are established.

Since only the *in situ* biomass is required for these methods, the results project a relatively unbiased picture of the community. Because the analysis is a phylogenetic one, population members are incisively related to known organisms in terms of their fundamental biochemical properties. The experimental approaches are generally applicable and promise new insights into the microbial world.

In the following sections we consider some of the technical details of these experimental approaches and review the results of some analyses of natural microbial populations. First, we discuss the properties of the ribosome and the rRNAs in necessary detail and argue that rRNA nucleotide sequences are particularly (possibly uniquely) useful for broad, phylogenetic characterizations. We then detail the methods used to determine 5S rRNA sequences and discuss their use and limitations in analyzing mixed populations. Next we examine cloning strategies for rRNA genes, taking into account the necessity for detecting novel organisms, and discuss new tools for rapidly sequencing portions of the rRNA genes. Then we consider the theoretical and technical aspects of evaluating the derived sequence data and relating organisms by rRNA sequence comparisons.

2. Molecular Phylogeny and Microorganisms

There is a need for the general application of molecular phylogenetic analysis to microorganisms, both procaryote and eucaryote. Broadly speaking, there is no more straightforward way to classify and relate microorganisms. Their simple morphologies often provide few clues for identification; even conspicuous traits may require subjective interpretations and an experienced eye. Physiological properties often are ambiguous and, in any case, demand extensive (and expensive) testing for adequate characterization. The microbial ecologist is particularly hindered by these constraints, since many, perhaps most, organisms resist the cultivation that is so important to laboratory testing.

The use of macromolecular sequence comparisons to define phylogenetic relationships is now well established (Zuckerkandl and Pauling, 1965). Direct determination of nucleotide or amino acid sequences is the best approach to this. Comparisons of organisms by DNA–DNA or DNA–RNA hybridization tests have been useful, but require pairwise comparison of all organisms considered, and there is no accumulation of a "data base" of reference information that can be related to new findings. Macromolecular sequences, on the other hand, are readily referred to and provide for quantitiative interpretation of phylogenetic relationships.

Phylogenetic determinations by protein sequences have most often been employed in the past, because nucleic acid sequences could not be approached until very recently. Studies comparing C-type cytochromes, ribonucleases, globins, etc., have been rewarding, although they have proven most useful with higher eucaryotes (Goodman, 1982). Among the microbes, the phylogenetic and biochemical diversity is such that even the identification of homologous proteins is not a straightforward task (Demoulin, 1979). It was not until the remarkable investigations of Carl Woese and his colleagues, using partial ribosomal RNA sequences, that a comprehensive phylogeny of procaryotes, and indeed of life, began to emerge (Fox *et al.*, 1980). Their findings, still not widely understood, changed the way that we must think of evolutionary descent.

Prior to the Woese studies, the consensus thought was that the eucaryotes were relative late-comers on the evolutionary scene, generated by the fusion of procaryote cosymbionts. The studies with 16S rRNA confirmed that the major organelles, the mitochondria and chloroplasts, are of procaryotic origins. However, it emerged that the eucaryotic line of descent (that is, the nuclear genotype) is as ancient as the bacterial genotype. To distinguish the predecessor of the eucaryotic cell from more modern, organelle-containing eucaryotes, Woese termed it "urcaryote" (Woese and Fox, 1977). Even more striking, the Woese studies revealed that extant life consists not of two primary lines of evolutionary descent, procaryotes and eucaryotes, but three. There proved to be two phylogenetically distinct groups of procaryotes; these were termed the "eubacteria" and the "archaebacteria." This latter name was coined to signify that the archaebacteria (the extreme halophiles, thermoacidophiles, and methanogens) diverged from the eubacteria (e.g., the enterics, *Bacillus, Pseudomonas,* the cyanobacteria, etc.) at an ancient stage in evolution. Initially from rRNA sequences, but now from many other lines of evidence, it seems demonstrated that the archaebacteria are as phylogenetically distinct from the eubacteria as either is from the eucaryotes (Kandler, 1982). These relationships are summarized in Fig. 1.

Figure 1 also indicates two of many important questions: Are modern versions of the urcaryote still to be found? Will other primary king-

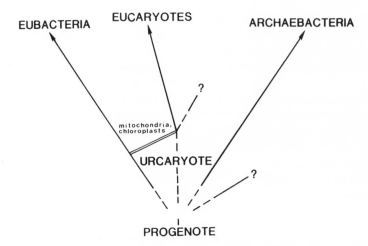

Figure 1. The primary lines of descent. Partial 16S rRNA sequences define three large domains of phylogenetic diversity, the modern-day three primary kingdoms. The rRNA sequences also establish the origins of mitochondria and chloroplasts. The predecessors of the eucaryotes, the "urcaryote," and of the three lines of descent, the "progenote," have been discussed by Woese and his colleagues (Fox *et al.*, 1980; Woese and Fox, 1977).

doms be recognized? We do not yet know the answers; the phylogenetic standings of too few organisms have been inspected. Moreover, in general it has only been possible to characterize organisms in pure laboratory cultures. Naturally occurring microbial populations that have never been cultivated certainly contain a wealth of unknown phylogenetic and hence biochemical diversity. One key to exploring that diversity is molecular phylogeny, using the appropriate macromolecular markers. Because of the universal requirement for protein synthesis, the elements of the translation apparatus, the protein-synthesizing machinery, seem best suited for broad phylogenetic analyses, where unknown breadth of diversity must be anticipated.

We now briefly review the components of the translation apparatus and argue that the ribosomal RNAs offer the best markers for molecular phylogenetic studies.

3. The Translation Apparatus and Its RNAs

3.1. Introduction

The ribosome, the seat of protein synthesis, probably is best viewed as a multienzyme machine, with protein ensembles providing binding and catalytic functions and RNA serving to coordinate the complex in space. The rRNA molecules likely provide machine action to drive the translation apparatus, and conceivably act in catalytic functions as well (Woese *et al.*, 1983). Each rapidly growing *Escherichia coli* cell contains ~20,000 ribosomes, which amount to some 40% of the dry weight of the cell. The rRNAs comprise about 50–60% of the ribosome mass, so about 20% of the cell dry weight is rRNA. Considering in addition the tRNAs, protein factors, and aminoacyl-tRNA synthetases, one finds that the translation apparatus may constitute more than half the dry weight of the bacterial cell (Ingraham *et al,*, 1983).

The ribosome and its RNA components have been well reviewed (Garrett, 1979; Chambliss *et al.*, 1980). The ribosomes of all cells and genome-containing organelles consist of two subunits, termed "small" and "large" subunits. The small subunit of the procaryotic ribosome contains one 16S rRNA molecule (~1500 nucleotides) plus about 20 proteins; the large subunit contains 23S (~2900 nucleotides) and 5S (~120 nucleotides) rRNAs plus about 30 proteins. The structural homologues of these three RNAs are present in all organisms and organelles so far inspected, with the exception of the animal and fungal mitochondria, which seem to lack 5S rRNA. There is some size variation in the ribosome and its constituents from different organisms* (Woese *et al.*, 1983). Eucaryotic ribosomes tend to be ~20–40% larger than their procaryotic counterparts and more protein-rich. Fungal and vertebrate mitochondrial ribosomes generally are ~70% the size of the typical bacterial ribosome, whereas the higher plant mitochondrial ribosomes tend to be more bacterial in aspect. Chloroplast ribosomes clearly are cyanobacterial in type.

Eucaryotic ribosomes contain a fourth rRNA, 5.8S rRNA, which is not found in the procaryotic ribosomes so far investigated (Walker and Pace, 1983). The 5.8S rRNA is not an evolutionary novelty of the eucaryotes, however, since the 5.8S rRNA corresponds in nucleotide sequence and in secondary structure to the 5′ end of the procaryotic 23S rRNA. The 5.8S rRNA is specifically cleaved from the large rRNA by an RNA processing event during ribosome biosynthesis. The reason that the 5.8S rRNA is disjoined from the 23S-equivalent molecule is not known. Specific fragmentation of the conventional rRNAs is proving to be common. *Drosophila,* for instance, cleaves its 5.8S rRNA even further, generating a "2S" rRNA (Pavlakis *et al.*, 1979); flowering plant chloroplasts sever a 4.5S (~100 nucleotides) segment from the 3′ end of the 23S rRNA, and *Rhodopseudomonas* and *Anacystis* specifically cleave their 23S rRNAs (Maars and Kaplan, 1970; Doolittle, 1973). Conceivably, these fragmen-

---

*Because of this size variation, even within rather closely related groups, designation of exact sedimentation coefficients for the rRNAs is impractical, and, in fact, they are seldom determined. We use the terms 16S, 23S, and 5S as generic ones, to indicate homologous rRNAs.

tations enhance conformational mobility important to ribosome function.

3.2. The Translation Apparatus and Phylogeny

The translation apparatus, the ribosome with the tRNAs, is an ancient mechanism. The similarity in architecture of the ribosomes and tRNAs from the three primary kingdoms—the eubacteria, the archaebacteria, and the eucaryotes—means that the translation apparatus emerged pretty much in its modern form even before the phylogenetic radiation of the three kingdoms. Thus, phylogenetic analyses employing components of the translation apparatus allow in principle, the tracing of relationships among organisms to nearly the origins of life.

The RNA components of the translation apparatus have been the most useful for phylogenetic analysis. Although many sequences have been determined for the ribosomal proteins and the various other protein factors involved in ribosome function (Chambliss *et al.*, 1980), the data are from too few organisms for significant phylogenetic comparison. The available information makes it clear, however, that these various proteins generally will not be useful as evolutionary markers.

The purification and analysis of individual ribosomal proteins is tedious and difficult, and even the definition of homologous proteins between compared organisms sometimes involves more speculation than objectivity. For instance, ribosomes reconstituted from *Bacillus* rRNA and *E. coli* ribosomal proteins, or vice versa, are active in protein synthesis (Nomura *et al.*, 1968), but the assignment of homologous pairs of the ribosomal proteins has proven remarkably difficult (Wittmann, 1983). The identified homologous proteins are significantly variable in size and sequence, even between organisms not separated by great phylogenetic distance. Defining homologous ribosomal proteins across deep phylogenetic groups (for instance, the eubacteria and eucaryotes) seems not very practical, in the main.

The tRNAs are not very useful for phylogenetic characterizations because they are too constrained in structure. Several hundred tRNA species have now been sequenced. However, considering that each organism contains 50–60 different tRNA species and that most attention has fallen on rather few organisms, the phylogenetic breadth so far examined has not been great. There has been some success in casting generally credible phylogenetic trees with homologous tRNA species (those using a particular amino acid), but those molecules intrinsically cannot yield a highly reliable information base for phylogenetic analysis (Cedergren *et al.*, 1981). The tRNAs consist of too few nucleotides (70–90), of which only ~30–40% may change even across kingdom boundaries. Moreover, those nucleotides that can change probably cannot do so independently without perturbing other positions in the molecule. This is because the tRNA structure is tightly locked up in an intricate tertiary structure. Almost every residue in tRNA has contacts with at least one other residue, and the tight interlocking of the molecule imposes conformational constraints on all residues. The main problem, however, is the limited number of mutable residues in homologous tRNAs. The small number of changes in compared molecules means that the statistical error in calculated evolutionary distance is great.

We are left, then, with the rRNAs as the most useful tools for phylogenetic explorations. There are several explicit reasons for focusing on the rRNAs:

1. The rRNAs, as key elements of the protein-synthesizing machinery, are of profound importance to all organisms.
2. The rRNAs are ancient molecules and are extremely conservative in overall structure. Thus, the homologous forms of the rRNAs generally are readily identifiable simply by their sizes.
3. The conservative nature of rRNA structure carries over to the nucleotide sequence level. Some sequence islands are invariant across the biological kingdoms, while others vary to greater or lesser extent. The islands of conserved sequence and secondary structure allow the alignment of disparate sequence so that only homologous sequences are employed in any phylogenetic analysis. The highly conserved regions also provide convenient hybridization targets for cloning the rRNA genes and for primer-directed sequencing techniques (see below).
4. The rRNAs constitute a significant component of cellular mass, and they generally are easily recovered from all types of organisms. This assists the accumulation of a data base of reference sequences.
5. The rRNAs are sufficiently lengthy in nucleotide sequence that they provide for statistically significant comparisons.
6. The rRNA genes seem to be free from artifacts of lateral transfer between contemporary organisms. Thus, relationships established by rRNA sequence comparisons represent true evolutionary relationships (Stackebrandt and Woese, 1981).

Taken together, these features point to the rRNAs as being particularly suitable, probably uniquely suitable, for establishing phylogenetic relationships among very different organisms.

Of the three generic rRNAs, 5S, 16S, and 23S rRNA, the 5S and 16S molecules have received the most attention, largely for historical and technical reasons. The 5S rRNA, because of its relatively small size, was amenable to sequence analysis by the late 1960s. It is larger than tRNA ($\sim$120 nucleotides versus $\sim$70) and also is unmodified and significantly less constrained in secondary structure. Thus, 5S rRNA is easier to sequence than tRNA and contains more phylogenetically useful information (i.e., independently varying positions) than simple size difference would imply. As a practical matter, however, 5S rRNA is still sufficiently small that it cannot be used indiscriminantly for phylogenetic inferences.

The 16S rRNA, in contrast, contains a wealth of useful phylogenetic information, but is too large for traditional RNA sequencing methods. This required the development of the DNA sequencing protocols; the first 16S rRNA gene sequence, that of *E. coli,* was not available until 1978 (Brosius *et al.*, 1978). However, the 16S rRNA was subject to partial sequence analysis, so-called "oligonucleotide cataloging," the approach used by Woese and his collaborators (see Section 6.3). Similarly, the 23S rRNA required DNA sequencing; it is twice the size of 16S rRNA (Brosius *et al.*, 1980). The 23S rRNA has not been used for phylogenetic analysis because attention has been directed to the smaller, hence more tractable, 5S and 16S rRNAs. The 23S rRNA should be an excellent phylogenetic measure, however, if the sequence collection is expanded to include many organisms.

4. Population Analysis Using 5S rRNA

When we first undertook the characterization of mixed, natural microbial populations using *in situ* samples we had not yet developed procedures for rapidly gaining access to 16S rRNA sequences. Moreover, there was little hope in being able to separate mixtures of 16S rRNAs, a prerequisite for their direct sequence characterization. We therefore focused initially on 5S rRNA. By virtue of its small size, minor structural differences in sequence or secondary structure of mixed 5S rRNAs are reflected in differing mobilities during high-resolution polyacrylamide gel electrophoresis. Thus, it proved possible to sort out mixtures of 5S rRNAs for nucleotide sequencing and the definition of the phylogeny of the contributing organisms.

4.1. The 5S rRNA

About 250 5S rRNA sequences now are available. With few exceptions, all are $\sim$115–120 nucleotides in chain length and all can be folded into a common secondary structure. In consideration of RNA structures, not only the primary nucleotide sequences, but also permissible secondary structure is important. The secondary structure is formed by pairing complementary nucleotide sequences. Complementary pairings are mostly defined by the canonical Watson–Crick pairs, A–U, and G–C, although it now is clear that noncanonical pairs, such as G–U, A–G, A–C, and others, contribute to the stability of duplex stems (Papanicolaou *et al.*, 1984). The folded structures of 5S rRNAs from representatives of the three primary kingdoms are shown in Fig. 2. Each of the sequences theoretically could be folded in several other ways, using different local complementary sequences, but the folded structure shown is the only one that accommodates all 5S rRNA sequences. The fact that all 5S rRNAs can adopt this folding is considered phylogenetic "proof" that the secondary structure actually is of that form. Many other sorts of data are fully consistent with that conclusion, but the phylogenetic omparisons pointed to this compact 5S rRNA structure (Stahl *et al.*, 1981).

The 5S rRNA is seen to consist essentially of three helical domains, two of which are subdivided into shorter helical segments, as indicated by the designation of helices I–V (Fig. 2). Various tertiary interactions between the three main helical domains have been suggested (Pieler and Erdmann, 1982), but there is no real evidence for any of these. Delihas and Andersen have recently summarized variation in the known 5S rRNA sequences (Delihas and Andersen, 1982). It is possible to group 5S RNA structures (and hence their donor organisms) by a variety of "signatures," including local sequences, nucleotide chain lengths between landmarks, helix lengths, presence or absence of "bulged" residues in helices, and so on. For instance, almost without exception, the procaryotes, both eubacteria and archaebacteria, have the sequence CGAAC at positions corresponding to residues 43–47. This is frequently replaced in the eucaryotic 5S by something like UGAUG, as seen in the human 5S

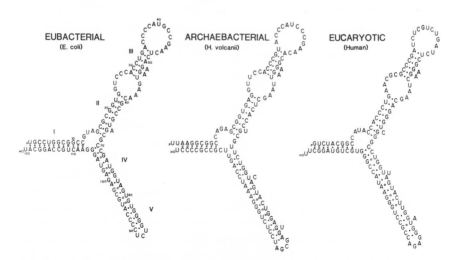

Figure 2. The 5S rRNAs from the three primary kingdoms. The sequences are folded into compatible structures according to Stahl *et al.* (1981).

rRNA (Fig. 2). Most regions of more slowly changing sequences are found in loops or "single-strand" portions of the molecule. Less conservative sequences are associated with duplex regions. Changes in duplex stems are compensated by a change in the complementary pairing partner. Presumably the conservative, "single-strand" regions have sequence-specific structure or function, whereas base-paired sequences generally are not so constrained, their roles being to provide duplex character, the exact sequences being less important.

Most chain-length variation in 5S rRNA occurs in helices IV and V. The inspected mycoplasma 5S rRNAs, for example, are only ~107 nucleotides in length, the deletion truncating helix V. The most bizarre 5S rRNA so far encountered, that of the archaebacterium *Halococcus morrhuae*, contains a 108-nucleotide insertion on the 3' side of the helix IV (Luehrsen *et al.*, 1981). The various group-specific anomalies or signatures are useful to some extent for phylogenetic definitions, but they are most useful for aligning compared sequences such that only truly homologous positions are considered in homology calculations using all or most of the molecule.

The tRNAs and high-molecular-weight rRNAs contain, in addition to the four usual nucleotides, many modified residues, most commonly base and sugar (2'-O) methylations. Modifications in the 5S rRNA are rare; however, they occasionally occur. *Sulfolobus* 5S rRNA, for example, probably contains a 2'-O-methyl C residue; that of *Saccharomyces* has a pseudouridine. Probably other modified residues occur that have not been detected during sequencing, but it does seem that modifications in 5S rRNA are so rare as to have little importance for phylogenetic considerations.

4.2. 5S rRNA and Phylogeny

The 5S rRNAs are highly conservative molecules, but they can change over almost their entire lengths. The frequencies of variation at the different nucleotide positions are not uniform, however; it is evident the evolutionary "clock speed" is not the same for all positions. Some positions in the 5S rRNA drift freely; some are constrained to universality, or nearly so.

Because of its long evolutionary history, 5S rRNA is applicable to broad phylogenetic comparisons (Hori and Osawa, 1979). The relatively small size of the molecule (~120 nucleotides) dictates that it be used cautiously, however. In the last analysis, phylogenetic conclusions based on sequence divergences are at the mercy of statistics; the reliability of relationships depends on the number of nucleotides compared. The upshot of this consideration is that there are boundaries upon the extent of sequence homology for which 5S rRNA is useful. Phylogenetic branching orders based on 5S rRNA are of limited accuracy if the compared sequences are too close in homology (i.e., vary at too few positions) or are too dissimilar (i.e., are similar at too few positions, those being highly conserved).

The higher metazoa, for instance, are a grouping which is phylogenetically so compact that 5S rRNA cannot provide detailed relationships, but only general outlines. The 5S rRNA sequence homology between human and *Xenopus* 5S rRNA is ~91%, and between human and sea urchin 84%. Human 5S rRNA is identical to marsupial 5S rRNA. Any relationships drawn within these groups, are based on very few nucleotide changes at relatively highly mutable positions, so statistical uncertainties become large. At a homology level of 0.90 for two compared sequences, the estimated error is ±0.03, 30% of the observed positional differences.

On the other hand, 5S rRNA is generally useful among the procaryotes, or even the entire grouping of eucaryotes, since great phylogenetic breadth is being inspected. Establishing relationships among similar organisms generally is not the goal. For comparison with the above-cited metazoan homologies, it is noteworthy that the 5S rRNA homology spanning the Enterobacteriaceae is ~90%; the enteric–*Bacillus* homology is ~75%; the enteric–cyanobacteria homology is ~60%. Kingdom-level 5S rRNA homologies between eubacteria, eucaryotes, and archaebacteria are at the 40–50% level. Although we are accustomed to thinking of the metazoa as a diverse lot, they really represent limited phylogenetic diversity compared to microorganisms. The phylogenetic (hence biochemical) diversity of life lies with the microbes, both procaryotic and eucaryotic.

Problems among the higher metazoa with the small size of 5S rRNA are compounded by the fact that many of those organisms produce a 5S rRNA population that is heterogeneous in sequence. Usually the heterogeneity is minor, variation occurring at only a few positions, so there is little impact on phylogenetic analyses. Minor heterogeneity likely arises because organisms contain multiple genes for 5S rRNA. Eubacteria commonly have 5–10 5S rRNA genes; metazoa commonly contain hundreds or thousands of 5S rRNA genes. A small amount of independent structural drift among the many genes seems tolerable and could account for minor sequence variations in the same organism. Over the long term, such sequence variation probably is buffered by unequal crossing-over events within multigene regions of the chromosome during recombination. The "optimal" 5S rRNA structure then would be maintained by selection.

In some cases, however, the heterogeneity in 5S rRNA sequence from a given organism is extreme, involving 10–20% of the residues, and it can confuse phylogenetic interpretations. The physiological significance of quite different 5S rRNA sequences in the same organism is not known. Sometimes it may reflect developmental processes. For instance, *Xenopus* somatic cells and oocytes produce dominant 5S rRNAs that differ at nine positions. The oocyte type of 5S rRNA conceivably has some special role during development, but the diversity also may (more likely) reflect different genetic organization of 5S genes expressed during oogenesis and during somatic cell growth. If the two batteries of 5S rRNA genes were segregated long ago and not continuously homogenized by recombination and selection, then the oocyte and somatic types of 5S sequence would have diverged much as do the genes of independently evolving organisms. Other cases of quite diverse 5S sequences from the same organism are not obviously related to development. Chicken somatic cells, for example, produce two (at least) types of 5S rRNA, differing at nine positions. One ribbon worm collection yielded two types of 5S rRNA, differing at an astounding 22 positions; that of a different ribbon worm collection was essentially homogeneous (Erdmann *et al.*, 1984). It is possible that the ribbon worm population with such 5S heterogeneity was parasite-ridden, and so constituted a mixed population of organisms. However, the overall phenomenon of 5S sequence variation within a given organism is another reason for caution in the use of the molecule for establishing relationships among organisms.

4.3. Populations Characterized by 5S rRNA

So far we have inspected three different types of mixed, natural populations using 5S rRNA-based phylogenetic characterizations. These are: (1) the 91°C source pool of Octopus Spring in Yellowstone National Park; (2) the unusual symbioses discovered around submarine hydrothermal vents, in which procaryotic endosymbionts confer sulfur-based chemoautotrophy upon the invertebrates *Solemya velum, Calyptogena magnifica,* and *Riftia pachyptila;* and (3) a leaching pond atop a copper recovery dump in the Chino Mine in southern New Mexico.

One of the more studied thermal habitats in Yellowstone National Park is Octopus Spring (Brock, 1978). Although *in situ* microbial accumulation in the slightly alkaline, 91°C source waters is easily demonstrated, these organisms have repeatedly eluded cultivation. Sequence analysis of 5S rRNA from *in situ* samplings established the presence of a complex community with three dominant members, two representatives of eubacteria, distantly related to *Thermus* spp., and one representative of the archaebacteria. The latter comprises about 50% of the extractable 5S rRNA and defines a deep phylogenetic division within the archaebacteria (Stahl *et al.*, 1985).

The submarine hydrothermal vent systems are associated with crustal spreading centers along the Mid-Ocean Ridge, which wanders more than 70,000 km over the earth's surface (Edmond and Von Damm, 1983). Ocean water, circulating by convection through magma-heated zones of the crust, is charged with soluble volcanic exudates (metallic sulfides, H_2S, CH_4, CO_2, and CO) before exiting the fissure system. Dense populations of chemolithotrophic bacteria develop in and around the vents and they in turn support rich animal communities through grazing and filter-feeding (Jannasch and Wirsen, 1979). In some cases, sulfur-oxidizing bacteria evidently have established symbiotic associations with the macrobiota (zoocoenotic symbioses), producing "chemolithotrophic animals."

The symbionts were first identified histologically and by the presence of high levels of certain Calvin cycle and sulfur-oxidative enzymes in the vestimentiferan tube worm, *Riftia pachyptila,* in which the bacteria fill a specialized organ, the trophosome (Cavanaugh *et al.*, 1981). Similar associations were noted in the gill tissues of vent molluscs, including the giant vent clam (*Calyptogena* sp.), and also in *Solemya,* a bivalve which inhabits sulfide-rich tidal flats (Cavanaugh, 1983; Felbeck *et al.*, 1981). All attempted isolations of the symbiotic bacteria have been negative or equivocal (Jannasch and Nelson, 1984). The phylogenetic status of the bacteria symbiotically associated with the worm *Riftia pachyptila,* the clam *Calyptogena magnifica,* and the bivalve *Solemya velum* have been deduced from the characterization of the respective 5S rRNAs (Stahl *et al.*, 1984).

Commercial copper recovery processing relies to a considerable extent upon microbial leaching (Brierly, 1982). Copper ores commonly are of high sulfide content. Ponds constructed on ore dumps consequently develop rich populations of sulfur-oxidizing chemoautotrophs. Sulfuric acid, resulting from microbial metabolism, percolates through the ore stack, solubilizing the copper as the sulfate salt. *Thiobacillus* spp. traditionally are considered primarily responsible for copper leaching, since they are the most abundant in cultivations from leaching sites. However, other sulfur-oxidizing acidophiles (e.g., *Sulfolobus*) are recovered from leaching fluids, and chemoautotrophs often are difficult to culture, so we considered the microbial contents of leaching sites rather open to question. Although not yet completed, the analysis of 5S rRNAs from one leaching pond at Chino Mine, New Mexico, reveals the major population constituents as *Thiobacillus ferrooxidans* and an organism not closely related to any of our data base.

4.4. Overview of the Analyses

The technical aspects of population analysis by sequencing 5S rRNAs and 16S rRNA genes are summarized in Fig. 3. The reader may find it useful to refer to this overview for (or in lieu of) the following

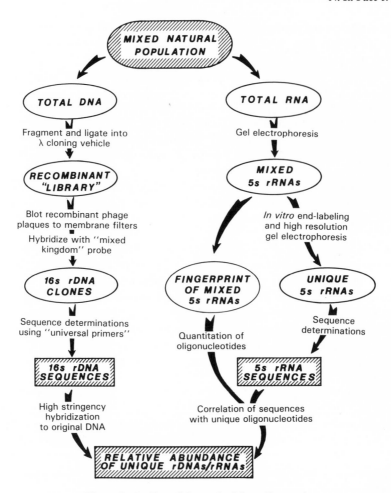

Figure 3. Outline of the methodology discussed.

discussions. Studies using 5S rRNA are discussed first, although, as will be evident, many of the considerations apply equally to population analyses using 16S rRNA genes.

4.5. Biomass Collection

This is an important step, requiring consideration of the type(s) of analysis to be performed and the nature of the environment. Points to consider include: (1) the density and potential diversity of cell types; (2) the potential for coisolation, with the biomass, of materials that interfere with nucleic acid recovery or manipulation; and (3) the potential introduction of sampling bias into the composition of the biomass collected, if quantitative analysis is important.

Sufficient biomass must be collected that subsequent nucleic acid manipulations are feasible. As a practical matter, the sequencing and fingerprinting steps described here require trivial (radioactive) amounts of rRNA: on the order of 10^9 cells, collected in reasonably clean form, will usually suffice for the full analysis depicted in Fig. 3. The environment under consideration (and the sampling biases one is willing to accept) ultimately shape the experimental strategy.

In the case of Octopus Spring, there is significant microbial accumulation in the cooler effluent channels and side pools (e.g., *Synechococcus* and *Thermus*). Additionally, abudant "pink tufts" of fibrous nature, presumably microbial depositions, are present on the walls of the clearwater source pool and immediate runoff channel. We were unable to recover significant amounts of nucleic acids from the pink tufts, apparently because the bulk of this material is dead and leached by the hot waters. Probably only a surface lamella of the fibrous deposits is composed of viable mass. However, contact slides immersed in the 91°C source showed vigorous growth. Providing suitable *in situ* growth is observed, sampling of surfaces thus seemed a reasonable approach to gaining sufficient material for analysis. Cotton and fiberglass batting sewn into a nylon screen for support yielded generous amounts of biomass following immersion in the source for 7–10 days. Thus, in this instance, sampling is biased toward organisms capable of growth on surfaces. However, these are probably representative of organisms persisting in a flowing environment. There also was concern that the cotton might itself provide a metabolizable growth substrate, thus enriching for organisms able to utilize it. In this case, the cotton and fiberglass batting appeared to support identical populations.

At the Chino mine, samples of the surface mud underlying the pH 2.5, iron-rich leaching pond were transported frozen on dry içe to the laboratory. Upon thawing, the mud was vortexed vigorously with two changes of buffer and sedimented by a brief, low-speed centrifugation. Cells were then concentrated from the supernatant by higher speed centrifugation and stored as frozen pellets. From 10^9 to 10^{10} cells were recovered from ∼1 kg of mud, as estimated by microscopic observation.

Frozen, dissected tissue samples served as starting material for analysis of the sulfur-oxidizing invertebrate/procaryote symbioses. In the two bivalves, *Calyptogena* (giant clam) and *Solemya* (small mussel), éndosymbionts had been detected only in gill tissue, whereas in the tube worm, *Riftia,* they colonize a tubular organ, the trophosome. Trunk wall and trophosome tissue from two sections of *Riftia* were examined. Gill tissue from *Calyptogena* and gill and foot tissue from *Solemya* were inspected as well. In each case, only a small amount of tissue (1 to a few g) was more than adequate for the analyses.

4.6. Nucleic Acid Isolation

Total RNA is prepared by fairly standard protocols. The invertebrate tissues were first homogenized with buffer in a Dounce homogenizer, then treated in essentially the same way as the bacterial cell pellets collected from Octopus Spring and the Chino leaching pond. Cell lysis is accomplished by any of a variety of means, including lysozyme and/or pronase treatment or passage through a French pressure cell. The simplest method, which proved adequate in the cases discussed here, involved suspending the cells in a modest volume of 0.05 M NaOAc, 0.01 M EDTA, pH 5.1 (extraction buffer), to which an equal volume of the same buffer containing 1% sodium dodecyl sulfate (SDS) is added. With unfamiliar samples, aliquots should be monitored for complete lysis microscopically (concentrating if necessary). Two or three freeze/thaw (dry ice–ethanol/60°C) cycles may assist lysis, and passage of the lysate through an 18- or 21-gauge needle (not recommended for DNA extraction) is useful to reduce the viscosity of concentrated samples. The lysed suspensions were then extracted several times at 60°C against phenol and the total nucleic acids were recovered by precipitation from 70% ethanol.

The 5S rRNA was purified from total RNA preparations by polyacrylamide gel electrophoresis, as illustrated in Fig. 4A. As is evident, 5S rRNA is cleanly resolved from high-molecular-weight RNA (primarily 16S and 23S rRNAs) and tRNA, although individual 5S rRNA species are only poorly resolved. The 5S regions of the gels (identified by UV shadow) were excised and the 5S rRNA was eluted and recovered by precipitation from ethanol.

An aliquot of the unfractionated 5S rRNA was reserved for the relative abundance measurements and the remainder radioactively end-labeled at the 3' termini using [5'-^{32}P]cytidine bisphosphate (pCp) and RNA ligase, or at the 5' termini (following treatment with alkaline phosphatase to remove 5'-terminal phosphate residues) using [γ-^{32}P]ATP and polynucleotide kinase (Stahl *et al.*, 1984). Individual 5S rRNA species (now either 5' or 3' terminally labeled with ^{32}P) were then fractionated on a high-resolution (80 cm long) "sequencing"-type polyacrylamide gel (Fig. 4B).

When very small amounts (suboptical quantities) of total RNA were recovered (as with the Chino leaching pond samples), the total RNA was labeled using RNA ligase and [5'-^{32}P]pCp. Native rRNAs generally contain a free 3' hydroxyl group, and thus are good RNA ligase substrates,

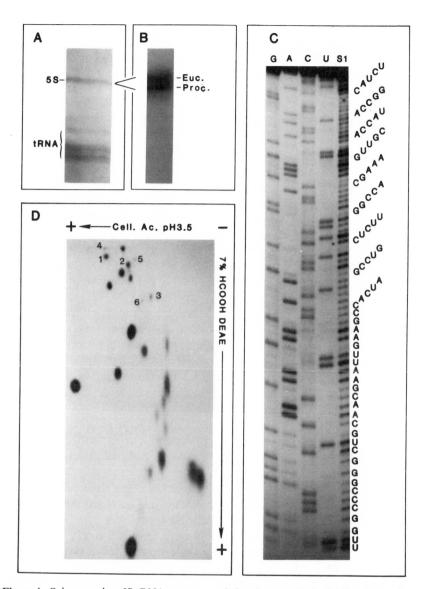

Figure 4. *Solemya velum* 5S rRNA sequence and abundance analysis. (A) Total RNA from *S. velum* gill tissue was fractionated on a preparative, 8% polyacrylamide gel, and RNA bands were visualized by UV shadow. (B) The 5S rRNA zone from the preparative gel was eluted, end-labeled, resolved by electrophoresis on an 80-cm, 8% sequencing gel, and the bands detected by autoradiography. (C) A portion of an autoradiogram of an RNA sequencing gel. (D) The mixture of 5S rRNAs from panel A was digested completely with RNase A. The resulting oligonucleotides were labeled using $[\gamma\text{-}^{32}\text{P}]\text{ATP}$ and polynucleotide kinase, then resolved by two-dimensional electrophoresis according to Sanger *et al.* (1965). Oligonucleotides specific to the eucaryotic host (1–3) and procaryotic symbiont (4–6) were located by autoradiography and quantitated by scintillation counting. The details of these experiments have been presented (Stahl *et al.*, 1984).

whereas degradation products generally have 3′ phosphates and do not serve as RNA ligase substrates. Thus, 3′ end-labeled 5S rRNAs can be isolated by gel electrophoresis directly. On the other hand, 5′-^{32}P end-labeling of 5S rRNA is not very efficient, and degraded RNA in a total RNA sample is a formidable competing substrate for the kinase. Unlabeled 5S rRNA for 5′ end-labeling or fingerprint analysis is isolated using the 3′-labeled 5S rRNA as a marker on preparative gels.

4.7. Nucleotide Sequence Determination

RNA sequencing procedures are well documented (Donis-Keller *et al.*, 1977; Peattie, 1979), are only outlined here, with a few cautionary comments. Purified, end-labeled, 5S rRNA is cleaved in separate reactions with base-specific enzymes or reagents such that, on average, one break per RNA molecule is produced. The partial digestion products then are resolved on high-resolution polyacrylamide gels. When G-, C-, A-, and U-specific reactions are electrophoresed in adjacent gel tracks, a "ladder" is produced on an autoradiograph of the gel from which the nucleotide sequence can be read directly. Generally, we have employed enzymatic protocols on both 5′ and 3′ end-labeled 5S rRNAs, and chemical digestion protocols (which offer superior discrimination between C and U residues) on 3′ end-labeled RNA. An example of a portion of a sequencing gel is shown in Fig. 4C.

A significant problem in gel sequencing of RNA, less so with DNA, is band compression or even rearrangement due to secondary structural effects. Highly stable secondary structures are not completely denatured in the 7 or 8 M urea-containing gels commonly employed, so anomalous banding patterns of partial digestion products often result. Certain portions of 5S rRNA sequences, such as helix V in Fig. 2, reproducibly yield anomalous banding patterns; however, such artifacts may arise in any part of the molecule. Sequencing from both termini usually permits the nucleotide sequences through such stable duplex features to be interpreted unambiguously.

Another problem encountered in 5S rRNA sequencing, which is aggravated in mixed populations, is that of terminal length heterogeneity. In many instances, multiple forms of 5S rRNA, differing only by the presence of additional nucleotides at the 5′ and/or 3′ termini, are isolated from a single cell. These are considered to be unimportant with respect to ribosome function and are not useful for phylogenetic analysis. The different forms must be separated from one another for nucleotide sequence analysis, however, because when the heterogeneity is at the labeled terminus, multiple, overlapping sequencing patterns, often too complex to be interpreted, appear upon electrophoresis of the sequencing reactions.

4.8. Incorporation of New Sequences into the Existing Data Base— Sequence Alignment

This is a critical step in the overall analysis, since the accuracy of all subsequent phylogenetic evaluation hinges on proper nucleotide sequence alignments. The accumulating knowledge of 5S rRNA structure and an understanding of the evolutionary models underlying "mutational distance" analysis now come into play. A further treatment of the alignment problem and a discussion of the theoretical and technical aspects of evaluating phylogenetic relatedness by molecular sequence comparisons are presented below. Here we simply point out that 5S rRNA sequences from the different organisms in the data collection are to be compared on a position-by-position basis. From the number of positional differences between each pair of sequences, the average number of mutational events separating each pair is estimated. Thus, proper sequence alignment is required to ensure that only evolutionarily homologous positions are compared.

Sequence alignment is necessarily a compromise between obtaining a maximum amount of homology (i.e., matching residues) and introducing the smallest number of gaps (i.e., postulating insertion or deletion events). Because 5S rRNA is highly conservative with respect to certain features of both primary and secondary structure (see Fig. 2), sequence alignment is a relatively straightforward matter. The process involves first aligning highly conserved residues, which occur with few or no exceptions at the same positions in all 5S rRNA sequences, followed by alignment of residues conserved among related groups of organisms. Further alignment then is made on the basis of conserved secondary structural features, as are evident in Fig. 2.

The most problematic region of the 5S rRNA molecule with regard to sequence alignment is that including helices IV and V in Fig. 2. This portion of the molecule is considerably more variable in structure than either helix I or the segment containing helices II and III. The helix IV–V region of the molecule also clearly distinguishes eubacterial from eucaryotic and archaebacterial 5S rRNAs.

The evolutionary relationships among organisms are best interpreted in the form of a phylogenetic tree, as discussed in more detail in Section 5.1. This is constructed by finding a tree geometry and branch lengths that best fit the pairwise homologies of all the aligned sequences considered.

4.9. Relative Abundance of Unique 5S rRNAs

Some appreciation of the relative abundances of the resident micro-organisms may be established by quantitation of the relative abundances of their representative 5S rRNAs. However, because of differences in secondary structures, different 5S rRNAs label at their termini with varying efficiencies, so the relative incorporation of radioactivity into individual, intact 5S rRNA species does not necessarily correlate well with their *in situ* abundances. To relieve this bias, the total collection of 5S-sized material is digested to completion with RNase T_1 (which cleaves specifically after guanosine residues) and the resultant oligonucleotides then are end-labeled at their 5′ termini with $[\gamma\text{-}^{32}P]ATP$ and polynucleotide kinase and fractionated by two-dimensional high-voltage paper electrophoresis (Fig. 4D). The derived oligonucleotides label with uniformly good efficiencies, thus eliminating the uncertainty of differential incorporation into the intact molecules. The evaluation of label incorporation into oligonucleotides unique to a particular 5S rRNA sequence scores its abundance in the original mixed population and provides an estimate of the abundance of the organisms that contributed that rRNA. This can only be an estimate, however, since rRNA content per cell is known to vary with metabolic status and between different cell types.

5. The Populations Revealed by 5S rRNA

5.1. Overview

The relationships of the procaryotic 5S rRNAs recovered from Octopus Spring, the marine invertebrates, and the Chino Mine pond are summarized in the phylogenetic trees shown in Fig. 5.

The eubacteria considered fall into the "purple photosynthetic bacteria" grouping, so named because the deepest branchings in the group include the purple photosynthetic phenotype (Gibson *et al.*, 1979). The inset summarizes the lines of eubacterial descent so far defined by partial 16S rRNA sequence characterization (Fox *et al.*, 1980), and is offered as reference to more familiar organisms. Notable in this phylogenetic tree is the distribution of sulfur-based energy metabolism throughout the array shown, and the close evolutionary associations of well-known heterotrophs with chemoautotrophic phenotypes. The traditional microbial taxonomic hierarchies based on autotrophy and heterotrophy appear to correlate poorly with phylogenetic associations. Also, our and other analyses indicate that many organisms classified as *Pseudomonas* spp. are phylo-

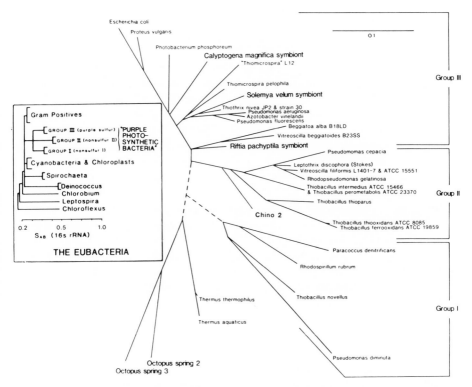

Figure 5. 5S rRNA-derived relationships among the purple photosynthetic bacteria. The natural population constituents so far characterized by 5S rRNA analysis (bold lettering) mostly fall into the "purple photosynthetic" group of the eubacteria [see inset (adapted from Stackebrandt and Woese, 1981) and text]. The sequences were aligned and the tree inferred as described by Stahl *et al.* (1984) (also see text). The 5S rRNA sequences are from Erdmann *et al.* (1984), Stahl *et al.* (1984), Lane *et al.* (1985), and N. R. Pace, D. A. Stahl, D. J. Lane, and G. J. Olsen (unpublished data). The scale bar represents an evolutionary distance of 0.1 nucleotide changes per sequence position. The root of the tree likely falls within the dashed segments.

genetically disparate (Palleroni, 1981); those included are scattered throughout the purple photosynthetic line of descent.

5.2. The Octopus Spring Population

We selected Octopus Spring, in Yellowstone National Park, for a first analysis of a mixed microbial habitat by rRNA sequences because it offers a vigorous community of apparently limited complexity. Also, although Brock and colleagues had characterized to some extent the *in situ* activity of the community, characteristic microorganisms had not been obtained in culture (Brock, 1978). The temperature of the source waters is nearly constant at 91°C, matching the temperature optimum for incorporation of radioisotopically labeled substrates (leucine, lactate, asparatate, phenylalanine, thymidine). The resident microbial population therefore appears to be optimally adapted for growth at the ambient temperature. The slightly alkaline waters are low ($\sim$5 μM) in sulfide, and addition of sulfide inhibits the uptake of labeled substances, so it seems unlikely that sulfide serves the metabolic needs of that community. These earlier studies also demonstrated that bubbling with CH_4, N_2, CO_2, or air did not stimulate uptake of labeled substrates (Brock, 1978).

Three different 5S rRNAs constituted more than 90% of the 5S rRNA retrieved from the Octopus Spring samplings. Two of these were eubacterial and one archaebacterial.

The two characterized eubacterial 5S rRNAs obtained from Octopus Spring, about half that recovered, most closely resemble the 5S rRNAs of the two representatives of the genus *Thermus* (*T. aquaticus* and *T. thermophilus*) in our reference collection. Although *T. aquaticus* is unable to grow at temperatures in excess of 85°C, the occurrence of *Thermus*-like organisms at temperatures greater than 90°C has been noted (Brock, 1978). The rudiments of the *Thermus* phylogeny emerging from 5S sequence comparisons suggest that it is of relatively ancient origin. The 5S rRNA analysis also suggests that the represented *Thermus* spp. and the eubacterial Octopus Spring organisms radiated from a common ancestor near the origin of the purple photosynthetic line of descent and should possibly be included in it (Fig. 5). No 16S rRNA sequence information is yet available for *Thermus,* and because phylogenetic analysis based on 5S rRNA does not satisfactorily bridge longer phylogenetic distances (Section 6.7), the suggested placement of *Thermus* with the purple photosynthetic bacteria should be considered tentative until the data base of reference sequences in this region of the phylogenetic "map" is more complete. The phylogenetic affinities of the Octopus Spring organisms for these *Thermus* spp., however, is clear.

Members of the genus *Thermus* are formally described as filamentous, nonmotile, obligately aerobic, oligotrophic heterotrophs. *Thermus*-like filaments have been observed in flowing thermal environments with organic concentrations as low as 2 ppm, and are conspicuous on contact slides immersed in the Octopus Spring source waters. Although we have no measured value for the organic contents of Octopus Spring, dense microbial mats (primarily *Chloroflexus* and *Synechococcus*), thriving in the cooler shallows surrounding the source waters, probably supply adequate organics to sustain oligotrophic growth. Therefore, the *Thermus* relatives encountered in the Octopus Spring source may follow the physiological profile of this genus.

The Octopus Spring archaebacterium, which accounts for about 50% of the isolatable 5S rRNA, is not closely related to any organism in our sequence collection; the list of available archaebacterial 5S rRNA sequences is scant. As shown in Fig. 6, however, it is affiliated with the

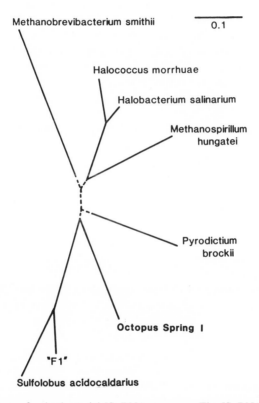

Figure 6. A phylogeny of archaebacterial 5S rRNA sequences. The 5S rRNA sequences representing several major lineages of the archaebacterial kingdom (two major divisions of the methanogens, the halophiles, the cell wall-less thermoacidophile *Thermoplasma acidophilum,* and the extreme thermoacidophile *Sulfolobus*) and the archaebacterial 5S rRNA isolated from Octopus Spring were compared and a phylogeny inferred as described by Stahl *et al.* (1984). The sequences are from Erdmann *et al.* (1984) and N. R. Pace, D. A. Stahl, D. J. Lane, and G. J. Olsen (unpublished data). The scale bar represents an evolutionary distance of 0.1 nucleotide changes per sequence position. The root of the tree probably falls within the dashed region.

so-called "sulfur-metabolizing (dependent)" branch of the archaebacteria, a physiologically diverse assemblage (Stetter and Zillig, 1985). Representatives of the genus *Sulfolobus* grow heterotrophically or as sulfur-oxidizing autotrophs. Other members of this branch grow by sulfur-dependent respiration of hydrogen or organic compounds, either heterotrophically or autotrophically.

The discovery of a representative of this assemblage in an alkaline source is noteworthy. There so far has been no successful isolation of a phylogenetically close relative of the sulfur-metabolizing archaebacteria from an alkaline environment and, until recently, this line of descent was termed the thermoacidophilic branch of the archaebacteria. With the recognition that many representatives of this collection continue to grow or grow optimally near neutrality, the alternative designation proposed by Setter and Zillig, based on sulfur utilization, might seem a more comprehensive description. However, we point out that most if not all methanogenic archaebacteria so far characterized have the capacity to substitute sulfur as an electron acceptor for the oxidation of hydrogen (Stetter and Gaag, 1983); *Sulfolobus* grows well heterotrophically (Brock, 1978). Thus, the designation of sulfur utilization may not address a fundamental biochemical difference between these two deep branchings of archaebacteria. Instead, the broad distribution of sulfur-metabolizing capabilities may reflect the ancient origin of sulfur-based energetics. The Octopus Hot Spring is a low-sulfide environment, suggesting that this archaebacterium, like the *Thermus* relatives, is growing heterotrophically.

5.3. The Symbionts and Their Invertebrate Hosts

Although microscopic examinations have indicated that the invertebrates we studied might harbor more than one type of symbiont (J. Tuttle, personal communication), we recovered only one procaryotic 5S rRNA from each; each animal is associated with a unique (or nearly so) eubacterial symbiont. As shown in Fig. 5, the symbionts are among the group III purple photosynthetic bacteria, which also include *Escherichia coli* and *Pseudomonas fluorescens* (Gibson *et al.*, 1979). Because of these phylogenetic affinities, the fundamental biochemical properties of the symbionts must be similar to those of the enterics and classic pseudomonads. The closest relatives of the symbionts among the free-living, sulfur-oxdizing chemoautotrophs in our 5S sequence collection are *Thiomicrospira pelophila* and L-12 (a *Thiomicrospira*-like hydrothermal vent isolate; Ruby and Jannasch, 1982).

The marine invertebrates inspected, *Riftia*, *Calyptogena*, and *Solemya*, each of course yielded a eucaryotic (host) 5S rRNA in addition to that of the symbionts. The 5S rRNA sequence homologies of the host animals with some other closely related invertebrates (a mussel, *Mytilus;* a brachiopod, *Lingula;* a snail, *Arion;* and two annelids, *Sabellastarte* and *Perinereis*), and *Drosophila* are presented as a homology matrix (Table I). We do not cast phylogenetic trees with these data because we are not confident in the precision of metazoan phylogenies based on 5S rRNA sequences. As emphasized above, the small size of this RNA, coupled with the statistical uncertainty (indicated in Table I) implicit in the fact that so few nucleotide changes are found between these eucaryotic 5S rRNAs, renders such affinities difficult to resolve unambiguously.

The symbiont hosts examined here are all molluscan in aspect, as indicated in the table. This was clearly to be expected for *Calyptogena* and *Solemya,* since both are conspicuously bivalves. Such affiliation for *Riftia,* however, was unexpected. It is clearly a representative of the phylum Pogonophora by morphological criteria; however, the phylogenetic status of that phylum has been in some dispute (Jones, 1981). Alternatively, they have been declared relatives of the hemichordates (deuterostomes) or annelids (protostomes). Our analysis clearly weighs strongly in favor of a protostome identity for *Riftia,* with a preference, however, for a molluscan rather than an annelid affiliation.

One other noteworthy inference can be drawn from this 5S rRNA analysis. The symbioses involving *Riftia, Calyptogena,* and *Solemya* appear to have been established independently by free-living procaryotes. Given the close evolutionary associations among the invertebrates relative to those among their procaryotic symbionts, it seems unlikely that any one host–procaryotic partnership (e.g., *Riftia*) coevolved to give rise to the others.

5.4. The Chino Mine Pond

The copper leaching pond atop the Chino dump number 1 has so far yielded two distinct 5S rRNA sequences. Their phylogenetic position also is indicated in Fig. 5; one of these is identical to the *Thiobacillus ferrooxidans* 5S rRNA sequence already in our sequence collection (ATCC19859). Since prior observations at the Chino site (J. Brierly, personal communication) had indicated that *T. ferrooxidans* was the predominant bacterial species present, this result was not unexpected. The second 5S rRNA sequence, indicated in Fig. 5 as Chino 2, is more difficult to "identify" in a physiologically relevant fashion; no close relatives of this organism are represented in our data collection. It may represent another iron-oxdizing organism, such as the facultatively thermophilic iron oxidizers TH1 and TH3, which have been isolated from this same environment (Norris *et al.,* 1980). Chino 2 may instead represent a sulfur oxidizer related in physiology to *T. thiooxidans.* Based on colony counts,

Table I. Homologies between Symbiont Host 5S rRNA Sequences and Those of Other Invertebrates[a]

Organism	Fractional homology with given organism								
	R. pa.	C. ma.	M. ed.	S. ve.	L. an.	A. ru.	S. ja.	P. br.	D. me.
Riftia pachyptila[b]	—	0.98 ± 0.01	0.96 ± 0.02	0.95 ± 0.02	0.92 ± 0.03	0.87 ± 0.03	0.90 ± 0.03	0.92 ± 0.02	0.83 ± 0.03
Calyptogena magnifica	0.99 ± 0.01	—	0.98 ± 0.01	0.97 ± 0.02	0.93 ± 0.02	0.89 ± 0.03	0.88 ± 0.03	0.92 ± 0.02	0.83 ± 0.03
Mytilus edulis	0.96 ± 0.02	0.98 ± 0.02	—	0.94 ± 0.02	0.94 ± 0.02	0.90 ± 0.03	0.86 ± 0.03	0.90 ± 0.03	0.85 ± 0.03
Solemya velum	0.96 ± 0.02	0.97 ± 0.02	0.95 ± 0.03	—	0.96 ± 0.02	0.89 ± 0.03	0.90 ± 0.03	0.92 ± 0.02	0.83 ± 0.03
Lingula anatina	0.93 ± 0.03	0.94 ± 0.03	0.95 ± 0.02	0.96 ± 0.02	—	0.89 ± 0.03	0.86 ± 0.03	0.91 ± 0.03	0.82 ± 0.04
Arion rufus	0.89 ± 0.04	0.90 ± 0.03	0.91 ± 0.03	0.89 ± 0.03	0.90 ± 0.03	—	0.82 ± 0.04	0.88 ± 0.03	0.87 ± 0.03
Sabellastarte japonica	0.92 ± 0.03	0.91 ± 0.03	0.89 ± 0.04	0.92 ± 0.03	0.88 ± 0.04	0.84 ± 0.04	—	0.87 ± 0.03	0.85 ± 0.03
Perinereis brevicirris	0.94 ± 0.03	0.94 ± 0.03	0.92 ± 0.03	0.93 ± 0.03	0.92 ± 0.03	0.89 ± 0.04	0.90 ± 0.03	—	0.82 ± 0.04
Drosophila melanogaster	0.86 ± 0.04	0.86 ± 0.04	0.87 ± 0.04	0.85 ± 0.04	0.84 ± 0.04	0.87 ± 0.04	0.87 + 0.04	0.84 + 0.04	—

[a]The homologies and one sigma uncertainties are given for pairwise combinations of the host sequences and those of related invertebrates, omitting the region of 3'-terminal length variation. The uncertainties are based upon the binomial counting uncertainty in the number of sequence differences (Hori and Osawa, 1979). The values in the upper right half of the table are calculated by giving all sequence positions equal weighting. The values in the lower left half of the table assign base paired positions (see Fig. 2) one-half of the weight of unpaired positions (see Fig. 3 legend). The generally higher homologies in the lower left half of the table reflect the preferential conservation of single-stranded regions (Mackay et al., 1982), while the larger uncertainties reflect the smaller number of positions that are assumed to be indepently varying and hence poorer (though more realistic) counting statistics.

[b]Based on the dominant nucleotide at each heterogeneous position.

T. ferrooxidans and *T. thiooxidans* are present in the Chino leaching liquor in about a 10:1 ratio. *Thiobacillus thiooxidans* (two strains, ATCC 808S and Sulfur Spring, having essentially the same 5S rRNA sequences) and *T. ferrooxidans* (ATCC 19859) are very closely related at the 5S rRNA level. DNA homology comparisons (Harrison, 1982) indicate that certain strains of *T. ferrooxidans* are actually more closely related to *T. thiooxidans* strains than they are to other so-called *T. ferrooxidans*. Thus, we are reluctant to assign phylogenetic significance to the trait of iron oxidation. In any case, the Chino 2 5S rRNA sequence is not closely related to any of the other thiobacilli whose 5S rRNA sequences are known.

6. Population Analysis Using 16S rRNA Genes

6.1. Overview

The communities discussed above are relatively simple in the number of their major constituents, and thus they are amenable to direct isolation and fractionation of 5S rRNAs. The practical limitation on this methodology in terms of community complexity has not yet been established, although the fractionation of ten or so unique species of 5S rRNA is easily within the range of the analysis. However, the 5S rRNA method is not sufficiently sensitive to detect all minor components of mixed populations. We therefore are developing an alternative approach, which seems to have no limitations due to either population complexity or the abundance of community members. This approach involves the direct cloning of 16S rRNA genes (rDNA) from DNA collected *in situ*.

The characterization of natural populations by cloned DNA sequences has several practical advantages over the analysis using 5S rRNA:

1. The 16S rRNA (or its gene) is much larger than the 5S rRNA ($\sim$1500 versus $\sim$120 nucleotides), permitting a more precise statistical determination of evolutionary affiliation over both short- and long-range evolutionary spans (close and distant relatives).
2. DNA is more stable than RNA to nuclease and chemical degradations that might be encountered during isolation from natural samples.
3. Mixtures of different rRNA genes are readily separated as recombinant phage clones.
4. In principle, there is no limit on the complexity of the population examined. Retrieval of minor population members only requires screening more clones.

5. Rapid sequencing methods for cloned DNA are substantially easier and more accurate than sequencing end-labeled RNA.
6. Since cloned rDNAs are amplified as recombinant phage prior to sequencing, smaller amounts of initial biomass are required in principle than for recovering isolated 5S rRNA.
7. Given a cloned rRNA gene, any other gene from the organism contributing that 16S rDNA may be retrieved from the original mixed-organism recombinant pool by "chromosome walking."

The overall approach using cloned 16S rRNA genes is outlined in Fig. 2.

6.2. The 16S rRNA

The full nucleotide sequences of the 16S-like rRNAs of about 25 diverse organisms, representative of each of the kingdoms and organelles, are now known. This information, coupled with the many partial 16S sequences accumulated by Woese and his collaborators over the past decade, provides a reasonably detailed picture of the molecule, at least in terms of its primary and secondary structures. Although we use the term "16S" as a generic one, the small-subunit rRNAs vary somewhat in size, say 15S–18S, corresponding to ~1500–1900 nucleotides in chain length. Mitochondrial small-subunit rRNAs are quite degenerate: some are shorter than 1000 nucleotides. There is substantial variation in the primary structures, but all of these are compatible with a common secondary structure (Woese *et al.,* 1983). This is illustrated in Fig. 7, which shows folded schema for the 16S rRNAs from the three primary kingdoms, as well as a "minimal" 16S rRNA, which includes only structures contained in all 16S rRNAs, including the mitochondrial and chloroplast versions. The folded structures from the three kingdoms are essentially superimposable over much of their lengths. This uniformity of higher order structure in 16S rRNA for all organisms is powerful testimony to its antiquity and functional constancy.

The 16S rRNA, like 5S rRNA, except more so, is seen to consist of multiple helix–loop domains. The larger "single-strand" segments or loops appear devoid of structure in Fig. 7, but these regions likely are engaged in dense tertiary structure, either local or more long range. Little is known of the molecular functions of these various domains, indeed of the entire molecule. Some of the domains probably function individually during protein synthesis; others may act in concert. Length variation in compared 16S sequences generally does not involve scattered, small insertions or deletions. Rather, entire domains, large sequence blocks presumably responsible for discrete functions, are added or eliminated (cf. the eucaryotic folding in Fig. 7). Phylogenetic comparisons based on

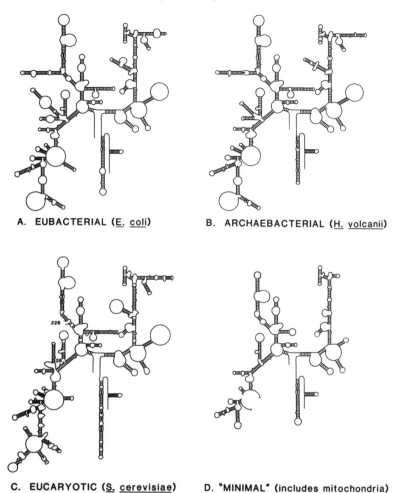

A. EUBACTERIAL (E. coli)   B. ARCHAEBACTERIAL (H. volcanii)

C. EUCARYOTIC (S. cerevisiae)   D. "MINIMAL" (includes mitochondria)

Figure 7. The 16S rRNA foldings for the three primary kingdoms, and a "minimum" small-subunit rRNA. The small-subunit rRNA sequences for the organisms indicated are schematically shown folded into their common pattern. Two sections of the *S. cerevisiae* structure, containing 25 and 226 nucleotides as indicated, were omitted. These regions vary extensively among the eucaryotes. The "minimal" folding includes only structures common to all three lines of descent, as well as the mitochondria. [Adapted from Woese *et al.* (1983), with permission.]

16S rRNA nucleotide sequences therefore must not be considered as single-marker comparisons, but as comparisons using numerous functional domains. This would be analogous to comparisons simultaneously using amino acid sequences of several different proteins from the same metabolic pathway.

Although there is substantial variation in homologous stretches of compared 16S rRNAs, the distribution of changes is not random. Some segments are perfectly conserved, or nearly so, while others drift freely. The constant sequences are the keys to defining reasonably readily phylogenetic affiliations by 16S rRNA structure. We have synthesized a series of DNA oligomers complementary to these constant sequences for service as initiating sites for enzymatic primer extension methods of sequencing. As discussed below, DNA polymerase is used to sequence 16S rRNA genes cloned from mixed, natural populations. Reverse transcriptase is used for the analysis with 16S rRNA templates from pure cultures at hand. Although in principle the "universal primers" can be used to sequence the entirety of a 16S rRNA, complete sequence information is not necessary for phylogenetic analysis.

6.3. 16S rRNA and Phylogeny

Comparisons of partial 16S rRNA sequences already have provided the most far-reaching of any phylogenetic explorations. These are the studies of Woese and his colleagues, who began in the late 1960s to accumulate partial 16S rRNA sequence data, specifically for defining evolutionary relationships. At that time it was not realistic to sequence 16S rRNA in entirety, and even 5S rRNA was not a light undertaking. They therefore characterized the 16S rRNAs by "oligonucleotide cataloging." In this procedure, uniformly ^{32}P-labeled 16S rRNA is isolated from cells grown in the presence of [^{32}P]orthophosphate, then digested exhaustively with RNase T_1. This enzyme cuts at the 3' side of all G residues, and so results in a collection of oligonucleotides which consist of variable amounts (and sequences) of A, C, and U and a single G residue at the 3' end of each oligonucleotide; e.g., AUCACCG or UUCAUAG. The collection of oligonucleotides in such digests can be separated by two-dimensional electrophoresis according to size, composition, and sequence, and individual oligonucleotides recovered and sequenced by digestions using RNases with base specificities different than RNase T_1 (Sanger *et al.*, 1965). The collection of sequences of oligonucleotides from a given organism (the "catalog") is diagnostic of that organism and it provides criteria for relating different organisms.

Woese and colleagues have now characterized the 16S-like rRNAs of over 300 organisms and organelles by oligonucleotide catalogs. They have focused on oligonucleotides of length six or more residues, which almost certainly are unique in a given molecule and likely derived from homologous sequences in compared molecules. In sum, these larger oligomers span ~25% of the entire 16S rRNAs. Organisms can be numerically related, in a pairwise manner, by an association coefficient, a so-called S_{AB} value, which is calculated according to

$$S_{AB} = 2(\overline{AB})/(A + B) \tag{1}$$

where $\overline{AB}$ is the number of residues in oligonucleotides of six or greater length that are *common* to the pair of organisms, and A and B are the numbers of residues in *all* oligonucleotides of length six or more for the respective organisms. The use of S_{AB} values to define phylogenetic relationships have been well reviewed. As already pointed out, the work resulted in the first credible phylogeny of procaryotes and the recognition of the archaebacteria as a primary line of evolutionary descent, as discrete in rRNA homology grouping as either the eubacteria or the eucaryotes (Fox *et al.*, 1980).

The evaluation of phylogenetic associations by S_{AB} analysis has been enormously useful, but establishing an oligonucleotide catalog for a 16S rRNA is difficult experimentally. Moreover, the analysis sacrifices substantial sequence information. Because of the frequency of G occurrence, hence RNase T_1 cuts, most oligonucleotides considered are in the six- to ten-nucleotide class. The evolutionary alteration of most residues to G therefore results in the loss of the oligonucleotide from the data collection because the oligonucleotide then is cast into the class of less than six residues in length, which is not scored in the analysis. Thus, most single position changes to G cost six or more positions of potentially comparable sequence.

The S_{AB} analysis focused upon more highly conserved oligonucleotide sequences, a focus which, to some extent, is convenient for the straightforward assignment of organisms to discrete groupings. However, now that continuous sequence collection is experimentally tractable—more tractable in fact than catalog analysis—it is expected that the primer extension sequencing methods discussed here will be used most in the foreseeable future. Nonetheless, the Woese collection of oligonucleotide catalogs will remain an important resource in the mapping of procaryotic phylogeny.

6.4. Isolating and Cloning DNA from Environmental Samples

The difficulties of extracting DNA from many microorganisms (cyanobacteria, filamentous fungi, etc.) are well documented (e.g., Garber and Yoder, 1983). The application of any generalized extraction protocol to a mixed, natural population must surely bias to some extent the species composition of the DNA isolated. Differential isolations of DNA are a much more severe problem than with 5S rRNA. In fact, 5S rRNA, because of its small size, generally can be isolated from microorganisms even without formal breakage: direct extraction with phenol and sodium dodecyl sulfate at elevated temperatures usually yields low-molecular-weight RNAs. Also, because 5S rRNA is not susceptible to shear, vigorous disruption procedures, such as repeated passage through a French pressure cell or grinding with alumina, are applicable. In the case of preparing DNA for recombinant phage construction, however, cell lysis must be effected and the retrieved DNA must be of relatively high molecular weight; i.e., it must undergo minimum shearing forces.

There are many DNA extraction protocols in the literature (Parish, 1972), most of which are applicable to microbial masses obtained from environmental samples. However, it is critical that the effectiveness of any disruption procedure be determined in pilot experiments, using microscopic examination for lysis. A review of cell lysis protocols is not the province of this chapter. We note, however, that in the case of cloning DNA from Octopus Spring, treatment of biomass with high levels of lysozyme, followed by sodium dodecyl sulfate at evaluated temperature, proved effective. A technique we are developing (D. Ward, unpublished results) which may prove generally applicable is a single pass through a French pressure cell at high salt concentrations and reduced pressures, conditions which minimize DNA shearing.

To ensure equal representation of all portions of the mixed-population genomes in recombinant phage, it is necessary to produce a more or less random population of DNA fragments of appropriate size and with termini complementary to an existing cloning site in the phage vector DNA. This can be realized in several ways. The usual approach is to partially digest the DNA with a restriction endonuclease common to one of the phage cloning sites. This derived population of fragmented DNA is sized, either on sucrose gradients or on agarose gels. Fragments of the desired size range (usually as large as the phage can accommodate) are covalently coupled to flanking halves (arms) of the phage genome by the action of DNA ligase. Our initial cloning efforts with natural population DNA have used a phage λ L47 derivative, one of the first vectors designed to accommodate large DNA inserts at several useful (i.e., common) restriction endonuclease sites.

Concerns with using restriction endonucleases to fragment DNA isolated from the environment relate to modifications, such as methylation, or coisolated factors that prevent the restriction of a fraction of the isolated DNA. To alleviate part of this concern, the restriction endonuclease used is refractory to the common restriction modification of methylation within its tetranucleotide recognition sequence. We have had good success using the restriction enzyme *Sau*3A, which cleaves both methylated and unmethylated DNA. The termini generated are compatible with the termini generated in the phage vector DNA by cleavage with the hexanucleotide recognition restriction endonuclease *Bam*H1. Since *Sau*3A cleaves DNA at a tetranucleotide recognition sequence, and thus cuts on average every 256 nucleotides, the products of *partial Sau*3A digestion represent a near-random collection of overlapping genomic fragments. In the case of the vector currently in use, partial digestion fragments ~15 kb in size are convenient.

For the maximum recovery of recombinant molecules, the products of ligating natural population DNA with the vector arms are packaged *in vitro* into phage λ heads prior to infecting cells. This technique, in general terms, makes use of complementary phage mutants, each maintained as a lysogen and deficient in an element required for phage head assembly or DNA packaging. Thus, each lysogenic mutant, when induced, accumulates phage components and also fails to excise and replicate its own DNA. When extracts of induced cells are combined, the complementation of components allows the completion of phage head assembly and the packaging of added recombinant phage DNA. We have used the Sternberg strains (NS433 and NS428) with good success (Enquist and Sternberg, 1979). About 0.5% of input recombinant phage genomes are expressed as plaques. Although this value may appear low, it is superior by at least an order of magnitude in efficiency to cloning protocols that rely upon transformation for the retrieval of recombinant phage.

6.5. Detecting and Sequencing rDNA from Natural Populations

Recombinant phage clones carrying an inserted "passenger" DNA of interest are normally detected by hybridization of a radioactive probe complementary to the insert. The hybridization is carried out against a membrane filter replica of a plate containing many recombinant phage plaques. A potential problem in examining natural populations, since we do not necessarily know the evolutionary diversity of extant life, is the identification of rDNA-containing clones derived from uncharacterized microbial forms. As has been emphasized, the ribosomal RNAs are highly conservative molecules with respect to both higher order structure and nucleotide sequence. They contain sequences that have remained unchanged, or nearly so, since the divergence of the three primary lines of evolutionary descent. This conservation of primary structure is used in the determination, by DNA–RNA hybridization, of those phage clones that contain rDNA.

A probe we have inspected in some detail for general applicability is a "mixed-kingdom" probe, a mixture of 16S rRNA derived from a representative of each kingdom. Our initial characterizations have used a fragmented mixture of the 16S rRNAs of *Escherichia coli* (eubacterial), *Sulfolobus solfataricus* (archaebacterial), and *Dictyostelium discoideum* (eucaryotic). The rRNAs are reduced by mild alkaline hydrolysis to fragments a few hundred nucleotides long to enhance incorporation of label introduced with polynucleotide kinase and $[\gamma\text{-}^{32}P]ATP$ at available 5′ termini. Hybridization conditions of sufficiently low stringency (conditions that stabilize hybrids that contain some mispaired bases) can detect 16S rRNA sequences over broad phylogenetic distances. Minimally, it is essential that the rRNA representative of one kingdom cross-hybridize with all other representatives of that kingdom. Since we cannot exhaustively inspect all representatives of a specific kingdom, this criterion is best demonstrated by the the ability to detect hybridization to homologous sequences across kingdom-level divisions. An examination of this is shown in Fig. 8. Here, base-fragmented and labeled 16S rRNAs from the aforementioned kingdom representatives are hybridized individually, under conditions of low stringency, to cloned DNA containing the small-subunit rRNA genes of *Paramecium, Trypanosoma, Rattus, Dictyostelium, Escherichia coli,* and *Sulfolobus solfataricus.* Conditions for low-stringency hybridization have been optimized to reduce background hybridization to nonribosomal DNA; this is represented by the combined "plus strand" DNA target. [The cloned rRNA genes here examined are carried in one of the single-strand filamentous phage M13 vectors (Messing, 1983), and so can be expressed as either "plus" or "minus" strands. The minus strand serves as the template for transcription of the rRNA and therefore is complementary to the fragmented rRNA.] The noncomplementary, plus strand provides a convenient control for adjusting conditions of hybridization stringency. Hybridization conditions are described in the legend to Fig. 8.

The *E. coli* small-subunit rRNA demonstrates cross-kingdom hybridization only with *Sulfolobus* rDNA and therefore is apparently restricted to the identification of procaryotic 16S rDNAs. However, both the *Sulfolobus* and the eucaryotic rRNAs demonstrate hybridization to all three kingdoms. With good confidence, then, the "mixed-kingdom" probe will identify the small-subunit rRNA genes from the representatives of all three lines of descent.

The characterization of rDNA sequences from mixed populations first requires sorting among the many recombinant phage that hybridize the mixed-kingdom probe to select those that are truly rDNA-containing and are unique in the mixed population. In principle, recombinants containing identical cloned rDNA segments can be identified by comparing restriction enzyme digests of recombinant phage DNAs. Since scores of clones must be inspected, however, this approach is unacceptably expensive and cumbersome.

rDNA

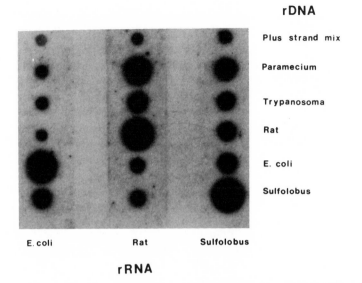

Plus strand mix

Paramecium

Trypanosoma

Rat

E. coli

Sulfolobus

E. coli   Rat   Sulfolobus

rRNA

Figure 8. Cross-kingdom hybridization tests. The small ribosomal subunit rRNA genes from the indicated organisms were cloned and propagated in bacteriophage M13mp8 or M13mp9 and the single-strand DNAs isolated (Messing, 1983). Fifty ng of each "minus"-strand (rRNA complement) was spotted onto a nitrocellulose sheet in $2\times$ SSC (Maniatis *et al.*, 1982); 50 ng of each of the respective "plus"-strand clones was mixed and applied to the nitrocellulose (the combined plus-strand control). The filters were washed in $2\times$ SSC, baked ~2 hr at 65°C, wetted in $6\times$ SSC, and prehybridized for 3 hr at 60°C in 0.25 ml/cm^2 of 50% formamide/0.9 M NaCl/10 mM Na phosphate, pH 6.8/1 mM EDTA/0.5% sodium dodecyl sulfate/$10\times$ Denhardt's solution lacking bovine serum albumin/10 μg/ml polyadenylic acid. The small-subunit rRNAs from the indicated organisms were fragmented to ~100 nucleotides by limited alkali treatment (Donis-Keller *et al.*, 1977) and 5′ end-labeled using [γ-^{32}P]ATP and polynucleotide kinase. About 1×10^7 cpm of labeled RNA/cm^2 at $\sim2\times10^8$ cpm/μg was added to each prehybridization solution following the prehybridization step. Hybridization occurred during slow cooling from 60 to 28°C over a period of ~20 hr. The membrane filters were washed at room temperature with several changes of $5\times$ SSC/0.1% sodium dodecyl sulfate, then dried and exposed to film for autoradiography.

The sorting and initial characterization of the 16S-containing clones from natural populations are done most conveniently by sequencing. The sequence determinations make use of well-described dideoxynucleotide chain termination protocols (Sanger *et al.*, 1977; Biggin *et al.*, 1983), in which synthesis is initiated by DNA polymerase or reverse transcriptase from a specific "priming" oligonucleotide complementary to a unique segment of DNA adjacent to the region sequenced. Chain elongation is terminated at either A, G, T, or C residues by inclusion of the respective dideoxynucleotides in the reaction mixes, and the products are resolved on high-resolution polyacrylamide gels, as described for the 5S RNA sequence determination (Stahl *et al.,* 1984).

As mentioned above, to facilitate sequencing rRNA genes we have synthesized a series of DNA oligonucleotide "primers" (15–20 nucleotides long) that are complementary to 16S rRNA sequences that are constant, or nearly so, across the primary kingdoms. Three of these regions in the 16S rRNA are indicated in Fig. 9. Thus, sequence blocks of 300–400 nucleotides are readily accessed from each primer. Figure 10 shows a series of dideoxynucleotide-terminated sequencing reactions using one of these "universal rRNA primers" with templates containing the 16S rDNA from one representative of each of the three kingdoms.

It has proven possible, using end-labeled primers rather than labeled dNTPs, to sequence directly from recombinant λ DNA rather than subcloning into single-strand phage M13, the conventional approach (N. R. Pace, D. A. Stahl, D. J. Lane, and G. J. Olsen, unpublished observation). In the case of sorting among clones for unique representatives, sequencing reactions need be carried out with only one of the chain-terminating nucleotides, using only one of the primers. Since the specific primers ensure that only homologous stretches of 16S rDNA are inspected, the patterns (e.g., of T residues on sequencing gels) readily identify clones containing different 16S rRNA genes. These are then submitted to more detailed analysis with each of the chain-terminating nucleotides and other primers. In general, complete 16S sequence determinations are not necessary. Use of three primers yields ~1000 nucleotides of sequence information, quite adequate for even detailed phylogenetic analysis. In fact, a sequence of ~300–400 nucleotides, obtainable from a single priming site, is sufficient for organism identification.

As with the 5S rRNA analysis, the derived 16S rDNA sequences are cast into phylogenetic trees, using available complete and partial 16S rRNA sequences. At this time we are limited in the available 16S rRNA reference sequences. Only about 20 full 16S rRNA sequences are available, but we are engaged in accumulating partial sequences from diverse organisms in culture for reference purposes. For this we again are using the above-mentioned "universal" primers and dideoxynucleotide sequencing, but using reverse transcriptase and the 16S rRNA as template (D. J. Lane, D. A. Stahl, G. J. Olsen, and B. Pace, unpublished results). Thus, the same sequences are read off the rRNA as are derived from the cloned rDNAs from natural populations. This effort to expand the reference data base will go on over the next years. We anticipate that deter-

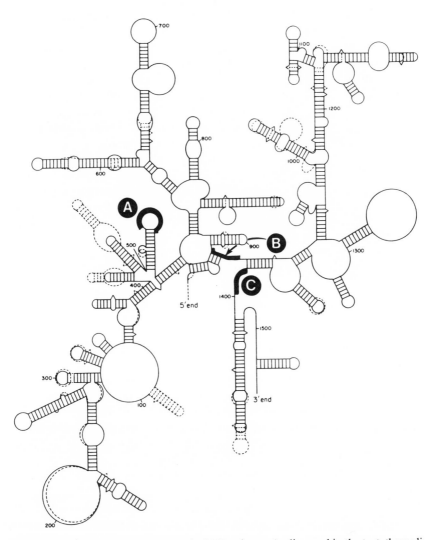

Figure 9. The "universal" small-subunit rRNA primers. As discussed in the text, three oligonucleotide primers are suitable for dideoxynucleotide sequence analysis, using as templates 16S–18S rRNAs or their genes, from all organisms. The positions of the complements of these primers in the folded 16S rRNA schema for (—) *Halobacterium volcanii* and (- - -) *E. coli* are shown. The three primers are complementary to (A) residues 519–536, (B) 906–920, and (C) 1392–1406, using the *E. coli* 16S rRNA nucleotide position numbering. The folded 16S rRNA structures are adapted from Gupta *et al.* (1983), with permission.

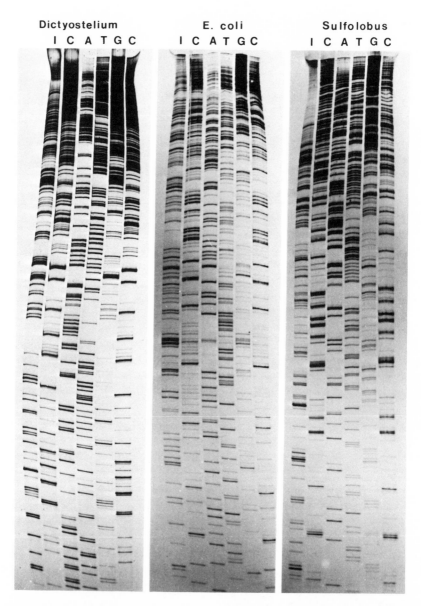

Figure 10. Sequence analysis with a "universal" rRNA primer. Primer C (Fig. 9) was used for dideoxynucleotide-terminated primer extension sequencing essentially as described by Biggin *et al.* (1983), using 16S–18S rRNA genes from each of the primary lines of descent, cloned into single-strand phage M13 (Messing, 1983). The organisms contributing the rRNA gene clones are indicated. *Solfolobus* is an archaebacterium, *E. coli* a eubacterium, and *Dictyostelium* a eucaryote. The letters C, A, T, and G indicate the dideoxynucleotide used to terminate chains in the reactions analyzed in that gel lane. The lane labeled I contained deoxyinosine triphosphate instead of deoxyguanosine triphosphate, but used dideoxyguanosine triphosphate as a chain-terminating analogue. Use of I instead of G alleviates band compression phonomena, which are sometimes problematic with the natural nucleotides.

mination of partial 16S rRNA sequences will become an important aspect of characterizing microbes, both procaryotes and eucaryotes, so reference sequences may accumulate quite rapidly.

Beyond the available partial and complete 16S rRNA sequences, an enormously useful fund of reference information is available in the RNase T_1 catalogs accumulated by Woese and his colleagues (Section 6.3). Since oligonucleotide sequences are implicit in the continuous sequence determined by the methods discussed here, the data collections are compatible for the correlation of organisms.

6.6. Estimation of Organism Abundance by 16S rRNA Gene Contents

We have not yet carried out quantitative analyses of mixed-population contents by 16S rRNA gene contents, but the approach is sufficiently straightforward that success seems promised. The notion is that, following sequence analysis of population contents, oligonucleotides *specific* for given 16S genes will be synthesized or recovered as restriction fragments from the cloned genes. Hybridization of an organism-specific sequence to bulk DNA derived from the natural sample then is a measure of the abundance of that gene in the original population. Hybridization of one of the above-mentioned oligonucleotide primers, complementary to all 16S genes so far as we know, provides an estimate of *total* rDNA present. The specific/total rDNA ratio really is only an approximate estimate of relative cell numbers, since rRNA gene copy numbers vary in different organisms. The inspected archaebacteria, for instance, have single copies of the rRNA genes; eubacteria commonly have 5–10 rDNA copies; and eucaryotes may have hundreds. Nonetheless, the amount of a specific rDNA in a mixed population likely reflects the contribution of that organism to the local metabolic potential.

6.7. 16S rRNA versus 5S rRNA as a Phylogenetic Tool

The question arises as to whether phylogenetic relationships inferred using 5S rRNA are the same as those derived using 16S rRNA sequences. Since phylogenetic relationships are defined by sequence homologies, this question reduces to one of whether 5S and 16S rRNA sequences diverge concomitantly. A useful span of complete 16S rRNA sequences is not yet available for this comparison. However, oligonucleotide catalogs of 16S rRNA overlap the 5S sequence collection to a useful extent. Figure 11 shows a semilog plot of 16S rRNA S_{AB} values versus 5S rRNA mutational distance K_{nuc} for several pairs of organisms (the plot is logarithmic because the calculated mutational distance for 5S rRNA, K_{nuc}, is an exponential function of mutational differences and hence of S_{AB} value). It indi-

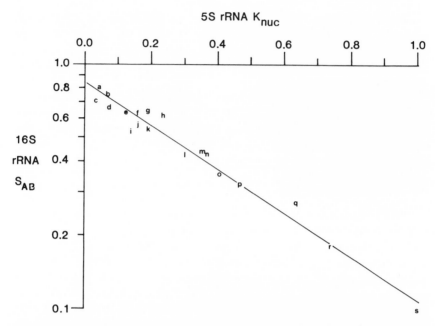

Figure 11. Empirical comparison of 16S rRNA catalog (S_{AB}) and 5S rRNA sequence (K_{nuc}) values. The S_{AB} values for organism–organism or group–group 16S rRNA, RNase T_1–oligonucleotide catalog comparisons have been extracted from the literature for which identical or comparable 5S rRNA sequence comparisons were also available. The comparisons, selected to give as broad a range of S_{AB} and K_{nuc} values as possible, are: (a) *Escherichia coli* vs. *Yersinia pestis;* (b) *Bacillus subtilis* vs. *B. megaterium;* (c) *E. coli* vs. *Proteus mirabilis;* (d) *B. subtilis* vs. *B. pasteurianum;* (e) *Methanococcus voltae* vs. *Mc. vannielii;* (f) *Methanobrevibacter smithii* vs. *Mb. ruminatium;* (g) *B. subtilis* vs. *B. stearothermophilus;* (h) *Methanogenium cariaci* vs. *Mg. marisnigri;* (i) purple photosynthetic bacteria, group II intragroup average; (j) enteric and *Vibrio* intragroup average; (k) *Bacillus* intragroup average; (l) *Bacillus* group vs *Lactobacillus* group; (m) purple photosynthetic bacteria, group I intragroup average; (n) purple photosynthetic bacteria, group III intragroup average; (o) *Clostridium thermosaccharolyticum* vs. *C. pasteurianum;* (p) purple photosynthetic bacteria intragroup average; (q) chloroplast (including cyanobacteria and *Prochloron*) intragroup average; (r) eubacterial intrakingdom average; (s) eubacterial vs. archaebacterial interkingdom average [Adapted from Lane (1983).]

cates that the S_{AB} values for 16S rRNA and sequence homologies for 5S rRNA yield consistent estimates of evolutionary distance, but with the peculiarity that the best-fit line does not extrapolate to the origin. In principle it must do so, since for a mutational distance K_{nuc} of 0, the S_{AB} value is 1.0. The most straightforward explanation for the deviation is that a small fraction of the 16S rRNA sequence is subject to very rapid accumulation of mutations, the remainder changing in concert with the 5S rRNA. These "fast clock" regions of the 16S sequence tend to be localized, and may prove useful for fine-structure phylogenetic mapping.

As emphasized, 16S rRNA is preferable to 5S rRNA because of the larger number of nucleotide positions used in homology evaluations. This means that the statistical error using 16S rRNA is substantially narrower than with 5S rRNA sequences. A plot of evolutionary distance versus sequence homology, indicating the error implicit in the use of both molecules, is shown in Fig. 12.

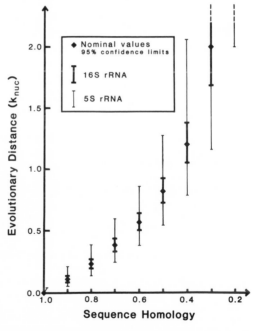

Figure 12. Relationship between observed homologies and the ability to estimate evolutionary distance. The inference of phylogenies from sequence data is dependent upon the accuracy with which the evolutionary divergence of the sequences can be estimated from the contemporary data. The accuracy with which the decay in sequence homology, and hence evolutionary distance, can be evaluated is dependent upon the number of independent sites over which the random fluctuations of the mutational process can be averaged. The figure illustrates the most probable evolutionary distance with its 95% confidence limits for several values of observed sequence homology between 5S rRNAs or 16S rRNAs. Because base pairing prevents many sequence positions from varying independently of one another (if one position is a C, its partner must be a G, etc.), the 5S rRNA has been taken to represent 80 independent positions and the 16S to be 1000 independent positions, about two-thirds of their actual lengths in each case. Although the confidence limits exceed the actual uncertainty in tree branching order, they illustrate the advantages of using a larger molecule for the inference of phylogenies. The figure also indicates the increase in distance uncertainty as the sequence homology falls (due either to a rapidly evolving molecule or to deep phylogenetic divisions).

7. Data Treatment

7.1. The Phylogeny of Organisms versus the Phylogeny of Molecules

The objective of the sequence determinations discussed above is to permit the inference of the relationships between the organisms from which the molecules were isolated. This procedure has several components, each with its own assumptions.

Initially, one must assume that the phylogeny of the molecules under study is the same as the phylogeny of the organisms possessing them. Two conditions must be met to satisfy this assumption. First, it is essential to study molecular species that are evolutionary and functional homologues in each of the organisms. If homologous molecules are to be found in all organisms, then ideally one should study a molecule that performs a universal function, such as DNA replication, transcription, or translation. The identification of homologous molecules can be difficult when examining molecules that are products of a multigene family; that is, when it is necessary to identify and select for use one specific member from among several similar molecules. Failure to accomplish this results in confusion of evolution within the gene family with the evolution of the organisms. Of the rRNAs, only the 5S rRNA appears to present this difficulty; as noted above, significantly different sequences can be found in a single organism. Minor heterogeneity in the high-molecular-weight rRNAs commonly occurs, but the number of sequence position changes does not significantly disturb phylogenetic calculations. However, as the rRNAs from additional organisms are characterized, one must remain aware of the possibility of this complication.

If a molecular phylogeny is to be extrapolated to the source organisms, it is also necessary that the genes for the molecules not be exchanged between species; i.e., the lateral transfer of the genes under study must be insignificant. The existence of transducing viruses and conjunctive plasmids with wide host ranges make this appear a substantial problem. It might seem easy for genetic information to cross species lines. The empirical fact is that this has not been observed to occur with the ribosomal RNA genes. The occurrence of lateral transfer can be inferred by comparing multiple, independent molecular phylogenies based on different cellular macromolecules. If they are consistent with a single phylogeny of source organisms, then it is highly unlikely that there has been significant lateral transfer of genetic information (Goodman, 1982).

With respect to rRNAs, one can rationalize the lack of observed transfer in terms of there being no selective pressure to pick up a new translation system; every viable organism (potential recipient) already has one of its own. Indeed, there is probably a selective disadvantage to replacing an important component of an existing translation system with one from a heterologous system (Woese, 1972). It is noteworthy here that organelles do not appear to share translation system components with their urcaryotic host, in spite of over a billion years of coexistence and an ability to exchange genetic information.

7.2. Sequence Alignments

After concluding that a molecular phylogeny can be taken to represent the phylogeny of the source organisms, the problem is reduced to inferring the molecular phylogeny. Two distinct elements are required for this: (1) a method for assessing the agreement between the sequence data and a given phylogenetic tree; and (2) a method for finding the phylogenetic tree that is most consistent, as defined by (1), with the sequence data. These elements are not unique to molecular phylogeny; they are required for any numerical taxonomy.

To construct a taxonomy, one evaluates a collection of "characters" by tabulating the "state" of each character displayed by each organism. The different characters from all the organisms are placed in one-to-one correspondence and compared; those characters that take similar states in various organisms are taken as evidence of specific affiliation of those organisms. To analyze nucleotide sequence data, each sequence position is taken to be a single character, which has four possible states (A, C, G, or U/T), one for each of the possible nucleotides at the sequence position. The occurrence of identical nucleotides at corresponding positions is taken as evidence of specific affiliation.

In general, when analyzing morphological or physiological characters, the correspondence of characters between the various organisms is obvious; there is little chance of accidently comparing the length of one rod-shaped bacterial species with the width of a second rod-shaped species. However, this is not straightforward with sequence data. Here, the characters are not the homologous molecules *per se,* but rather the homologous nucleotides, a fact not fully appreciated by those who consider rRNA homologies to be simply another (singular) character. Thus, it is necessary to define a one-to-one correspondence of sequence positions between the homologous molecules from each organism, a process referred to as "aligning" the sequences. This would be trivial if there were no length variations among macromolecular sequences. There are, however, two very different sources of sequence length variation: terminal length variation and internal length variation.

Terminal length variation is probably the least understood of the two and is the simplest to deal with. RNA and protein molecules are defined by corresponding regions of the genome, their genes. The precise extent of the DNA sequence represented in a mature molecule is defined by the details of the transcriptional, translational (in the case of proteins), and processing events giving rise to the molecule. These events, which define the endpoints of the molecules, are subject to variations that are independent of the DNA sequence of the gene itself. Therefore, we omit regions of terminal length variation from our analyses.

Internal sequence length variation is a consequence of nucleotide insertions or deletions in the course of evolution. At sites of internal length variation, one or more sequences will have a nucleotide for which there is *no* corresponding nucleotide in the other sequence(s). To establish a one-to-one correspondence of sequence positions it is necessary to introduce a fifth character state, an alignment gap, which is commonly represented as a hyphen (-) bridging the site of the alignment gap and symbolizing the covalent continuity of the sequence. Sequences are "aligned" by inserting alignment gaps into them as required to complete the one-to-one correspondence of positions (character traits) among the sequences being compared. It is important, however, to minimize the use of alignment gaps; unrestrained usage can "align" even random sequences.

The methodologies of sequence alignment are presently in the gray area between art and science. Although systematic procedures are increasingly available, most sequence alignment is still the product of experience, intuition, and many hours of staring at the data. The manual alignment of sequences is initiated by scanning the collection of homologous sequences for regions of similar or identical primary structure in roughly homologous positions. Then, alignment gaps are introduced to juxtapose the regions of homology in the various sequences. The procedure is repeated, and as the alignment becomes better defined it limits the regions that must be searched for homologies, permitting the identification of less obvious homologies. With nucleotide sequences that are identical in greater than 60% of their positions, this procedure is relatively straightforward.

When several sequences are aligned simultaneously by this procedure, it is fruitful to look for even very short regions (even a single nucleotide) that are present in similar positions in all sequences. The rationale for this is that the most important sequence features are highly conserved, and therefore provide landmarks in the alignment. This ability to consider preferentially sequence regions that are highly conserved is quite limited in automated alignment procedures.

After obtaining a preliminary sequence alignment based solely upon primary structure, it is possible to consider (in the case of RNA sequences) refining the alignment on the basis of conserved secondary structural features. Transfer RNA is the classic example of aligning sequences on the basis of their secondary structure. It has been standard practice to present tRNA sequences in the "cloverleaf" folding, and to present tRNA sequence alignments in the light of the folding pattern (Gauss and Sprinzl, 1984). Although few investigators would hesitate to infer the folding of a new tRNA sequence, the extension of secondary structural alignments to other molecules has been much slower in coming.

Increased acceptance of a universal 5S rRNA secondary structure has led to a greater confidence in the alignment of these sequences, as pointed out above, but this currently is the largest molecule for which a secondary structural model has received general acceptance. It has been our experience that the 16S rRNA secondary structure proposed by Woese, Noller, and their coworkers (Woese *et al.,* 1983) is now sufficiently detailed to permit rapid secondary structural alignment of eubacterial and archaebacterial small-subunit rRNA sequences as well. Thus, we find that both of the molecules emphasized in this chapter as phylogenetic tools are amenable to secondary structural alignment on existing models.

With the rRNAs, the alignment of a new sequence with an existing secondary structural model proceeds in much the same way as does a tRNA alignment. Nucleotide sequences are progressively superimposed onto the model structure until one comes to a region of length variation (most often a hairpin loop). Another landmark of conserved nucleotide sequence or universal pairing region is sought; then one works backward from the new landmark until reaching the region of length variation from the other side. Finally, one again works forward from the landmark. Generally, the overall alignment proceeds in an iterative manner. As the alignment improves, additional regions of conserved primary or secondary structure can be recognized.

Substantial work has been done on automated methods of sequence alignment. In particular, numerous authors have contributed to a family of algorithms often referred to as NWS (Needleman–Wunsch–Sellers) analyses (Smith *et al.,* 1981; Goad and Kanehisa, 1982; Fitch and Smith, 1983; Waterman, 1983). These algorithms provide a method of finding the "optimal" alignment between two sequences, where the quality of the alignment is defined by a series of coefficients which assign favorable scores to the occurrence of the same nucleotide in corresponding positions of the two sequences, and unfavorable scores to the occurrence of differing nucleotides in corresponding positions of the two sequences or to any alignment gaps. It has been our general experience, however, that automated alignments should be only a first approach, and any final result must be modified by subjective considerations, such as keying on secondary structure or other conserved features.

7.3. Estimating Evolutionary Distance

Methods of evaluating phylogenies from aligned sequences generally fall into one of two categories [see Penny (1976) and Felsenstein (1982) for more detailed lists]: methods that seek a phylogeny consistent with the smallest total number of mutational events (minimum change or maximum parsimony methods), and methods that estimate the total number of mutational events separating pairs of sequences and then seek the phylogeny most consistent with these pairwise distances (matrix or evolutionary distance methods). We have chosen to use an evolutionary distance method for the inference of molecular phylogenies. We find that this method avoids certain systematic errors in the reconstruction of phylogenies of sequences with very different rates of evolution (Olsen, 1983).

The first step of a matrix-type analysis is the estimation of the evolutionary distance that separates each pair of contemporary sequences. In this context, the evolutionary distance separating two sequences will be defined as the average number of independent, fixed mutations per sequence position that have occurred since the divergence of the sequences. Since the actual evolutionary history is unknown, this value must be estimated from the amount of homology between the contemporary sequences.

After sequences are aligned with one another, their "fractional homology," defined as the fraction of the positions at which the two aligned sequences have identical nucleotides, is computed. That is,

$$H = M/(M + U) \tag{2}$$

where H is the fractional homology, M is the number of sequence positions with matching nucleotides in the two sequences, and U is the number of sequence positions with nonmatching nucleotides. However, this definition does not account for alignment gaps.

As noted above, due to differences in length, it generally is not possible to establish a strict one-to-one correspondence between the nucleotides of two sequences without the introduction of alignment gaps (Section 7.2). The treatment of these sequence gaps in the evaluation of sequence homology (or evolutionary distance) is somewhat aribtrary, since there is no intrinsic similarity between length changes (insertions and deletions) and nucleotide substitutions. It is most common to treat these positions as either one or one-half of a base substitution (mismatched corresponding nucleotides). We generally have chosen the latter weighting, since the "position" in question is only present in one of the two sequences, and therefore is effectively one-half of a position. However, the number of gaps in a credible alignment is usually small compared to the number of mismatched bases, so the exact choice of treatment is not critical. With this treatment of deletions, the fractional homology becomes

$$H = M/(M + U + G/2) \qquad (3)$$

where G is the number of sequence positions that have an alignment gap in one sequence opposite a nucleotide in the second sequence.

A special case of sequence length variation is the occurrence of large insertions and deletions. These are likely the result of single, rare events rather than the compounding of large numbers of single-nucleotide events. If they represent single events, then it is not appropriate to weight them overly much in the evaluation of sequence divergence. For this reason, only the first five sequence gaps in a long string of gaps are counted in our sequence comparisons. Alternatively, regions subject to substantial length variation may be simply excluded from analyses, since a "gap" cannot be viewed as the functional homologue of a region of sequence.

As molecular sequences acquire significant numbers of mutations, there is a finite probability of multiple mutations occurring at the same site. Thus, if an adenine in an ancestral sequence mutates to become a cytosine, then a second mutation changing it to a guanine would not add any additional sequence difference and thus would not decrease the sequence homology. Similarly, if the second mutation converted the cytosine back to adenine, then the homology actually would be increased. The net effect of multiple mutations is to lower the amount of sequence difference relative to the actual evolutionary distance. In the extreme case, when the number of mutations approaches infinity, the number of sequence differences will vary randomly about an average value of 75% of the sequence positions.

Jukes and Cantor (1969) proposed that this tendency of multiple mutations to obscure the total number of mutational events can be partially compensated. If it is assumed that mutations are independent events and that all of the positions of a nucleotide sequence are independent, equally mutable, and capable of accepting any of the four nucleotides with equal ease, then a formula can be derived to estimate the actual number of mutations giving rise to the number of observed sequence differences. Specifically,

$$x = -(3/4) \ln[(4/3)(H - 1/4)] \qquad (4)$$

where x is the estimated evolutionary distance, ln is the natural logarithm, and H is the fractional sequence homology [see Eq. (3)]. For very similar sequences, the evolutionary distance estimated with this formula is only slightly different from one minus the homology. However, as the homology between the sequences approaches 25% [i.e., random for sequences with a 1:1:1:1 (A:C:G:U) base composition], then this distance metric approaches infinity. Although there are clear oversimplifications in the assumptions used to derive this distance metric, they certainly are preferable to ignoring multiple mutations. A critical examination of the errors in the assumptions suggests that they undercompensate, rather than introduce random artifacts or overcompensation (Fitch, 1976). Therefore, all of our estimations of evolutionary distance are calculated from the sequence homologies by this formula.

7.4. Construction of Phylogenetic Trees

The second major step in constructing a molecular phylogeny is finding the tree geometry and branch lengths that best fit the sequence data. Computational approaches to this are abundant in the literature, and to some extent are controversial. The calculation of tree geometry is an optimization problem. Unfortunately, this particular problem falls into a category for which there is no systematic method known for finding the solution in less than an amount of time which grows exponentially with the number of organisms (Karp, 1972).

The number of distinct phylogenetic trees with N organisms can be shown by mathematical induction to be

$$(2n - 5)!/2N - 3(N - 3)! \tag{5}$$

where "!" is the factorial operator. Although the number of possible tree geometries is a reasonably tractable 2×10^6 with ten sequences, it rises rapidly to 8×10^{12} and 2×10^{20} trees with 15 and 20 sequences, respectively. Therefore, in general, it is not practical, to test all the possible trees and choose the best. It is necessary to use an algorithm that systematically seeks better solutions, in the hope that it will ultimately find the best. Most methods approach the problem by testing a systematic set of rearrangements of the current tree, seeking one that is better. If no better tree is found within this set of alternatives, the current tree is taken to be optimal (technically, it is locally optimal for the set of rearrangements tested). Unfortunately, there is no guaranteed method for finding the optimum with a reasonable amount of computation (testing of trees). Therefore, it is necessary to choose a method on the basis of relatively arbitrary criteria.

Given the somewhat arbitrary nature of the choice of tree optimization methods, we use a method of "steepest descent." That is, given a phylogenetic tree, the effects of a given set of rearrangements upon the tree are tested, and then the best of all the tested alternatives (i.e., the most improved tree) is maintained and is used as the starting point for another round of optimization. The principal difference between this method and some others is the exhaustive search of local alternatives prior to selecting one, as opposed to immediately taking any step that yields an improved tree. This approach is based on the rationale that following the "steepest downhill slope" is more apt to lead to the optimum than are random (i.e., first found) "downhill" steps.

A critical component of the optimization procedure is the definition of the tree rearrangements that will be tested at each step. The larger the set, the more likely it is that one of them will lead to a better solution. On the other hand, the computational difficulty of a single round of optimization grows linearly with the number of alternative trees tested. Two simple classes of tree rearrangements are tested by the current optimization algorithm. Both of these regard a current tree as sets of subtrees connected by segments. A subtree can range from a single sequence up to the entire tree except for three sequences. A subtree can be moved to a new location by removing its nearest node from the tree and inserting this node (with the subtree attached) into a different tree segment. The effect of moving each subtree (one at a time) to every possible alternative location in the tree is systematically tested.

The second class of rearrangements tested are those that interchange the locations of a pair of subtrees. That is, each subtree is removed from its nearest node and reattached at the node from which a second subtree was removed. The effect of every possible pairwise interchange of subtrees (one pair at a time) is tested. There is a slight redundancy in these two sets of rearrangements, but this consumes only a small fraction (which decreases as the number of organisms increases) of the total computation.

It is necessary to define a quantitative measure of the agreement between a calculated phylogenetic tree and the data upon which it is based. The measure chosen for our work is the agreement (or lack of it) between the estimated evolutionary distance separating pairs of organisms and the sum of the tree segment lengths joining the organisms in the tree. That is, every pair of organisms is joined by a unique path of two or more segments through the phylogenetic tree under test. The sum of these segment lengths is the tree distance between the pair of organisms.

Ideally, every pair of organisms should be joined by segments that add up to the estimated evolutionary distance separating the organisms. The departure from this ideal is measured by squaring the difference between the two values. This error is normalized to a standard deviation by weighting it according to the statistical standard deviation of the distance estimate (Hori and Osawa, 1979),

$$\sigma = \frac{3}{4} \left[\frac{H(1 - H)}{N(H - 1/4)^2} \right]^{1/2} \tag{6}$$

where σ is the statistical standard deviation, H is the sequence homology, and N is the number of sequence positions compared. This weighting prevents unreliable distances (in particular, very long distances, for which the number of mutations approaches the number of sites) from having undue influence on the tree. The resulting weighted errors are summed over all pairs of organisms, and this is defined as the tree error. For a given tree topology the tree segment lengths that minimize this error are determined (this can be performed analytically, and hence rapidly, with linear algebra) and the residual error is evaluated. In summary, the optimization seeks to find the tree topology and tree segment lengths that minimize the standard deviation of the organism-to-organism evolutionary distances relative to the tree path lengths.

A possible mathematical artifact in this procedure is the introduction of negative tree segment lengths. These are of no evolutionary meaning (they *do not* correspond to convergent or reverse evolution). In order to prevent the tree optimization from converging on such a solution, the tree errors are also weighted by a term that increases by a factor of two for every negative length segment in the tree. Fitch and Margoliash (1967) simply discarded these trees, whereas the present method allows the optimization to proceed even if the initial tree has a negative length segment.

7.5. Implementation of Phylogenetic Calculations

The inference of phylogenies from sequence data has been carried out with the aid of two computer programs written in FORTRAN IV for execution on the Digital VAX-11/750. The first is an interactive program designed to facilitate the input, editing, alignment, and comparative analysis of nucleotide sequence data. This program is used for the sequence alignments and the calculation of homologies and evolutionary distances.

The second program facilitates the entry, editing, evaluation, and optimization of phylogenetic trees. The performance of this program is satisfactory for trees with less than about 30 organisms. It appears to have successfully found the optimum of all sequence sets tested (as evaluated by the independence of the "optimal" tree of the initial tree). The time required for each round of optimization depends only on the number of organisms in the tree (due to the testing of all local rearrangements prior to choosing the step) and is roughly

$$2.8 \times 10^{-5} N^6 \text{ sec} \qquad (7)$$

of central processing unit time on the Digital VAX-11/750, where N is the number of sequences. This corresponds to 28, 450, 1800, 7000, and 20,000 sec for trees with 10, 15, 20, 25, and 30 sequences, respectively.

8. From Phylogeny to Physiology?

Focus on rRNA and the use of the methods discussed here make it possible to define the phylogenetic affilitations of any organism, even if it is only identified in mixed, naturally occurring populations, but is not cultivatable. This would seem a boon to the microbiologist. However, an important question for the microbial ecologist to pose is: What is the relationship of the phylogenetic status of an organism to its physiological attributes?

We cannot yet predict, from phylogenetic placement, the detailed physiological traits of an organism unless it is nearly identical to one of our reference sequences. This is not a fault of the approach; rather, it is because the available collection of reference sequences is slender and restricted in its phylogenetic breadth. As the reference sequence data base expands over the next years to include many more organisms with diverse physiologies, the phylogenetic approach toward predicting the properties of organisms should become powerful indeed.

Our understanding of correlations between phylogeny and physiology is still rudimentary. At this time, we can only assert that if organisms share close phylogenetic affiliation, they must be similar in their fundamental biochemical motifs. Even the definition of a fundamental trait is not straightforward, however. Certainly the character of the ribosome seems fundamental; perhaps the components of the DNA replication machinery or the ion pumps are. Although significant in environmental considerations, important physiological traits such as sulfur oxidation or nitrogen fixation superficially are not "fundamental biochemical motifs." Sulfur-oxidizing autotrophs and fastidious heterotrophs, for instance, may be close relatives by rRNA sequences (Section 5.1). Sulfur oxidation is an ancient character, distributed throughout the procaryotes (at least), but its presence is dictated by the environment. The oxidation of sulfur *per se* may be no more fundamental to cellular function than, say, oxidation of a particular carbohydrate. The electron transport chains used by closely related heterotrophs and autotrophs conceivably are close phylogenetic homologues, but the identity of the electron donors seems less constrained.

One fascinating aspect of exploring the phylogeny of natural microbial populations, and microorganisms in general, is that vast evolutionary diversity is outlined. Hence potential biochemical novelties are identified. We have plucked, from natural populations, rRNA molecules and their genes that are not related to any that we know of closer than nearly kingdom-level phylogenetic depth. A more careful inspection of these organisms now becomes interesting. They must contain novel biochemical features, because they are so diverse from anything so far characterized.

ACKNOWLEDGMENTS. The original work reported here was supported by a U.S. National Institutes of Health research grant to N. R. P. The M13 clones of *Trypanosome* and *E. coli* were provided by Drs. Mitchell Sogin and Al Dahlberg, respectively.

References

Biggin, M. D., Gibson, T. J., and Hong, G. F., 1983, Buffer gradient gels and ^{35}S label as an aid to rapid DNA sequence determination, *Proc. Natl. Acad. Sci. USA* **80**:3963–3965.

Brierley, C. L., 1982, Microbiological mining, *Sci. Am.* **2**:44–53.

Brock, T. D., 1978, *Thermophilic Microorganisms and Life at High Temperatures,* Springer-Verlag, New York.

Brosius, J., Palmer, M. L., Kennedy, R. J., and Noller, H. F., 1978, Complete nucleotide sequence of a 16S ribosomal RNA gene from *Escherichia coli, Proc. Natl. Acad. Sci. USA* **75**:4801–4805.

Brosius, J., Dull, T. J., and Noller, H. F., 1980, Complete nucleotide sequence of a 23S ribosomal RNA gene from *Escherichia coli, Proc. Natl. Acad. Sci. USA* **77**:201–204.

Cavanaugh, C. M., 1983, Symbiotic chemoautotrophic bacteria in marine invertebrates from sulfide-rich habitats, *Nature* **302**:58–61.

Cavanaugh, C. M., Gardiner, S., Jones, M. L., Jannasch, H. W., and Waterbury, J. B., 1981, Prokaryotic cells in the hydrothermal vent tubeworm *Riftia pachyptila* Jones: Possible chemoautotrophic symbionts, *Science* **213**:340–342.

Cedergren, R . J., LaRue, B., Sankoff, D., and Grosjean, H., 1981, The evolving tRNA molecule, *Crit. Rev. Biochem.* **11**:35–104.

Chambliss, G., Craven, G. R., Davies, J., Davis, K., Kahan, L., and Nomura, M. (eds.), 1980, *Ribosomes: Structure, Function, and Genetics,* University Park Press, Baltimore.

Delihas, N., and Andersen, J. 1982, Generalized structures of the 5S ribosomal RNAs, *Nucleic Acids Res.* **10**:7323–7344.

Demoulin, V., 1979, Protein and nucleic acid sequence data and phylogeny, *Science* **205**:1036–1039.

Donis-Keller, H., Maxam, A., and Gilbert, W., 1977, Mapping adenines, guanines, and pyrimidines in RNA, *Nucleic Acids Res.* **4**:2527–2538.

Doolittle, W. F., 1973, Postmaturational cleavage of 23S ribosomal ribonucleic acid and its metabolic control in the blue-green alga *Anacystis nidulans, J. Bacteriol.* **113**:1256–1263.

Edmond, J. M., and Von Damm, K., 1983, Hot springs on the ocean floor, *Sci. Am.* **248**:78–93.

Enquist, L., and Sternberg, N., 1979, *In vitro* packaging of λ Dam vectors and their use in cloning DNA fragments, in *Methods in Enzymology,* Vol. 68 (R. Wu, ed.), pp. 281–298, Academic Press, New York.

Erdmann, V A., Wolters, J., Huysmans, E., Vandenberghe, A., and De Wachter, R., 1984, Collection of published 5S and 5.8S ribosomal RNA sequences, *Nucleic Acids Res.* **12**:r133–r166.

Felbeck, H., Childress, J. J., and Somero, G. N., 1981, Calvin–Benson cycle and sulfide oxidation enzymes in animals from sulfide-rich habitats, *Nature* **293**:291–293.

Felsenstein, J., 1982, Numerical methods for inferring evolutionary trees, *Q. Rev. Biol.* **57**:379–404.

Fitch, W. M., 1976, The molecular evolution of cytochrome c in eukaryotes, *J. Mol. Evol.* **8**:13–40.

Fitch, W. M., and Margoliash, E., 1967, Construction of phylogenetic trees: A method based on mutational distances as estimated from cytochrome c sequences is of general applicability, *Science* **155**:279–284.

Fitch, W. M., and Smith, T. F., 1983, Optimal sequence alignments, *Proc. Natl. Acad. Sci. USA* **80**:1382–1386.

Fox, G. E., Stackebrandt, E., Hespell, R. B., Gibson, J., Maniloff, J., Dyer, T. A., Wolfe, R. S., Gupta, R., Bonen, L., Lewis, B. J., Stahl, D. A., Luehrson, K. R., Chen, K. N., and Woese, C. R., 1980, The phylogeny of prokaryotes, *Science* **209**:457–463.

Garber, R. C., and Yoder, O. C. 1983, Isolation of DNA from filamentous fungi and separation into nuclear, mitochondrial, ribosomal, and plasmid components, *Anal. Biochem.* **135**:416–422.

Garrett, R. A., 1979, The structure, assembly, and function of ribosomes, *Crit. Rev. Biochem.* **25**:121–177.

Gauss, D. H., and Sprinzl, M., 1984, Compilation of tRNA sequences, *Nucleic Acids Res.* **12**(suppl.):r1–r58.

Gibson, J., Stackebrandt, E., Zablen, L. B., Gupta, R., and Woese, C. R., 1979, A phylogenetic analysis of the purple photosynthetic bacteria, *Curr. Microbiol.* **3**:59–64.

Goad, W. B., and Kanehisa, M. I., 1982, Pattern recognition in nucleic acid sequences. I. A general method for finding local homologies and symmetries, *Nucleic Acids Res.* **10**:247–263.

Goodman, M. (ed.), 1982, *Macromolecular Sequences in Systematic and Evolutionary Biology,* Plenum Press, New York.

Gupta, R., Lanter, J. M., and Woese, C. R., 1983, Sequence of the 16S ribosomal RNA from *Halobacterium volcanii,* an archaebacterium, *Science* **221**:656–659.

Harrison, A. P., Jr., 1982, Genomic and physiological diversity amongst strains of *Thiobacillus ferrooxidans,* and genomic comparison with *Thiobacillus thiooxidans, Arch. Microbiol.* **131**:68–76.

Hori, H., and Osawa, S., 1979, Evolutionary change in 5S RNA secondary structure and a phylogenetic tree of 54 5S RNA species, *Proc. Natl. Acad. Sci. USA* **76**:381–385.

Ingraham, J. L., Maaløe, O., and Neidhardt, F. C., 1983, *Growth of the Bacterial Cell,* Sinauer, Sunderland, Massachusetts.

Jannasch, H. W., and Nelson, D. C., 1984, Recent progress in the microbiology of hydrothermal vents, in: *Current Perspectives in Microbial Ecology* (M. J. and C. A. Reddy, eds.), pp. 170–176, American Society for Microbiology, Washington, D. C.

Jannasch, H. W., and Wirsen, C. O., 1979, Chemosynthetic primary production at east Pacific sea floor spreading centers, *Bioscience* **29**:592–598.

Jones, M. L., 1981, *Riftia pachyptila*, new genus, new species, the vestimentiferan worm from the Galapagos rift geothermal vents (Pogonophora), *Proc. Biol. Soc. Wash.* **94**:1295–1313.

Jukes, T. H., and Cantor, C. R., 1969, Evolution of protein molecules, in: *Mammalian Protein Metabolism* (H. N. Munro, ed.), pp. 21–132, Academic Press, New York.

Kandler, O. (ed.), 1982, First workshop on archaebacteria, Munich, 1981, *Bakteriol. Zentralbl. Hug. I Abt. Orig. C3* **1982**:1–161.

Karp, R. M., 1972, Reducibility among combinatorial problems, in: *Complexity of Computer Computations,* (R. E. Miller and J. W. Thatcher eds.), pp. 85–103, Plenum Press, New York.

Lane, D. J., 1983, 5S rRNA phylogenetic analyses of certain free-living and symbiotic sulfur-oxidizing chemolithotrophs, Thesis, Health Sciences Center, University of Colorado.

Lane, D. J., Stahl, D. A., Olsen, G. J., Heller, D., and Pace, N. R., 1985, Phylogenetic analysis of the genera *Thiobacillus* and *Thiomicrospira* by 5S rRNA sequences, *J. Bacteriol.* **163**:75–81.

Luehrsen, K. R., Nicholson, D. E., Eubanks, D. C., and Fox, G. E., 1981, An archaebacterial 5S rRNA contains a long insertion sequence, *Nature* **293**:755–756.

Mackay, R. M., Spencer, D. F., Schnare, M. N., Doolittle, W. F., and Gray, M. W., 1982, Comparative sequence analysis as an approach to evolving structure, function, and evolution of 5S and 5.8S ribosomal RNAs, *Can. J. Biochem.* **60**:480–485.

Maniatis, T., Fritsch, E. F., and Sambrook, J., 1982, *Molecular Cloning: A Laboratory Manual,* Cold Spring Harbor Laboratory, Cold Spring Harbor, New York.

Marrs, B., and Kaplan, S., 1970, 23S precursor ribosomal RNA of *Rhodopseudomonas sphaeroides, J. Mol. Biol.* **49**:297–317.

Messing, J., 1983, New M13 vectors for cloning, in: *Methods in Enzymology,* Vol. 101, (R. Wu, L. Grossman, and K. Moldave, eds.), pp. 70–78.

Nomura, M., Traub, P., and Bechman, H., 1968, Hybrid 30S ribosomal particles reconstituted from components of different bacterial origins, *Nature* **219**:793–799.

Norris, P. R., Brierley, J. A., and Kelly, D. P., 1980, Physiological characteristics of two facultatively thermophilic mineral-oxidizing bacteria, *FEMS Microbiol. Lett.* **7**:119–122.

Olsen, G. J., 1983, Comparative analysis of nucleotide sequence data, Thesis, Health Sciences Center, University of Colorado.

Palleroni, N., 1981, Introduction to the family Pseudomonadaceae, in: *The Prokaryates* (M. P. Starr, H. Stolp, H. G. Truper, A. Balows, and H. G. Schlegel, eds.), pp. 655–669, Springer-Verlag, New York.

Papanicolaou, C., Gouy, M., and Ninio, J., 1984, An energy model that predicts the correct folding of both the tRNA and the 5S RNA molecules, *Nucleic Acids Res.* **12**:31–44.

Parish, J. H., 1972, *Principles and Practice of Experiments with Nucleic Acids,* pp. 104–111, Longman, London.

Pavlakis, G. N., Jordan, B. R., Wurst, R. M., and Vournakis, J. N., 1979, Sequence and secondary structure of *Drosophila melanogaster* 5.8S and 2S rRNAs and of the processing site between them, *Nucleic Acids Res.* **7**:2213–2238.

Peattie, D. A., 1979, Direct chemical method for sequencing RNA, *Proc. Natl. Acad. Sci. USA* **76**:1760–1764.

Penny, D., 1976, Criteria for optimizing phylogenetic trees and the problem of determining the root of a tree, *J. Mol. Evol.* **8**:95–116.

Pieler, T., and Erdmann, V. A., 1982, Three-dimensional structural model of eubacterial 5S rRNA that has functional implications, *Proc. Natl. Acad. Sci. USA* **15**:4599–4603.

Ruby, E. G., and Jannasch, H. W., 1982, Physiological characteristics of *Thiomicrospira* sp. strain L-12 isolated from deep-sea hydrothermal vents, *J. Bacteriol.* **149**:161-165.

Sanger, F., Brownlee, G. G., and Barrell, B. G., 1965, A two-dimensional fractionation procedure for radioactive nucleotides, *J. Mol. Biol.* **13**:373–398.

Sanger, F., Nicklen, S., and Coulson, A. R., 1977, DNA sequencing with chain-terminating inhibitors, *Proc. Natl. Acad. Sci. USA* **74**:5463–5467.

Smith, T. F., Waterman, M. S., and Fitch, W. M., 1981, Comparative biosequence metrics, *J. Mol. Evol.* **18**:38–46.

Stackebrandt, E., and Woese, C. R., 1981, The evolution of prokaryotes, in: *Molecular and Cellular Aspects of Microbial Evolution* (M. J. Carlisle, J. R. Collins, and B. E. B. Moseley, eds.), pp. 1–31, Cambridge University Press, Cambridge.

Stahl, D. A., Luehrsen, K. R., Woese, C. R., and Pace, N. R., 1981, An unusual 5S rRNA, from *Sulfolobus acidocaldarius,* and its implications for a general 5S rRNA structure, *Nucleic Acids Res.* **9**:6129–6137.

Stahl, D. A., Lane, D. J., Olsen, G. J., and Pace, N. R., 1984. Analysis of hydrothermal vent-associated symbionts by ribosomal RNA sequences, *Science* **224**:409–411.

Stahl, D. A., Lane, D. J., Olsen, G. J., and Pace, N. R., 1985, Characterization of a yellowstone hot spring microbial community by 5S rRNA sequences, *Appl. Environ. Microbiol.* **49**:1379–1384.

Stetter, K. O., and Gaag, G., 1983, Reduction of molecular sulfur by methanogenic bacteria, *Nature* **305**:309–311.

Stetter, K. O., and Zillig, W., 1985, *Thermoplasma* and the thermophilic sulfur-dependent Archaebacteria, in: *The Bacteria,* Vol. 8 (C. R. Woese and R. S. Wolfe eds.), pp. 85–170, Academic Press, New York.

Walker, T. A., and Pace, N. R., 1983, 5.8S ribosomal RNA, *Cell* **33**:320–322.

Waterman, M. S., 1983, Sequence alignments in the neighborhood of the optimum with general application to dynamic programming, *Proc. Natl. Acad. Sci. USA* **80**:3123–3124.

Wittmann, H. G., 1983, Architecture of prokaryotic ribosomes, *Annu. Rev. Biochem.* **52**:35–65.

Woese, C. R., 1972, The evolution of cellular tape reading processes and macromolecular complexity, in: *Evolution of Genetic Systems, Brookhaven Symp. Biol.* **23**:326–365.

Woese, C. R., and Fox, G. E., 1977, The concept of cellular evolution, *J. Mol. Evol.* **10**:1–6.

Woese, C. R., Gutell, R. R., Gupta, R., and Noller, H. F., 1983, A detailed analysis of the higher-order structure of 16S-like ribosomal RNAs, *Microbiol. Rev.* **47**:621–669.

Zuckerkandl, E., and Pauling, L., 1965, Molecules as documents of evolutionary history, *J. Theor. Biol.* **8**:357–366.

2

The Ecology of Heterotrophic Microflagellates

TOM FENCHEL

1. Introduction

The wide variety of unicellular, phagotrophic eucaryotes known collectively as heterotrophic microflagellates has recently attracted much attention particularly among biological oceanographers. Knowledge of the morphology, systematic affinities, and general biology of members of this heterogeneous assemblage of protists is still far from complete. Even so, literature spanning over more than a century gives evidence of the diversity of these forms and of their ubiquitous occurrence. Lohmann (1911, 1920) attempted to quantify these small protozoans in seawater and assess their ecological significance, and Griessmann (1914) isolated a variety of forms in culture and described aspects of their biology.

The current interest in these forms is not due to the recent discovery of their existence, but has other reasons. They are difficult to study; they lend themselves neither to classical bacteriological methods nor to those of the zoologist, botanist, or planktologist, and often the resolving power of the light microscope is insufficient for identification or for the observation of morphological details necessary to understand function. Hence, until recently they were mostly ignored by ecologists and left to protozoologists or phycologists with an inclination for taxonomy. Of interest to biological oceanography is the recent evidence that bacteria (including photosynthetic cyanobacteria) play a substantial role in pelagic food

TOM FENCHEL • Department of Ecology and Genetics, University of Aarhus, DK-8000 Aarhus C, Denmark.

chains [for recent reviews see Williams (1981), van Es and Meyer-Reil (1982), and Azam et al. (1983)]. It has been shown that as much as 40–50% of the primary production is utilized by bacteria, that bacterial production may represent as much as 20% of the primary production, and that bacterial generation times may be as short as 12 hr. Since bacterial numbers in natural seawater usually remain rather stable (varying from perhaps 10^6 to 5×10^6 ml^{-1} in coastal waters and estuaries and from 5×10^5 to 10^6 ml^{-1} in off-shore waters), there must be a sink for bacteria. It has been difficult to identify grazers of bacteria among the better known zooplankters (including larger protozoans, such as ciliates). Many small metazoans and larger protozoans are known to feed and grow on a bacterial diet and occur in habitats with a very high bacterial productivity, where bacteria are associated with surfaces, but the role of such animals in controlling bacterial populations in planktonic systems has been difficult to demonstrate and is doubtful (Fenchel, 1980a,b, 1984). On the other hand, evidence based on general considerations, pure culture experiments, and *in situ* quantification of microflagellates in the sea has shown that these organisms are capable of consuming the bacterial production (e.g., Haas and Webb, 1979; Sorokin, 1979, 1981; Fenchel, 1982b,d, 1984; Sieburth and Davis, 1982; Sherr et al., 1983, 1984; Laake et al., 1983; Andersen and Fenchel, 1985). Similar results have been obtained for the microbial successions during the degradation of particulate organic matter and for an organic lake sediment (Fenchel, 1975; Fenchel and Harrison, 1976; Fenchel and Jørgensen, 1977; Hänel, 1979; Linley et al., 1981; Newel et al., 1981; Robertson et al., 1982). In addition, in soils, microflagellates play a role as consumers of bacteria, although here similarly sized amebae play a relatively larger role (Stout, 1980; Clarholm, 1981). The microflagellates, therefore, play a key role in nature by controlling bacterial populations and by converting this resource into larger particles, which are avilable as food to larger protozoans and metazoan filter-feeders.

Few things, in fact, are common to the members of the heterogeneous assemblage of protists called microflagellates. Besides the fact that they are phagotrophs and possess one or more flagella, which serve for motility and food particle capture, what unites them is their small size: measuring from less than 3 μm to the (arbitrarily chosen) upper limit of 10 μm, they are the smallest free-living phagotrophs. It is, above all, their small size that is responsible for the functional properties that make them a necessary component of ecosystems. It is this discovery that has stimulated the interest in these organisms among aquatic ecologists.

An additional reason why these organisms are interesting is that they contain information on the earliest evolution and diversification of eucaryotes, and, with the systematic application of modern methods of study, such as electron microscopy and sequencing of proteins and nucleic acids, much of this information may become accessible.

This review is mainly concerned with the functional biology and ecological significance, in particular with respect to planktonic environments, of heterotrophic microflagellates. No reference is given to similar commensalistic gut faunas found in almost all metazoans, nore to more specialized flagellate parasites, although from some taxonomic, ecological, and evolutionary viewpoints, comparisons with free-living forms might be illuminating. Although the review in no way pretends to contribute to the clarification of the still rather chaotic and incomplete taxonomy of all levels, it will open with an overview of the diversity of microflagellates. Also somewhat outside the main objective of the review, some comments on the methodology available for the study of microprotozoans will be given.

2. The Diversity of Heterotrophic Microflagellates

As already indicated, the subject of this review does not in any way represent a taxonomic unit, but rather a collection of independent eucaryote lineages whose relationships, mutually or to other eucaryotes, are still largely unknown. Traditionally, protozoologists recognize the phylum Sarcomastigophora comprising flagellates and sarcodines (ameboid forms in a wide sense). The flagellates are again divided into two classes, the Phytomastigophorea and the Zoomastigophorea; the former group contains species with chloroplasts or nonpigmented species obviously related to pigmented ones, and the latter group contains forms that are always devoid of chloroplasts (Levine *et al.,* 1980). This high-level systematic classification does not reflect any phylogenetic reality whatsoever. The individual orders comprising the above classes seem mainly to represent natural groups. However, to the extent that any relationships between these orders are recognized, they more often than not cross the boundaries of the higher taxons. Thus, the kinetoplastid flagellates (Zoomastigophorea) show affinities to the euglenoids (Phytomastigophorea), and the choanoflagellates, while being a rather isolated group among other protists, seem to have a common ancestry with metazoa (Corliss, 1983). All this is further confounded by the fact that protozoologists and phycologists have both claimed a partly overlapping share of the protists, so that two parallel schemes of classification and nomenclature exist. [An extraordinary example of the pseudoproblem of what are animals and what are plants among the protists derived from a doubtful and rather anecdotal finding of a pigmented choanoflagellate. For a period, this led to a claim from phycologists that the group be removed to the plant kingdom under the name "Craspedophycea"; see Throndsen (1974) and Parke and Leadbeater (1977) for references.] Clearly, the presence of chloroplasts is not so profound a systematic character among the protists as previously believed and may have been acquired (and lost) several times during evolution. Also, the distinction between phyto- and zooflagellates is not always very clear-cut from a functional viewpoint, since some pigmented species are phagotrophs as well. It is evident that a modern revision of the systematics of the Sarcomastigophora is necessary.

Another problem derives from the fact that light microscopy is often insufficient for the study of taxonomically relevant characters (e.g., flagellar and mitochondrial fine structure). As a consequence, many of the earlier accounts (e.g., Griessmann, 1914; Ruinen, 1938; Hollande, 1942; Grassé, 1952, 1953; Skuja, 1939, 1948, 1956) gave names to a large number of forms for which the systematic position is still entirely or partly unknown. Due to often superb light-microscopic descriptions, some of these forms are frequently identified from water samples by modern workers, but their systematic position will remain obscure until they are brought into culture and studied with the aid of the electron microscope.

The best definition of heterotrophic microflagellates I can offer, then, is that they are 2- to 10-μm-long, flagellated, phagotrophic protists (including some pigmented forms). Taxonomically, most of them belong to well-defined, but mutually unrelated, groups, whereas some cannot at the moment be assigned to any particular group. None of the forms covered by this definition seem to display sexuality. In the following, some of the most important free-living forms will be discussed briefly.

2.1. Choanoflagellida

The choanoflagellates occur in all types of aquatic habitats. The spherical, usually 3- to 5-μm-diameter cells have one smooth flagellum, which drives water through a collar of tentacles (Figs. 1 and 3). The collar acts as a filter, and food particles caught on the outside are phagocytized by temporary pseudopodia arising from the external base of the collar (Laval, 1971; Leadbeater and Morton, 1974). The cells may be naked (Codonosigidae; Fig. 1A), encased in a membranous collar (Salpingoecidae; Fig. 2D), or possess a complicated siliceous skeleton [Acanthoecidae, Fig. 3A; see also Leadbeater and Manton (1974)]. This latter group is confined to seawater, whereas the two other families are commonly represented in freshwater as well. Choanoflagellates may be permanently or temporarily attached to solid substrates or be freely suspended in the water, and some forms are colonial. The choanoflagellates are never pig-

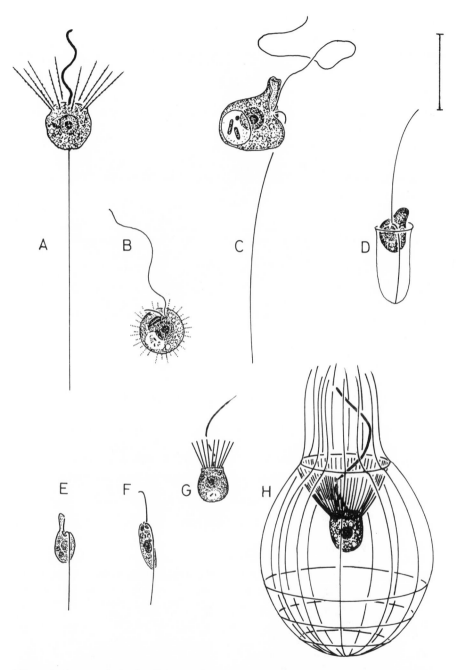

Figure 1. Some typical microflagellates from marine plankton. (A) *Pteridomonas* (helioflagellate); (B) *Paraphysomonas,* (C) *Pseudobodo,* and (D) *Bicoeca* (all chrysomonads or relatives); (E) *Rhynchomonas* and (F) *Bodo* (both kinetoplastids); (G) *Monosiga* and (H) *Diaphanoeca* (choanoflagellates). Scale: 10 μm.

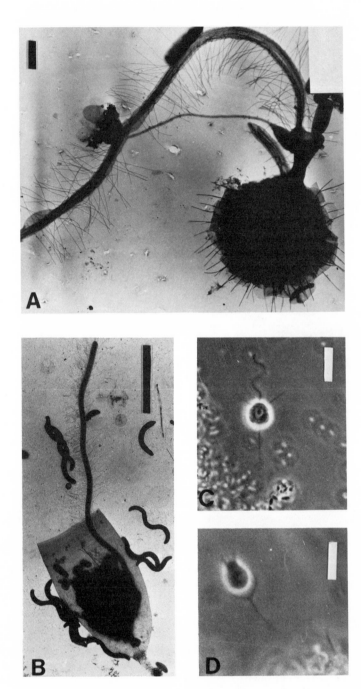

mented and seem without exception to feed on suspended bacteria. They are particularly important in marine plankton, where they often constitute more than 50% of the flagellate fauna (Sieburth, 1979; Fenchel, 1982d). This well-defined group has been studied in detail with respect to morphology (Laval, 1971; Leadbeater 1977, 1983) and their diversity and geographic distribution are also well documented. This applies in particular to the acanthoecids, since their skeleton makes easily recognizable (and esthetically pleasing) whole mounts under the transmission electron microscope (Fig. 3A) and specific identification can also be made with the light microscope (Boucaud-Camou, 1966; Leadbeater, 1972a–c, 1974; Thomsen, 1973, 1976, 1978, 1979, 1982; Thomsen and Boonruang, 1983a,b; Thomsen and Moestrup, 1983; Throndsen, 1974, which also give references to older literature).

2.2. Chrysomonadida

The chrysomonads show a continuum from phototrophic nutrition to obligatory phagotrophy and many species possess chloroplasts and ingest bacteria as well. Some forms cannot exist without particulate food in spite of the presence of chloroplasts and many are nonpigmented (Pringsheim, 1952; Swale, 1969; Aaronson, 1980; Fenchel, 1982a). The typical chrysomonad has two flagella, one of which is hispid. The structure of the flagellar hairs, the patterns of microtubules in the cells, and some other fine structural details characterize the group (Moestrup, 1982). The phagocytotic forms possess a cytostome supported by bundles of microtubules. The hispid flagellum drives water currents toward the cell, and suspended bacteria touching the cytostome are engulfed (Fig. 3B) (Fenchel, 1982a). Species of *Ochromonas,* which possess chloroplasts, but also depend on phagocytosis of bacteria, are abundant in limnic and marine habitats. Quite similar forms devoid of chloroplasts are assigned to the genus *Monas,* but the presence of a chloroplast is often difficult to establish in the light microscope. The genus *Paraphysomonas* comprises nonpigmented species with a characteristic covering of siliceous scales (Fig. 2A). These flagellates are extremely common in freshwater and seawater, and, due to the morphology of the scales as displayed by whole mounts under the electron microscope, a species taxonomy is well estab-

←———————————————————————————

Figure 2. (A) *Paraphysomonas vestita,* TEM, whole mount, culture isolated from Aarhus Bay. (B) *Bicoeca* sp., TEM, whole mount, freshwater enrichment culture. (C) *Pteridomonas* sp., living cell from culture isolated from Aarhus Bay. (D) *Codosiga* sp. attached to detrital particle, living cell, from centrifuged water sample, Limfjorden. Scales: (A) 1 μm, (B–D) 5 μm.

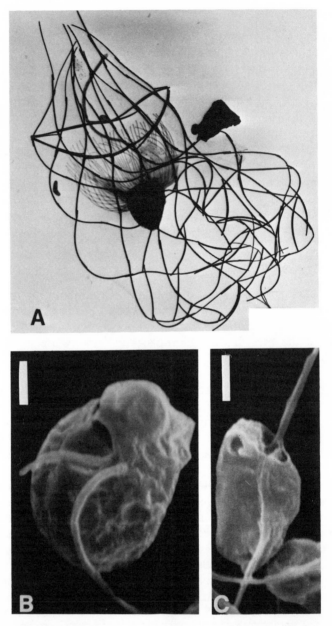

Figure 3. (A) *Diaphanoeca grandis,* TEM, whole mount, culture isolated from Limfjord. (B) *Ochromonas* sp. from culture, fixed immediately after engulfing a bacterium, the outline of which can be seen in the anterior end of the cell above the cytostome. Flagellar hairs are not resolved in this SEM picture. (C) *Pleuromonas jaculans,* SEM, from culture; the cell shows the cytostome and the flagellar pocket. Scales: (A) 5 μm, (B,C) 1 μm.

lished (Lucas, 1968; Pennick and Clarke, 1972; Leadbeater, 1972d; Thomsen, 1975, 1979; Thomsen *et al.*, 1981; Preisig and Hibberd, 1982).

The bicoecids, previously classified among the zooflagellates, show so many structural similarities to the chrysomonads that they are now placed within that group (Mignot, 1974; Moestrup and Thomsen, 1976; Moestrup, 1982). These organisms secrete a cup-shaped or conical test. The smooth flagellum attaches to the bottom of the test, which again is attached to water films or detrital material. When feeding, the oral part of the cell and the anterior, hispid flagellum protrude from the test (Figs. 1D and 2B). Several species are described from seawater and freshwater [see Moestrup and Thomsen (1976) for references]. A number of very common nonpigmented flagellates resemble the bicoecids closely, but are devoid of a test. Instead, they use the smooth flagellum for temporary attachment. These forms may have a superficial resemblance to kinetoplastids and many are presumably classified as such in the older literature. An example is *Pseudobodo tremulans,* an extremely common marine flagellate, originally described as a bodonine flagellate by Griessmann (1914), which has proven to be a nonloricate bicoecid (Fig. 1C) (Fenchel, 1982a).

2.3. Helioflagellida

These forms show a close affinity to the chrysomonads (in particular to the pigmented *Pedinella*) on the one hand and to the actinophryid heliozoans on the other (Davidson, 1982; Patterson and Fenchel 1985), and they are sometimes classified as heliozoans (e.g., Grassé, 1953). In *Pteridomonas* [described under the name *Actinomonas* in Fenchel (1982a,b,e)] and *Actinomonas* [which differs from *Pteridomonas* by possessing more than one ring of pseudopodia; see Patterson and Fenchel (1985) for taxonomic discussion] the hispid flagellum drives a water current toward the cell and through the pseudopodial collar, which acts as a filter for suspended food particles (Figs. 1A and 2C). The pseudopodia, which are supported by microtubules, carry "extrusomes" which cause the prey bacteria to stick to the pseudopodia. Phagocytosis takes place by temporary pseudopodia arising from the cell surface. In *Ciliophrys* (Fig. 4) feeding is more like that in a typical heliozoan. The flagellum is immobile and the bacteria are caught on the pseudopodia due to the motility of the prey. When disturbed, however, the pseudopodia contract and within minutes the cell swims off, looking like a ordinary flagellate (Griessmann, 1914; Davidson, 1982). All these forms seem to be omnipresent in seawater, from which cultures can easily be established; at least *Actinomonas* and *Pteridomonas* occur in freshwater as well.

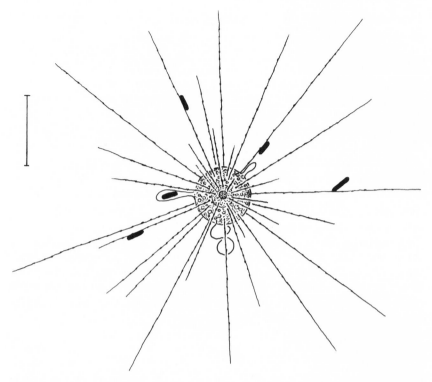

Figure 4. The helioflagellate *Ciliophrys marina* with bacteria attached to the pseudopodia. Scale: 10 μm.

2.4. Euglenida

The euglenoids comprise pigmented as well as colorless forms. Many of the latter are too large to be considered "microflagellates," and, indeed, choose prey larger than bacteria (e.g., *Peranema*). Typically, the euglenoids possess two flagella (but one may be absent) arising from a depression in the anterior end of the cell, the "cytopharynx." The flagella carry characteristic hairs (different from those found in chrysomonads and relatives) and are also characterized by a "paraxial rod" situated parallel to the axoneme (Moestrup, 1982). The phagotrophic forms possess a cytostome situated outside the cytopharynx and supported by microtubular rods, which play a role in the engulfment of prey cells. The forms to be considered here (e.g., *Entosiphon* and *Anisonema*) slide along the surfaces of solid substrates on their trailing flagellum and seem to depend on attached bacteria for food. While hardly of significance in the plankton, they are very numerous in sediments and in decomposing detritus.

Species descriptions may be found in Lackey (1936, 1962, 1963), Hollande (1942), and Skuja (1939, 1948, 1956).

2.5. Kinetoplastida

The kinetoplastids resemble the euglenoids in many important respects (e.g., with respect to the flagellar apparatus and the structure of the cytostome) and the two groups are presumably related. Kinetoplastids are always colorless. A peculiar feature of the kinetoplastids is the structure of the single mitochondrion. In the anterior end of the cell (close to the flagellar bases) it has a swelling filled with a large amount of extranuclear DNA, the function of which is unknown (Englund, 1981). This part of the organelle is referred to as the kinetoplast. Among the free-living forms, the genera *Bodo, Rhynchomonas* (with its very short anterior flagellum) and *Pleuromonas* are best known (Figs. 1E, 1F, and 3C). Representatives of the two former genera are mainly associated with surfaces and specialize in feeding on attached bacteria; representatives of the latter attach temporarily with the long, posterior flagellum and drive water with suspended bacteria past the cytostome with the aid of the anterior flagellum. Modern accounts of the structure and classification of free-living kinetoplastids are found in Swale (1973), Vickerman (1976), and Eyden (1977).

The species systematics is very difficult. The large number of species descriptions in the older literature (e.g., Griesmann, 1914; Ruinen, 1938; Hollande, 1942; Skuja, 1939, 1948, 1956), and even in some more recent accounts (e.g., Hänel, 1979), are confusing and of restricted value, since they are based on vague characters, such as size and general shape. It is also likely that a number of unrelated forms have previously been described as bodonines (Vickerman, 1976). A modern taxonomic revision of this group is strongly needed.

2.6. Cryptomonadida

Cyptomonad flagellates may be very abundant in plankton samples. Most forms are pigmented, but colorless, phagotrophic forms occur as well (Pratt, 1959). Most are of the *Chilomonas* type, but the tiny (5 μm) *Cyathomonas* (Mignot, 1965) also occurs in both freshwater and seawater (Skuja, 1939; T. Fenchel, unpublished observations).

2.7. Other Forms

The older literature gives the description of a large number of non-pigmented flagellates which cannot readily be assigned to any of the established orders of flagellates. Some of these seem to occur quite frequently in nature. In particular, Griessmann (1914), Ruinen (1938), and Skuja (1948, 1956) have given name to such species. Some of them resemble (and were assigned to) genera otherwise only known as commensals [e.g., *Cryptobia;* see Ruinen (1938)]. Also, a number of ameboid flagellates *(Cercobodo, Mastigamoeba)* have been described. Although frequent in some habitats, they lead an obscure existence from a taxonomic point of view. The characteristic organism described and placed in the genus *Tetramitus* by Ruinen (1938) has recently been isolated into culture from Danish waters (T. Fenchel, unpublished results). It bears little resemblance to the species *Tetramitus rostratus,* usually considered an amebae with a flagellated life stage (e.g., Hollande, 1942; Page, 1976).

Finally it should be mentioned that a number of other flagellate groups, ignored here, perhaps could qualify as "heterotrophic microflagellates." The dinoflagellates comprise many (pigmented and colorless) phagotrophic forms. However, they are usually larger (>10 μm) and in general depend on food items larger than bacteria. It is also possible that in addition to the chrysomonads and the euglenoids, some of the other nanoflagellates usually considered photosynthetic may comprise phagotrophic forms.

3. Methods of Study

3.1. Concentration of Cells in Natural Samples

Many types of samples (sediments, detrital material, sewage) contain such a high concentration of cells that no further concentration is needed in order to observe the living organisms. In samples from open water, however, this will generally be necessary. Two methods are used: reverse filtration (Sieburth, 1979) or centrifugation. While neither method can be used reliably for quantitative work, they do provide material for light microscopic observation or, following fixation in OsO_4 or glutaraldehyde solutions, for whole mounts for electron microscopy.

3.2. Pure Cultures

Progress in understanding the morphology and functional biology of these forms depends on pure (clone) cultures. To date, quite a large number of forms have been isolated and maintained in culture containing a mixture of bacteria (usually maintained on a boiled wheat grain, soil extract medium, or similar medium) or on a single species of bacterium (e.g., Swale, 1973; Eyden, 1977; Leadbeater, 1977, 1983; Fenchel,

1982a,b; T. Fenchel, unpublished results). The main problem always seems to be to isolate a single cell, since the microflagellates are too small to be picked up individually with a capillary pipette as is easily done with slightly larger protozoans. At least some bodonids can grow on agar plates with bacteria (T. Fenchel, unpublished results), but in most cases clonal cultures can only be obtained on the basis of serial dilution cultures using natural samples or enrichment cultures as a source. Consequently, it is difficult to isolate the less numerous forms. Axenic culture has apparently not been attempted [except for *Ochromonas,* see e.g., Aaronson (1980)], but should probably not pose any particular difficulty using methods established for other heterotrophic protozoans.

3.3. Quantification

Serial dilution cultures have been extensively used by soil biologists for the enumeration of protozoans (Darbyshire *et al.,* 1974; Stout *et al.,* 1982) and since this method is believed to be the only one available for microflagellates in soils, its reliability is not known. In marine environments, Lighthart (1969) and Throndsen (1969) used this method for the enumeration of heterotrophic flagellates and it is likely that their results underestimate the real number to a considerable degree (Fenchel, 1982d). In this latter paper, the method was used in parallel with direct counts with the epifluorescence microscope. It was found that the serial dilution method did crudely reflect the results from direct counts both with respect to total numbers and species composition. However, some forms did not turn up at all and some only appeared from more dilute inoculants suggesting that competitive exclusion in the cultures biases the results. Also, since many of the species tend to associate with suspended detrital particles, the assumption of random distribution is not met. Finally, the presence of cysts, which excyst during incubation, may give an unrealistic picture of the active organisms present in the original sample. The method is therefore unreliable for quantitative work.

Russian workers in the field (e.g., Sorokin, 1979) have used light microscopy with a low-power objective to count flagellates directly in water samples. The method would seem difficult, at least if any identification is attempted, and it would seem to exclude the possibility of distinguishing between pigmented and nonpigmented forms. The simplest and most reliable method employed so far is the epifluorescence method for total counts of bacteria as described by Hobbie *et al.* (1977). Various refinements of this method (mainly with respect ot the use of different fluorochromes) have been suggested in order to improve the identification of chloroplasts, recognition of bacteria in feeding vacuoles, and differentiation of cytological details (Davis and Sieburth, 1982; Haas, 1982;

Caron, 1983; Sherr and Sherr, 1983). Fenchel (1982d), using the method of Hobbie *et al.* (1977), was able to identify most of the important groups of flagellates, sometimes to generic level in marine samples.

Fenchel (1975) quantified microflagellates in a detrital lake sediment by counting the number of cells per unit surface area of sediment particles in live samples stained with the fluorochrome acridine orange. The number of cells per unit volume of sediment was then calculated from estimates of total particle surface. This method is hardly very accurate, but seems to be the only one available for sediments.

4. Functional Biology

4.1. Significance of Size

It was mentioned that the small size of the microflagellates is a key to the understanding of their ecological role as consumers of bacteria and that, in particular, planktologists hitherto by and large have ignored phagotrophic organisms smaller than about 100 μm, that is, 100–200 times larger than the average bacterium. Among larger organisms, we are accustomed to the fact that there are limits to the size ratio between a predator and its prey due to mechanical constraints on the efficiency of food particle capture. We are therefore astonished by the extremes in possibilities, such as a 15-m-long whale feeding on 10-cm-long krill, but if we scale krill down to bacterial dimensions, a 150-μm-long bacterivorous ciliate becomes a protozoan baleen whale. For similar mechanical reasons, the efficiency of a feeding apparatus decreases if the variance in particle size is very large; therefore, phagotrophs are usually specialized on a rather limited size range of prey. This fact contributes much to the diversity of living things.

The characteristic time scale of organisms also changes with body size. A predator population that most efficiently utilizes a fluctuating prey population will have a generation time not much longer than that of the resource. This is another reason to expect that the increment in size between a prey species and its predator will be limited.

Such constraints suggest the properties of an efficient bacterivorous phagotroph. We may consider the situation in marine plankton. Here we typically find about 10^6 bacteria ml^{-1}. Assuming they each have a volume of 1 μm^3 (this is an exaggeration, but will do for the present argument), one finds that the volume fraction of bacteria in seawater is 10^{-6}. Now, if the predator is to divide every 24 hr (and this is roughly the maximum generation time for balanced growth in the flagellates) and if its growth efficiency (in terms of volume/volume) is 50%, then it must clear 2×10^6

times its own body volume of water for bacteria per day or about 10^5 hr^{-1}. This value for volume-specific clearance is not found in larger suspension feeding animals, but as shown in Section 4.2, it is attainable and realized in organisms in the size range of 3–10 μm.

4.2. Bioenergetics of Balanced Growth

The parameters describing growth can be derived from batch culture experiments. This was done by Fenchel (1982b) for six species of suspension feeding microflagellates. The experiments were carried out by adding a small inoculum of flagellates to suspensions of bacteria of known concentrations. The media used (aged, filtered seawater or, in the case of freshwater species, a mineral solution) do not support bacterial growth. For each experiment, an exponential growth rate constant $\mu(x)$ and a yield $Y(x)$ = eventual number of flagellates/initial number of bacteria can be calculated as a function of the initial bacterial concentration x (Fig. 5A). Since Y is invariant at least within the range of growth rates studied (Fig. 5C), the uptake of bacteria per unit time is given by $U(x)$ = $\mu(x)/Y$ and the clearance is given by $F(x) = U(x)/x$. In practice, these calculations must be modified somewhat, since, when the flagellates start to starve toward the end of the experiment, they tend to divide once or twice, resulting in cells one-half or one-quarter the size of growing cells. Hence, the value of Y must be divided by 2 or 4 to give correct estimates for growing cells.

Since Y is invariant with growth rate, $\mu(x)$ and $U(x)$, the "functional response," must have similar functional forms. The data can be fitted closely to a hyperbolic function (Fig. 5B) and this can be rationalized as follows. If the cells clear F_m volumes of water per unit time, the capture of bacteria should be $U = F_m x$, where x is bacterial concentration and F_m is the maximum value of clearance, realized at very low values of x. How-

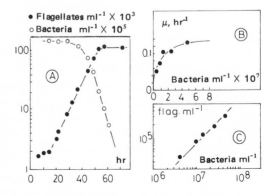

Figure 5. Growth data from batch cultures of *Pseudobodo tremulans*. (A) Growth in a culture initiated with 1.2×10^7 bacteria/ml. (B) Growth rate constants as a function of bacterial density. (C) Eventual flagellate concentration as a function of initial bacterial concentration in five batch cultures. [Data from Fenchel (1982b).]

ever, it takes a finite time τ to phagocytize a bacterium, during which another one cannot be engulfed (and in many forms, flagellar motion stops during phagocytosis). Hence, uptake should equal

$$U(x) = F_m X[1 - \tau U(x)] = U_m x/(x + K)$$

where $U_m = \tau^{-1}$ and the half-saturation constant $K = U_m/F_m$. When $x \to$ 0, $U \to F_m x$, and when $x \to \infty$, $U \to U_m$. This Michaelis–Menten analogue for describing food uptake and growth as a function of food particle concentration also has been shown to fit experimental data well for a number of other protozoans (Curds and Cockburn, 1968, 1971; Fenchel, 1980b).

These considerations, in conjunction with experimental data, allow the estimation of most parameters pertaining to balanced growth. Measurements of carbon contents of predator and prey cells and of flagellate respiration rates (which, as expected, are nearly proportional to growth rate) allow the calculation of gross growth efficiency (yield) and net growth efficiency in terms of carbon.

Such figures are compiled in Table I for 20°C. It must be pointed out that, strictly speaking, these values only refer to a specific food bacterium, in this case an ~ 1.3-μm-long *Pseudomonas* sp. As discussed in Section 4.3, the efficiency of particle retention is size dependent and is so in a different manner in different flagellates, so that clearance for a given species always refers to a given food particle size. Also, biochemical differences between bacteria may affect yield and growth rate for a given consumption of organic C. That different species of bacteria give different values of yield and growth was shown clearly in growth experiments with a *Monas* sp. carried out by Sherr *et al.* (1983). In other respects, their results are consistent with those in Table I.

To be noted in the table is that minimum generation times are within the range of 3–4 hr, that yield is within the range 30–50% in terms of cell volume (the food bacterium is about 0.6 μm^3) or in terms of C, and that the net growth efficiency in terms of C is about 60%, as is typical for growing eucaryote cells (Calow, 1977). It can be calculated from these data that some 30–40% of the ingested bacterial C is egested. This, at least in part, represents cell wall material, which is visible in electron micrographs of the egestive vacuole, but loss of dissolved organic material is also possible. Kopylov *et al.* (1980), using a more complicated experimental design than described here, found a yield of only 18–20%, but their estimate of the net growth efficiency was in the range of 60–70% in the flagellate *Parabodo*.

The flagellates can at most engulf 0.3–4 bacteria min^{-1}, according to

Table I. Parameters of Growth and Feeding[a]

Species	Volume (growing cells) (μm^3)	Clearance (ml hr^{-1} $\times 10^{-6}$)	Specific clearance (hr^{-1} $\times 10^5$)	U_m (bacteria hr^{-1})	μ_m (hr^{-1})	Yield (flagellates bacterium^{-1} $\times 10^{-3}$)	Net growth efficiency C:C (%)
Monosiga (choanoflagellate)	20	2	0.98	27	0.17	6.2	—
Paraphysomonas (chrysomonad)	190	17	0.91	254	0.23	0.91	—
Ochromonas (chrysomonad)	200	10	0.52	190	0.19	1.0	59
Pseudobodo (chrysomonad)	90	10	1.1	84	0.15	1.8	—
Pteridomonas (helioflagellate)	75	79	11.0	107	0.25	2.3	—
Pleuromonas (kinetoplastid)	50	1.4	0.55	54	0.16	3.0	60

[a] After Fenchel (1982b). Based on batch culture experiments at 20°C with a *Pseudomonas* sp. (volume 0.6 μm^3) as food.

the size of the species, and this corresponds to 60–80% of their body volume hr^{-1}. According to species, they can clear from 2×10^{-6} to 7×10^{-5} ml hr^{-1}, corresponding to a maximum volume-specific clearance of 6×10^4 to 10^6 hr^{-1}; the high value for *Pteridomonas* is undoubtedly related to its specialization to larger food particles (see Section 4.3), and otherwise the value 10^5 hr^{-1} is a reasonable generalization for the specific clearance of suspension feeding microflagellates.

Table I does not include information on the minimum balanced growth rate that can be sustained. This is experimentally much more difficult to answer. Fenchel (1982b) achieved exponential growth rates with generation times as long as about 24 hr in *Ochromonas;* at even lower food concentrations some growth could be detected, but no rate constant could be assigned to these populations. Based on indirect evidence from field populations and bottle incubations of natural water samples (see Section 5.1.2), it can be assumed that bacterial concentrations within the range of $(0.5–1.5) \times 10^6$ ml^{-1} represent the minimum that can sustain growth corresponding to generation times of 1–2 days, but these are crude estimates. It is likely that continuous culture techniques will give somewhat better answers to this question.

One aspect that often attracts interest among ecologists is the excretion of N and P from grazing organisms. Rates of NH_4^+ excretion were directly measured for growing populations of *Monas* sp. by Sherr *et al.* (1983). They found rates of 0.5–0.6 μmole NH_4^+/mg dry weight hr^{-1}. However, the magnitude of such rates must parallel the C flow of the population and is easily estimated from data such as those shown in Table I, with the reasonably robust assumption that bacteria and flagellates have similar C/N ratios. Clearly, the amount of N (and P) excreted must correspond to the bacterial C being respired. The cells studied by Sherr *et al.* doubled every 5 hr, so assuming a net growth efficiency of 60% and that the C content of cytoplasm is 50% of dry weight, one has that 1 mg dry weight of flagellates would need 0.83 mg C/5 hr or 0.166 mg C hr^{-1}, of which 0.066 is respired and that the remaining is incorporated. Setting the C/N ratio of bacteria to 6, we see that the flagellates should excrete 0.066/6 = 0.011 mg N or about 0.78 μmole NH_4^+ hr^{-1}, a reasonably close fit to the experimental findings.

The fact that yield is nearly invariant with growth rate (and consequently that respiration rate is proportional to the growth rate constant) shows that by far the largest part of the energy requirement is due to macromolecular synthesis during growth, while maintenance energy and power requirements for motility are slight. This is less of a surprise to a microbiologist than to a zoologist and is a general property of small organisms (Sections 4.3 and 4.4) (Fenchel and Finlay, 1983).

4.3. Mechanisms of Feeding

The external feeding apparatus of the flagellates serves to concentrate particles in suspension, which in case of the planktonic environment is indeed dilute. We have seen that the flagellates must and can clear about 10^5 times their cell volume of water hr^{-1}. As discussed in Fenchel (1984), three possible mechanisms that could achieve the necessary efficiency in concentrating suspended particles are: (1) sieving water through a pseudopodial filter; (2) directly intercepting particles following the flow lines in the vicinity of the cell; and (3) taking advantage of the motility of the food particles to bring them into contact with the protozoan, onto which they somehow stick. This last, heliozoan type of feeding is realized by *Ciliophrys* (Fig. 4); it has not been studied quantitatively and will not be discussed further. Fenchel (1984) applied scaling arguments to show that, for organisms in the size range of the microflagellates, all three mechanisms are about equally efficient and can attain the necessary values of clearance (although mechanism 3 requires motile bacteria). The efficiencies of the three mechanisms (in terms of specific clearance), however, change in a different manner as a function of prey and predator sizes. The efficiency of sieving varies inversely with predator size, whereas direct interception decreases with the second power of predator size and increases with the first power of food particle size. The efficiency of "diffusion feeding" also decreases with the second power of cell size. Hence, it is conceivable that among larger bacterivorous organisms (e.g., ciliates) only filtration is realized, whereas among the small flagellates "raptorial" feeding on suspended bacteria also occurs. A fourth mechanism is that of feeding on bacteria attached to surfaces; this will be discussed briefly at the end of this section.

Sieving and direct interception depend on motility and this is achieved by one or (rarely) more flagella. This organelle usually measures 5–10 μm in length. It functions by producing undulations or waves, which move along the flagellum and away from the cell. The undulations of the flagellum are characterized by the frequency (typically within the range 10–60 Hz as measured in stroboscopic light) and by the amplitude and wavelength, which can be recorded on photographs of living cells using an electronic flash (Fig. 2C). In smooth flagella, the water current moves in the direction of the waves, while in hispid flagella it moves in the opposite direction, that is, toward the cell. Maximum water velocities attained close to the center of the flagellum are usually within the range of 50–200 μm sec^{-1}, but in some filter feeding forms it is lower. The hydrodynamics of eucaryote flagella is described in detail in Holwill (1974) and Lighthill (1976).

Sleigh (1964) gave a qualitative account of the feeding currents produced by a choanoflagellate and two types of chrysomonads. Recently, I
have studied feeding currents quantitatively by recording the feeding of
different flagellates in suspensions of bacteria or latex beads on a video
recorder (T. Fenchel, unpublished results). From recordings of the position of individual particles in sequential frames, it is possible to recon-
struct the entire flow field generated by the flagellate. This also allows the
calculation of clearance by integration of the flow within "critical flow
lines" inside which food particles are intercepted. In filter feeding forms
(choanoflagellates, helioflagellates) this flow is, of course, delimited by the
area of the pseudopodial collars. In forms employing "direct interception" (e.g., chrysomonads) the critical flow lines are given by the distance
from the cytostome to flow lines in which passing particles will just be
intercepted. Some examples of flow fields are shown in Figs. 6 and 7 and
estimates of clearance based on hydrodynamic data and some other pertinent data are shown in Table II.

As found by Sleigh (1964) and in accordance with the fact that helioflagellates have hispid flagella, water passes toward the cell, and food particles are intercepted on the inside of the pseudopodial collar whereas the
converse is true for choanoflagellates. There is a considerable difference
between *Pteridomonas* and *Diaphanoeca* (and other choanoflagellates)
with respect ot the velocity of the feeding current (Table II). This
undoubtedly reflects the difference in the porosity of the filter in the two
types of flagellates. In *Pteridomonas* the spacing between adjacent pseudopodia is 1–3 μm, while in choanoflagellates it is only 0.1–0.3 μm; this
reflects different food niches with respect to particle sizes retained on the

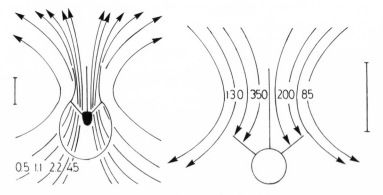

Figure 6. Feeding current flow fields of (left) the choanoflagellate *Diaphanoeca* and (left)
the helioflagellate *Pteridomonas* (right). Numbers refer to flow velocities (μm/sec) beneath
the lorica and at the level of the center of the flagellum, respectively. Scales: 10 μm.

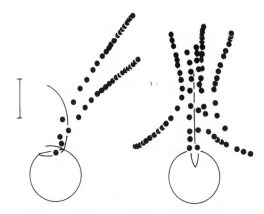

Figure 7. Seven flow lines recorded as the position of seven latex beads at time intervals of 0.016 sec in a video recording of two feeding *Paraphysomonas* cells. Scale: 5 μm.

filters. The low porosity of the choanoflagellate filter, however, means a relatively higher pressure drop across the filter for a given velocity. As discussed in Fenchel (1980a, 1984), the hydrostatic pressure that a flagellum can generate limits the velocity of the feeding currents in ciliary filter feeders. Theoretical calculations show that the pressure drop is always within the range 10–20 dynes cm^{-2} (0.1–0.2 mm H$_2$O) irrespective of the characteristic water velocity generated by different species (see Table II). Therefore, it is a tradeoff situation for the choanoflagellates: they are capable of retaining even the smallest procaryote cells, but in return they have a lower value of clearance than species feeding on larger particles.

Chrysomonads such as *Paraphysomonas* feed by a direct interception method. In these forms, the flow field is asymmetric due to the asymmetric position of the flagellar hairs (Fig. 2B) and due to the slight bending of the flagellum, which is superimposed on the undulations (Fig. 7).

Values for specific clearance based on hydrodynamic data (Table II)

Table II. Hydrodynamic Data Pertaining to Feeding[a]

Species	Porosity of filter (μm)	Average velocity through filter (μm sec^{-1})	Average velocity within critical flow line[b] (μm sec^{-1})	Pressure drop over filter[c] (dyne cm^{-2})	Specific clearance ($\times 10^5$ hr^{-1})
Diaphanoeca	0.1–0.3	9	20	12	1.7
Pteridomonas	1–3	90	350	9	5.0
Paraphysomonas	—	—	120	—	0.5

[a]T. Fenchel (unpublished data).
[b]At the middle of the flagellum.
[c]Calculated theoretically; see Fenchel (1980a).

agree reasonably well with those obtained from growth data (Table I), although the former seem to be systematically lower. This is likely to be due to the hydrodynamic drag imposed by the coverslip during the recording of water currents, a source of error that cannot be totally avoided, due to the short working distance of high-power objectives; nor is it easy to evaluate the magnitude of this effect.

Lighthill (1976) showed theoretically that, in the types of feeding described above, the hydrodynamic situation changes according to whether the flagellate swims freely or is anchored to some substrate (assuming that the flagellar beat parameters are identical in the two situations). In the former case, the flagellum creates practically no far-field flow and the flow experienced by the cell is the Stokes flow arising from the cell being pulled through the water by the flagellum. If the cell is anchored, on the other hand, it is the flow field created by the flagellum that the cell utilizes as a feeding current. It can be shown that this latter situation is the most favorable one and this explains the almost universal tendency for suspension feeding protozoans to attach to surfaces (in contrast to those feeding on bacteria attached to surfaces or to photosynthetic nanoflagellates). Some forms (e.g., bicoecids with a test, some choanoflagellates) are permanently attached and nearly all other forms do so temporarily. In *Pteridomonas* (Fig. 2C) and in many chrysomonads this is accomplished with a temporary stalk; in naked bicoecids and in kinetoplastids the posterior flagellum serves this purpose. In growing cultures of microflagellates, the largest fraction of the cells is usually found attached to the walls of the container or the surface film, while a smaller number is found swimming freely at any one moment. In planktonic environments a large fraction of the cells seems associated with detrital particles (Section 5.1.2).

Attachment to a solid surface, however, poses another problem. At the low Reynolds numbers experienced by the flagellates, the viscous drag associated with the surface will slow down the feeding currents considerably. This again explains the extremely long stalks or flagella employed for attachment, which often bound the cells ten cell diameters or more away from the surface to which they are anchored. This effect may also explain the complex loricas built by acanthoecid choanoflagellates. They often stick to surface films or detrital particles, but the lorica will keep the cell and its filtration apparatus at a considerable distance from the surface. This is quite speculative, however, and seems a rather pedestrian explanation for the adaptive significance of such extraordinary structures.

Flagellates such as species of *Bodo* and colorless euglenoids that graze surfaces for attached bacteria are important in many environments, such as detrital particles, surface films, and sediments (a niche which they share with small amebae). The only quantification of the feeding of such forms, of which I am aware is the experiment shown in Fig. 8, which

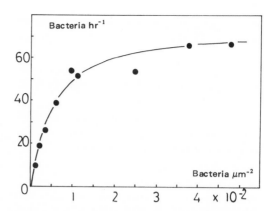

Figure 8. Ingestion of bacteria attached to a coverslip by a *Bodo* sp. as a function of bacterial density and fitted to a hyperbolic function.

depicts the functional response of a *Bodo* sp. grazing on coverslips with different densities of bacteria attached. The data are fitted to a hyperbolic function, the parameters of which are consistent with the fact that the flagellates spend about 45 sec ingesting a bacterium, during which time the flagellate does not move. The flagellate slides over its substrate with a velocity of 3.5 μm sec^{-1} and so only detects and ingests bacteria lying in an ~ 1-μm-wide band along the traveling path of the cell. This is consistent with a cytostome size of about 1 μm.

A final word should be given on the energy spent on particle capture. It is often assumed, and is intuitively supported by the impressive activity displayed under high magnification, that protozoans spend a substantial part of their energy budget on mechanical work. In fact, this is not so. The power requirements of protozoans for mechanical work (swimming and producing water currents for food particle capture) constitute an extremely small part of the total power production. This result can be achieved by assessing the hydrodynamic work in conjunction with reasonable estimates of the conversion efficiency from chemical to mechanical work or by using data in the literature on the power output of isolated cilia [for details, see Fenchel and Finlay (1983)].

4.4. Adaptations to Temporal Heterogeneity

No free-living organisms live in an entirely constant environment and among the factors that vary is food availability. For the organisms discussed in this review this is evident by the fact that they can display balanced growth with rate constants spanning at least a factor of ten. This plasticity in growth rate is certainly adaptive and no doubt the whole span of potential growth rates is realized in nature over a sufficiently long period of time. That this is so for forms living in environments characterized by the spatially and temporally patchy occurrence of resources (e.g., the mass occurrence of bacteria on particulate, decomposing plant, and animal tissue) is obvious. Even in marine plankton, however, the availability of bacteria is sufficiently patchy in time and space that "a feast and famine existence" in the sense of Koch (1971) is an appropriate term with which to describe the life of the microflagellates. The study of balanced growth is therefore not sufficient for a full description of the ecological bioenergetics of these (or other) microorganisms. Unfortunately, little work has so far been carried out on protozoans.

When food availability reaches a certain threshold below which balanced growth cannot be sustained, protozoans undergo a number of adaptive physiological changes. Many forms encyst as a response to adverse conditions and can survive for long periods in the encysted stage, excysting when conditions (e.g., food availability) improve (Corliss and Esser, 1974). Among the heterotrophic microflagellates, cyst formation is well established among the chrysomonads, the choanoflagellates, and the kinetoplastids, and such cysts have been observed in seawater samples and occur in abandoned enrichment cultures (Bourelly, 1963; Booth *et al.,* 1980; Silver *et al.,* 1980; Fenchel, 1982c).

Not all species or clones encyst as a response to starvation, but instead survive in an active state until they eventually die off, and this applies to the six forms studied in pure culture by Fenchel (1982b,c). It is conceivable that this behavior confers a higher fitness in some types of habitats than would encystment. This is because there are two conflicting fitness components involved when an organism faces starvation. One of these is survival for a maximum time period and the other is the ability to intitiate growth and division as rapidly as possible once food again becomes available. Cysts will display a longer lag before they can feed and divide than will active cells, so if the expected duration of famine is not too long (shorter than the life expectancy of active, starving cells) encystment will not be favored.

Fenchel (1982c) studied the physiological and cytological changes following starvation in some flagellates and in particular with respect to the chrysomonad *Ocromonas.* When exponentially growing cells are starved (by transfer to a bacterial-free medium or after exhaustion of food in a batch culture) some effects are immediately observed. One is that the cells continue to divide once or twice, producing cells one-half or one-quarter the size of the parent cells. The number of postgrowth divisions seems to be species dependent. It may also be a function of the previous growth rate, which determines the size of exponentially growing cells (slowly growing cells are very small), but this aspect has not been studied. In *Pseudobodo* and in *Monosiga* characteristic slender and more motile "swarmer cells" are produced by these postgrowth divisions (Fig. 9). The adaptive nature of this behavior is clear: the number of starvation-

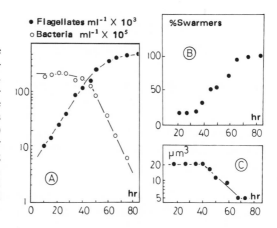

Figure 9. The choanoflagellate *Monosiga* in a batch culture initiated with 2.2 × 10⁷ bacteria/ml. (A) Growth curve and consumption of bacteria. (B) Percentage swarmer cells, which appear as the flagellates start to starve. (C) Mean cell volume; four swarmer cells are formed for each growing cell by two successive divisions, but without increase in total cell volume. [Data from Fenchel (1982b,c).]

swarmers produced per cell should just exceed the factor by which survival is decreased by becoming smaller and in this way the genome maximizes the probability that at least one copy arrives in a more favorable environment. The increased motility of the swarmers should also be understood in this light.

Another immediate effect is a decrease in respiratory rate, which eventually becomes only 2–5% of that of growing cells. This is, of course, crucial for survival. If starving cells maintained the respiratory rate of dividing cells, for example, every 4 hr, they would burn up entirely within about 6 hr. *Ochromonas* cells can starve for about 60 hr (20°C) without mortality; thereafter they die off exponentially with a half-life of about 20 hr. Viability remains 100% until the cells disintegrate. During starvation a number of cytological changes take place. Among the most obvious is the formation of autophagous vacuoles; in particular, mitochondria are digested so that eventually only about 10% of the original mitochondrial volume remains. The loss of mitochondrial volume and the loss of enzymatic activity associated with the electron transfer system does not quite parallel the decrease in respiratory rate, so that a somewhat higher respiratory potential is maintained (Fenchel, 1982c; Finlay *et al.*, 1983).

Previously starved cells immediately feed and increase in cell volume when offered bacteria. However, depending on the duration of the starvation period, the respiratory rate characteristic of growing cells is regained more slowly. Also, an increasing lag time prior to the initiation of the first cell division is evident. After 200 hr of starvation it takes more than 15 hr after refeeding before the first divisions occur. This corresponds to about three generations of balanced growth at the bacterial concentration offered to the cells following starvation.

4.5. Tolerance to Environmental Factors

It is still far from possible to give a comprehensive or even a reasonably informative treatment of the habitat niches of heterotrophic microflagellates. The reason for this is, at least in part, the poor species systematics, which makes comparisons of faunal compositions compiled by different authors for different habitats questionable, in conjunction with the virtual absence of experimental work. An example is the relation to salinity. Acanthoecid choanoflagellates are restricted to seawater, but many forms extend into dilute brackish water (Thomsen, 1973; P. Andersen, personal communication). Within many other groups of flagellates, however, apparently quite similar forms occur in both freshwater and in the sea, although others seem to be specific for one or the other type of habitat. The extent to which entirely euryhaline forms exist can only be determined by experimentation with clonal cultures. Further, the apparent global distribution of forms (see Section 5.1 for references) may in fact reflect morphologically identical but physiologically different forms.

It would seem that extreme environments (hyperhaline lakes, extreme temperatures, anoxia) will yield interesting faunas of microflagellates from a taxonomic as well as a physiological point of view (analogous to procaryotes and ciliated protozoa). Save for a very few notes in the older taxonomic literature, there is nothing to report, presumably because workers have not yet looked for these creatures in such environments.

5. The Habitats

This section discusses the role of microflagellates in different habitats, particularly with respect to their quantitative occurrence and their impact on other organisms. When the size scale of microprotozoans is considered, the traditional distinctions between plankton, sediments, and soils are not very clear-cut. Thus, flagellates associated with suspended particles may be forms adapted to life on solid surfaces and typically occur in sediments, whereas minute volumes of water such as found in the interstitia of coarse sediments may constitute a pelagic environment to a flagellate. Consequently, when flagellates are isolated or observed directly in samples from the environment, one may get the impression of a great diversity of eurytopic species while overlooking microhabitat specializations on a much smaller scale (Lackey, 1961). Therefore, the following subsections represent a somewhat arbitrary classification of the habitats. Most information on the role of microprotozoans derive from terrestrial soils and from marine plankton. The latter will receive the most detailed treatment, since the role of heterotrophic flagellates in marine plankton has drawn much recent attention and much new knowledge has accumulated. The significance of protozoans in soils has long been recognized and has recently been reviewed (Stout, 1980; Stout *et al.*, 1982).

5.1. Marine Plankton

Heterotrophic microflagellates have been recorded from coastal as well as oceanic waters and are known from polar regions, including sea ice (Booth *et al.*, 1980; Silver *et al.*, 1980; Takahashi, 1981; Thomsen, 1982), as well as from tropical regions (Thomsen, 1978; Thomsen and Boonruang, 1983a,b; Landry *et al.*, 1984). The fact that these forms have a substantial impact on the carbon flow of the sea mainly in terms of bacterial consumption was first suggested by work of Pomeroy and Johannes (1968), Sorokin (1977), and Sorokin and Mikheev (1979) on the basis of quantitative occurrence and by, for example, Kopylov and Moiseev (1980) and Haas and Webb (1979) on the basis of their behavior in enrichment cultures.

5.1.1. Quantitative Occurrence and Bacterial Grazing

During the last 5 years, several quantitative studies on microflagellates in marine plankton have been carried out. These results are compiled in Table III together with other pertinent information when available. As a crude generalization, one can say that there is about one flagellate for every 10^3 bacteria or roughly 10^3 flagellates/ml^{-1}. However, there is a considerable variation in flagellate numbers both in terms of temporal fluctuations within localities and in terms of variation from one place to another, and this variation far exceeds that of bacterial populations. Thus, in the Limfjord, the ratio between maximum and minimum population sizes for flagellates is about 70, but only about 7 for bacteria. Similarly, when average values are considered for different localities, the ratio spans a factor of perhaps 30 for the flagellates and at most 10 for the bacteria as based on the data in Table III.

To the extent that flagellates control bacterial populations, this could be expected, because differences in the resource availability for the prey (bacteria) will mainly show up as changes in predator numbers. The number of bacteria will roughly equilibrate at the level that can just support flagellate growth and this seems to be within the range $(0.5-2) \times 10^6$ ml^{-1}. The population fluctuations will be discussed in more detail in Section 5.1.2.

On the basis of values for clearance estimated in laboratory batch

Table III. Quantitative Occurrence and Clearance of Microflagellates in Marine Plankton

Locality	Heterotrophic flagellates		Bacteria (Number ml⁻¹ × 10⁶)	Estimated clearance (% day⁻¹)	Remarks	Reference
	Number ml⁻¹ × 10³	µg C l⁻¹ × 10³				
Limfjord	0.2–3	—	1.4–3.3	10–70 (av. 20)	0–7 m, August	Fenchel (1982d)
Limfjord	0.2–14	0.4–30	2–14	5–365 (av. 45)	0.7 m, March–July	P. Andersen (unpublished results)
Off Peru	0.2–13	—	—	—	0–100 m, most cells, 0–30 m	Sorokin and Mikheev (1979)
Japan Sea	0.7–9.5	—	0.5–1.4	—	Heterotrophic succession phase	Sorokin (1977)
Black Sea	0.6–1.5	—	—	—	Aerobic zone	Moiseev (1980)
Atlantic, off Georgia	0.3–3.3	0.6–9.5	2–10	30–50	Shore to 15 km off shore	Sherr et al. (1984)
Atlantic off Georgia	0.1–0.5	—	—	—	25–120 km off shore	Sherr and Sherr (1983)
Around Woods Hole	1–2.7	—	—	—	—	Caron (1983)
Sargasso Sea	0.2–0.7	—	—	—	—	Caron (1983)
Narragansett Bay	0.9–37	—	—	—	—	Davis and Sieburth (1982)
Atlantic shelf	0.9–5.1	—	—	—	—	Davis and Sieburth (1982)
Gulf Stream	0.9–1.4	—	—	—	—	Davis and Sieburth (1982)
Sargasso Sea	0.9–1.3	—	—	—	—	Davis and Sieburth (1982)
Bering Sea	0.01–0.1	—	0.6–2.4	0.2–2.4	0–30 m, August	P. Andersen (unpublished results)
English Channel	—	0.2–3.3	0.3–2.4	—	0–60 m	Linley et al. (1983)
Kaneohe Bay, Hawaii	0.9	—	0.8	—	—	Landry et al. (1984)

cultures of a representative number of species and in conjunction with quantitative data from the field, it is possible to calculate the fraction of the water column that is cleared for bacteria per day. Fenchel (1982d) and P. Andersen (unpublished results) did this for the Limfjord, a wide sound connecting the North Sea with the Kattegat. Fenchel's data from the month of August suggested that on the average 20% (range 10–70%) of the water column is cleared for bacteria per day. Andersen's data, extending over the period March–July 1983, suggest an average of 45%, ranging from 5 to 365% of the water column per day. If there are no causes of bacterial mortality other than flagellate grazing, these results suggest that bacterial generation times range from 6 hr to 20 days and average 2.2 days Sherr et al. (1984) and Andersen and Fenchel (1985) estimated flagellate clearance from changes in population sizes in incubations of freshly collected seawater samples; these results agree reasonably well with those from laboratory cultures. Using their own data and counts of flagellates from Georgia off-shore localities, Sherr et al. (1984) calculated that the protozoans clear 30–50% of the water column for bacteria per day.

These results must be taken with a certain amount of reservation. They are more or less based on the extrapolation of in vitro experiments to in situ conditions. One of the problems is that the laboratory-grown bacteria used in experiments are larger than those in natural seawater and this may lead to an overestimate of in situ grazing rates. The fact that some pigmented microflagellates also consume bacteria may result in an underestimate of flagellate grazing, since they have been excluded from the number of flagellates on which calculations are based. Therefore, the methods and assumptions on which these grazing estimates are based warrant more detailed and critical studies. Also, such studies must be extended to different regions before very solid generalizations can be made. Still, the average values and range of the available data are quite consistent with contemporary estimates of bacterial generation times in seawater (Azam et al., 1983), and the evidence that heterotrophic microflagellates are the main consumers of bacteria on the water column is well supported.

Another demonstration of the role of microflagellates was given by Linley et al. (1981), Newell et al. (1981), and Stewart et al. (1981). These authors incubated fragmented kelp, kelp mucilage, or phytoplankton debris with seawater and followed the microbial successions. In these laboratory simulations of a hetertrophic plankton succession, bacterial numbers quickly increased. This increase was followed by an increase in flagellate numbers, which caused a decline in the bacterial populations. When the flagellate populations peaked, the bacterial numbers had fallen to about 10^6 ml^{-1}. Ciliates and amebae occurred later in the successions. Similar results were obtained by Robertson et al. (1982) in experiments on microbial flocculation based on dissolved organic matter.

Whereas the importance of microflagellates as bacterial grazers in marine plankton is well supported, other groups of organisms may also play a role in this respect. Small bacterivorous amebae are omnipresent (Davis *et al.*, 1978); these forms are presumably important on suspended particles and at the water–air interface. Suspended particles and "marine snow" also harbor ciliates, some of which are bacterivorous (Caron *et al.*, 1982). In oligotrophic waters the great majority of ciliates are feeders on particles larger than bacteria (Fenchel, 1980a), but in eutrophic estuaries, bacterivorous ciliates seem to be important as well (Burkill, 1982). There are very few convincing reports on metazoan suspension feeders that can retain and utilize bacteria (King *et al.*, 1980; Fenchel, 1984).

Not much is known about the role of microflagellates as food for other organisms. Their size range means that many protozoan and metazoan suspension feeders will retain them, and many ciliates feed and grow on these flagellates in culture (Jørgensen, 1975; Fenchel, 1980b), but their role in the planktonic food web remains to be studied in detail.

5.1.2. Population Oscillations

Fenchel (1982d) found during a 5-week period in the Limfjord that bacterial and flagellate populations oscillated in size, and interpreted this in terms of coupled prey–predator oscillations (Fig. 10). In this area, tidal as well as net flow of water is so low that spatial heterogeneity cannot explain the data. Laake *et al.* (1983) found a similar phenomenon in experimental seawater enclosures. In both papers, attempts to model the

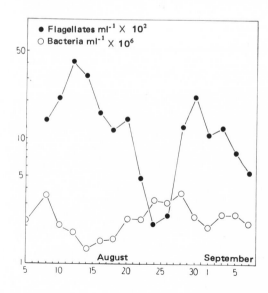

Figure 10. Numbers of heterotrophic microflagellates and of bacteria in the surface waters of the Limfjord during late summer 1981. [Data from Fenchel (1982d).]

situation were made and a reasonable agreement between amplitude and cycle periods on the one hand and parameters of growth and feeding on the other was found. More recently, P. Andersen (unpublished results) followed bacterial and flagellate populations in the Limfjord over a 5-month period. The population curves show many irregularities, which are probably explained by changes in the availability of resources for the bacteria and in the predation on the flagellates by larger plankton organisms. However, six peaks in bacterial density each followed by a peak in flagellate density could be demonstrated. The cycle period of the oscillations was about 10 days during summer, increasing to about 18 days during spring and autumn, when the water was cooler. The flagellate peaks lagged 1/4–1/3 cycle period behind the bacterial peaks (Fig. 11).

Such oscillations will be initiated by an increase in bacterial resources; e.g., the decay of a phytoplankton bloom as observed by Sorokin (1977) and Sherr *et al.* (1982). Certain characteristics of these oscillations, however, contain more general information about the system. The cycle period of the oscillations is mainly a function of the potential growth rate of the bacteria and the death rate of the flagellates. The stability of the system (that is, whether it is damped or whether it enters a stable limit cycle of smaller or greater amplitude) is determined by the ratio of the carrying capacity of the prey (the level of bacterial resources) to the equilibrium or average prey population size (which is determined by the minimum bacterial density that supports flagellate growth). According to May (1973), the ratio of the minimum and mean predator population sizes is roughly given by $\exp[-c(K/\bar{x})^2]$, where K is the carrying capacity of the prey, $\bar{x}$ is its mean population size, and c is a con-

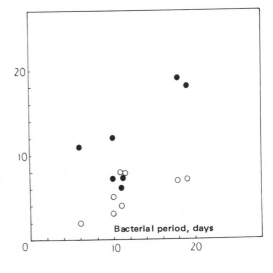

Figure 11. Periods between successive peaks in flagellate abundance plotted against corresponding periods between bacterial peaks for the Limfjord, summer 1983 (●). Also shown is the lag time between a bacterial peak and the following flagellate peak (○). The data strongly suggest a coupling between these events. [P. Anderson (unpublished data).]

stant of order unity. Hence, one should expect greater fluctuations in flagellate numbers in eutrophic waters, but rather more stability in oligotrophic situations. This is shown in Fig. 12, based on data from Table III; the fluctuations in flagellate numbers are expressed as the logarithm of the ratio between the maximum and minimum population densities recorded. It is seen that, while in estuarine and upwelling areas flagellate numbers may fluctuate by a factor of about 60, they show nearly constant numbers in the oligotrophic Sargasso Sea. This shows in a general way that in the latter area the flagellates just manage to keep the bacterial numbers slightly below their carrying capacity (the numbers they would attain in the absence of predation). In more eutrophic areas the mean number of bacteria is kept well below the carrying capacity by grazing.

5.1.3. Spatial Distribution in the Water Column

It was previously mentioned (Section 4.3) that many flagellates tend to anchor themselves to solid surfaces, and that this behavior facilitates suspension feeding. In addition, some forms (and amebae) are specialized to feed on attached bacteria. There is evidence that many "pelagic" flagellates are in fact associated with suspended particles in nature. This was directly observed by Caron et al. (1982) on suspended particles from an oceanic environment. Fenchel (1982d) found that filtration of water samples through 20-μm nylon gauze reduced flagellate numbers by about

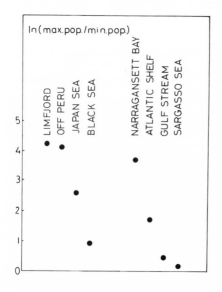

Figure 12. Logarithm of the ratio between maximum and minimum flagellate densities recorded from eight marine areas, based on the data in Table III.

60%, whereas numbers of bacteria and of photosynthetic nanoflagellates were not affected; it was assumed that the missing cells were attached to particles too large to pass through the filter.

None of the microflagellates cultured in the laboratory so far exhibit negative geotaxis. The sinking velocity of immobilized cells is within the range 0.3–0.6 μm sec^{-1}. In unstirred cultures, free-swimming cells therefore tend to concentrate close to the bottom (T. Fenchel, unpublished results). In the open water, however, turbulent diffusion would totally override the effect of gravitational sinking, and, in the absence of growth, one would expect a fairly even distribution through the water column. (This argument does not hold, of course, if the cells are attached to larger detrital particles.) Fenchel (1982d and unpublished results) found the flagellates evenly distributed with depth in the relatively shallow (8–10 m) Limfjord during windy periods. During calm weather there was a clear maximum in the upper 2–3 m of the water column. This distribution probably reflects higher growth rates close to the surface, which overrides the effect of turbulence. Sorokin's (1977, 1979) data from the Japan Sea and off the coast of Peru clearly show peak abundances, coinciding with maximum bacterial densities and growth rates. Presumably, growth and turbulent diffusion are the dominating factors determining the vertical distribution, whereas the motility of the cells themselves and gravitational sinking play a diminishing role.

5.2. Freshwater Plankton

While the presence of heterotrophic microflagellates is well established in freshwater (for references, see Section 2), there are very few studies on the quantitative occurrence in lake water. It is reasonable to expect a role similar to that in the sea. Sorokin and Paveljeva (1972) give some data from a lake in Kamchatka. Sherr *et al.* (1982) found a mass occurrence of heterotrophic microflagellates following the decay of a dinoflagellate bloom in Lake Tiberias, Israel.

5.3. Aquatic Sediments

As in the case of freshwater plankton, there is evidence of the presence of microflagellates in aquatic sediments, but very little knowledge on their ecological significance. In particular, the paper by Lackey (1961) illustrates the diversity and microhabitat specializations of colorless euglenoids in a marine sediment.

In marine organic detritus derived from seagrasses, mangrove leaves, and similar materials, between 10^7 and 10^8 flagellates/g dry weight is found, thus making up a biomass comparable to that of bacteria (Fenchel,

1970; Fenchel and Jørgensen, 1977), and similar numbers were found in an organic sediment in an arctic tundra pond (Fenchel, 1975). As in the plankton, the ratio of numbers of flagellates to bacteria is roughly $1:10^3$. An indication of the role of microflagellates in controlling bacteria in sediments derives from observations on the microbial succession on previously sterilized sediment or detrital material incubated in seawater or freshwater (Fenchel, 1970; Fenchel and Harrison, 1976; Fenchel and Jørgensen, 1977). Initially, bacteria increase in numbers, but then decrease as microflagellate populations develop. Control systems devoid of microprotozoans show much higher bacterial levels than do grazed ones. As discussed in some detail by Fenchel and Harrison (1976), protozoan grazing somehow accelerates bacterial degradation of particulate organic matter, but the exact mechanism involved is not clear.

5.4. Sewage

There is much literature to show that heterotrophic microflagellates play a considerable quantitative role in sewage purification systems. Curds (1965), for example, found that during the first days, microflagellates constitute nearly 100% of the protozoan fauna of activated sludge, only then to be followed by ciliates, which have so far attracted much more attention in the microbial ecology of sewage treatment. A memoir on flagellates associated with sewage treatment has been published by Hänel (1979), to which the reader is referred for older references. This paper advocates the use of microflagellates for monitoring waste water and the purification process; in practice, the usefulness of this approach is probably limited by the difficulty of species identification.

5.5. Terrestrial Soils

In contrast to the situation for aquatic environments, the role of microprotozoans in the turnover of bacterial populations has long been recognized and was demonstrated experimentally over 60 years ago (Cutler and Crump, 1920; Cutler et al., 1922; Cutler, 1927), including the demonstration of coupled oscillations of bacterial and protozoan populations. The literature on soil protozoans has been reviewed by Stout (1980) and Stout et al., (1982). In arable and unplanted soils, from 6×10^2 to 6.4×10^4 flagellates g^{-1} dry weight has been recorded (Darbyshire and Greaves, 1967), but this is about one order of magnitude lower than numbers found for small amebae, which play an accordingly larger role in such soils (Clarholm, 1981). In swamps and forest litter, however, flagellate numbers and diversity are higher, up to 3×10^6 g^{-1} dry weight of soil (Bamforth, 1976). The fauna is dominated by colorless chrysomonads and euglenoids, bodonids, and ameboid flagellates (Bamforth, 1967).

6. Concluding Remarks

This review has been concerned with a wide, if not nebulous, topic, that is, all aspects of the ecology and functional biology of a group of organisms that do not form a natural taxonomic group, but rather constitute one life form among the eucaryotes and are mainly characterized by their small size and bacterivory. It is becoming clear that they play a key role in most ecosystems as consumers of bacteria and as a link in the food chains between bacteria and larger phagotrophic organisms. In spite of the very long reference list cited, there is still much to be learnt about the biology of microflagellates. Most effort so far has been concerned with naming species and with counting them in nature. This approach will still be necessary for some time to come. However, the greatest scope for new insight lies, I believe, in experimental approaches to problems on the borderline between cell physiology and ecology. This is still a very neglected area, but will be the only way to obtain a more complete understanding of the diversity and biology of these organisms.

ACKNOWLEDGMENTS. I thank Dr. P. Andersen for allowing me to use some of his unpublished data. I am also grateful to Annie Sølling for technical assistance and photographic work and to Marianne Szygenda for typing the manuscript. My studies on microprotozoans have been supported by grants 20949, 11-3650, and 11-4688 from the Danish Natural Science Research Council.

References

Aaronson, S., 1980, Descriptive biochemistry and physiology of the Chrysophyceae (with some comparisons to Prymnesiophyceae), in: *Biochemistry and Physiology of Protozoa,* Vol. 3 (M. Levandowsky and S. H. Hutner, eds.), pp. 117–169, Academic Press, New York.

Andersen, P., and Fenchel T., 1985, Bacterivory by microheterotrophic flagellates in seawater samples, *Limnol. Oceanogr.* **30:**198–202.

Azam, F., Fenchel, T., Field, J. G., Gray, J. S., Meyer-Reil, L. A., and Thingstad, F., 1983, The ecological role of water-column microbes in the sea, *Mar. Ecol. Prog. Ser.* **10:**257–263.

Bamforth, S. S., 1967, A microbial comparison of two forest soils of southeastern Louisiana, *Proc. La. Acad. Sci.* **30:**7–16.

Bamforth, S. S. 1976, Rhizosphere–soil microbial comparisons in sub-tropical forests of southeastern Louisiana, *Trans. Am. Microsc. Soc.* **95**:613–621.

Booth, B. C., Lewin, J., and Norris, R. E., 1980, Siliceous nanoplankton. I. Newly discovered cysts from the Gulf of Alaska, *Mar. Biol.* **58**:205–209.

Boucaud-Camou, E., 1966, Le choanoflagellés des cotes de la Manche: I. Systematique, *Bull. Soc. Linn. Normandie Ser. 10* **7**:191–209.

Bourrelly, P., 1963, Loricae and cysts in the Chrysophyceae, *Ann. N. Y. Acad. Sci.* **108**:421–429.

Burkill, P. H., 1982, Ciliates and other microplankton components of a nearshore food-web: Standing stocks and production processes, *Ann. Inst. Oceanogr. Paris* **58**:335–350.

Calow, P., 1977, Conversion efficiencies in heterotrophic organisms, *Biol. Rev.* **52**:385–409.

Caron, D. A., 1983, Technique for enumeration of heterotrophic and phototrophic nanoplankton, using epifluorescence microscopy, and comparison with other procedures, *Appl. Environ. Microbiol.* **46**:491–498.

Caron, D. A., Davis, P. G. Madin, L. P., and Sieburth J. McN., 1982, Heterotrophic bacteria and bacivorous protozoa in oçeanic macroaggregates, *Science* **218**:795–797.

Clarholm, M., 1981, Protozoan grazing of bacteria in soil—Impact and importance, *Microb. Ecol.* **7**:343–350.

Corliss, J. O., 1983, A puddle of protists, *Sciences* **23**(3):35–39.

Corliss, J. O., and Esser, S. C., 1974, Comments on the role of the cyst in the life cycle and survival of free-living protozoa, *Trans. Am. Microc. Soc.* **93**:578–593.

Curds, C. R., 1965, An ecological study of the ciliated protozoa in activated sludge, *Oikos* **15**:282–289.

Curds, C. R., and Cockburn, A., 1968, Studies on the growth and feeding of *Tetrahymena pyriformis* in axenic and monoxenic culture, *J. Gen. Microbiol.* **54**:343–358.

Curds, C. R., and Cockburn, A., 1971, Continuous monoxenic culture of *Tetrahymena pyriformis, J. Gen. Microbiol.* **66**:95–108.

Cutler, D. W., 1927, Soil protozoa and bacteria in relation to their environment, *J. Queckett. Microsc. Club* **15**:309–330.

Cutler, D. W., and Crump, L. M., 1920, Daily periodicity in the numbers of active soil flagellates: With a brief note on the relation of trophic amoebae and bacterial numbers, *Ann. Appl. Biol.* **7**:11–24.

Cutler, D. W., Crump, L. M., and Sandon, H., 1922, A quantitative investigation of the bacterial and protozoan population of the soil, with an account of the protozoan fauna, *Phil. Trans. R. Soc. Lond. B* **211**:317–350.

Darbyshire, J. F., and Greaves, M. P., 1967, Protozoa and bacteria in the rhizosphere of *Sinapis alba* L., *Trifolium repens* L., and *Lolium perenne* L., *Can. J. Microbiol.* **13**:1057–1068.

Darbyshire, J. F., Wheatley, R. E., Greaves, M. P., and Inkson, R. H., 1974, A rapid micromethod for estimating bacterial and protozoan populations in soil, *Rev. Ecol. Biol. Sol.* **11**:465–475.

Davidson, L. A., 1982, Ultrastructure, behavior, and algal flagellate affinities of the helioflaggellate *Ciliophrys marina,* and the classification of the helioflagellates (Protista, Actinopoda, Heliozoea), *J. Protozool.* **29**:19–29.

Davis, P. G., and Sieburth, J. McN., 1982, Differentiation of phototrophic and heterotrophic nanoplankton populations in marine waters by epifluorescence microscopy, *Ann. Inst. oceanogr. Paris* **58**:249–260.

Davis, P. G., Caron, D. A., and Sieburth, J. McN., 1978, Oceanic amoebae from North Atlantic: Culture, distribution and taxonomy, *Trans. Am. Microsc. Soc.* **96**:73–88.

Englund, P. T., 1981, Kinetoplast DNA, in: *Biochemistry and Physiology of Protozoa,* Vol. 4 (M. Levandowsky and S. H. Hutner, eds.), pp. 334–383, Academic Press, New York.

Eyden, B. P., 1977, Morphology and ultrastructure of *Bodo designis* Skuja 1948, *Protistologica* **13**:169–179.

Fenchel, T., 1970, Studies on the decomposition of organic detritus derived from the turtle grass *Thalassia testudinum*, *Limnol. Oceanogr.* **15**:14–20.

Fenchel, T., 1975, The quantitative importance of the benthic microfauna of an arctic tundra pond, *Hydrobiologia* **46**:445–464.

Fenchel, T., 1980a, Relation between particle size selection and clearance in suspension feeding ciliates, *Limnol. Oceanogr.* **25**:733–738.

Fenchel, T., 1980b, Suspension feeding in ciliated protozoa: Functional response and particle size selection, *Microb. Ecol.* **6**:1–11.

Fenchel, T., 1982a, Ecology of heterotrophic microflagellates. I. Some important forms and their functional morphology, *Mar. Ecol. Prog. Ser.* **8**:211–223.

Fenchel, T., 1982b, Ecology of heterotrophic microflagellates. II. Bioenergetics and growth, *Mar. Ecol. Prog. Ser.* **8**:225–231.

Fenchel, T., 1982c, Ecology of heterotrophic microflagellates. III. Adaptations to heterogenous environments, *Mar. Ecol. Prog. Ser.* **9**:25–33.

Fenchel, T., 1982d, Ecology of heterotrophic microflagellates. IV. Quantitative occurrence and importance as consumers of bacteria, *Mar. Ecol. Prog. Ser.* **9**:35–42.

Fenchel, T., 1982e, The bioenergetics of a heterotrophic microflagellate, *Ann. Inst. Oceanogr. Paris* **58**:55–60.

Fenchel, T., 1984, Suspended marine bacteria as food source, in: *Flow of Energy and Materials in Marine Ecosystems* (M. J. Fasham, ed.), pp. 301–314, Plenum Press, New York.

Fenchel, T., and Finlay, B. J., 1983, Respiration rates in heterotrophic, free-living Protozoa, *Microb. Ecol.* **9**:99–122.

Fenchel, T., and Harrison, P., 1976, The significance of bacterial grazing and mineral cycling for the decomposition of particulate detritus, in: *The Role of Terrestrial and Aquatic Organisms in Decomposition Processes* (J. M. Anderson and A. Macfayden, eds.), pp. 285–299, Blackwell, Oxford.

Fenchel, T., and Jørgensen, B. B., 1977, Detritus food chains of aquatic ecosystems: The role of bacteria, in: *Advances in Microbial Ecology*, Vol. 1 (M. Alexander, ed.), pp. 1–58, Plenum Press, New York.

Finlay, B. J., Span, A., and Ochsenbein-Gattlen, C, 1983, Influence of physiological states on indices of respiration rate in protozoa, *Comp. Biochem. Physiol.* **74A**:211–219.

Grassé, P.-P., 1952, *Traité de Zoologie*, 1.1. *Protozoaires*, Masson, Paris.

Grassé, P.-P., 1953, *Traité de Zoologie*, 1.2. *Protozoaires*, Masson, Paris.

Griessmann, K., 1914, Über marine Flagellaten, *Arch. Protistenkd.* **32**:1–78.

Haas, L. W., 1982, Improved epifluorescence microscopy for observing planktonic microorganisms, *Ann. Inst. Oceanogr. Paris* **58**:261–266.

Haas, L. W., and Webb, K. L., 1979, Nutritional mode of several non-pigmented microflagellates from the York River Estuary, Virginia, *J. Exp. Mar. Biol. Ecol.* **39**:125–134.

Hänel, K., 1979, Systematik und Ökologie der farblosen Flagellaten des Abwassers, *Arch. Protistenkd.* **121**:73–137.

Hobbie, J. E., Daley, R. J., and Jasper, S., 1977, Use of Nucleopore filters for counting bacteria by fluorescence microscopy, *Appl. Environ. Microbiol.* **33**:1225–1228.

Hollande, A., 1942, Étude cytologique et biologique de quelques flagellés libres, *Arch. Zool. Exp. Gen.* **83**:1–268.

Holwill, M. E. J., 1974, Hydrodynamic aspects of ciliary and flagellar movement, in: *Cilia and Flagella*, M. A. Sleigh; ed.), pp. 143–175, Academic Press, London.

Jørgensen, C. B., 1975, Comparative physiology of suspension feeding, *Annu. Rev. Physiol.* **37**:57–79.

King, K. R., Hollibaugh, J. T., and Azam, F., 1980, Predator–prey interactions between the larvacean *Oikopleura dioica* and bacterioplankton in enclosed water columns, *Mar. Biol.* **56**:49–57.

Koch, A. L., 1971, The adaptive responses of *Escherichia coli* to a feast and famine existence, *Adv. Microb. Physiol.* **6**:147–217.

Kopylov, A. I., and Moiseev, E. V., 1980, Effect of colorless infusoria on the estimation of bacterial production in seawater, *Doklady Akad. Nauk SSSR* **252**:503–505 (in Russian).

Kopylov, A. I., Mamayeva T. I., and Batsanin, S. F., 1980, Energy balance of the colorless flagellate *Parabodo attenuatus* (Zoomastigophora, Protozoa), *Oceanology* **20**:705–708.

Laake, M., Dahle, A. B., Eberlein, K., and Rein, K., 1983, A modelling approach to the interplay of carbohydrates, bacteria and non-pigmented flagellates in a controlled ecosystem experiment with *Skeletonema costatum, Mar. Ecol. Prog. Ser.* **14**:71–79.

Lackey, J. B., 1936, Occurrence and distribution of the marine protozoan species in the Woods Hole area, *Biol. Bull.* **70**:264–278.

Lackey, J. B., 1961, Bottom sampling and environmental niches, *Limnol. Oceanogr.* **6**:211–279.

Lackey, J. B., 1962, Three new colorless Euglenophyceae from marine situations, *Arch. Microbiol.* **42**:190–195.

Lackey, J. B., 1963, The microbiology of a Long Island bay in the summer of 1961, *Int. Rev. Gesamten Hydrobiol.* **48**:577–601.

Landry, M. R., Haas, L. W., and Fagerness, V. L., 1984, Dynamics of microbial plankton communities: Experiments in Kaneohe Bay, Hawaii, *Mar. Ecol. Prog. Ser.* **16**:127–133.

Laval, M., 1971, Ultrastructure et mode de nutrition du choanoflagellé *Salpingoeca pelagica* sp. nov. Comparaison avec les choanocytes des spongiaries, *Protistologica* **7**:325–336.

Leadbeater, B. S. C., 1972a, Fine-structural observations on some marine choanoflagellates from the coast of Norway, *J. Mar. Biol. Assoc. U. K.* **52**:67–79.

Leadbeater, B. S. C., 1972b, Identification, by means of electron microscopy, of flagellate nanoplankton from the coast of Norway, *Sarsia* **49**:107–124.

Leadbeater, B. S. C., 1972c, Ultrastructural observations on some marine choanoflagellates from the coast of Denmark, *Br. Phycol. J.* **7**:195–211.

Leadbeater, B. S. C., 1972d, *Paraphysomonas cyclocophora* sp. nov., a marine species from the coast of Norway, *Nor. J. Bot.* **19**:179–185.

Leadbeater, B. S. C., 1974, Ultrastructural observations on nanoplankton collected from the coast of Jugoslavia and the bay of Algiers, *J. Mar. Biol. Assoc. U. K.* **54**:179–196.

Leadbeater, B. S. C., 1977, Observations on the life-history and ultrastructure of the marine choanoflagellate *Choanoeca perplexa* Ellis, *J. Mar. Biol. Assoc. U. K.* **57**:285–301.

Leadbeater, B. S. C., 1983, Life-history and ultrastructure of a new marine species of Proterospongia (Choanoflagellida), *J. Mar. Biol. Assoc. U. K.* **63**:135–160.

Leadbeater, B. S. C., and Manton, I., 1974, Preliminary observations on the chemistry and biology of the lorica in a collared flagellate (*Stephanoeca diplocostata* Ellis), *J. Mar. Biol. Assoc. U. K.* **54**:269–276.

Leadbeater, B. S. C., and Morton, C., 1974, A microscopical study of a marine species of *Codosiga* James-Clark (Choanoflagellata) with special reference to the ingestion of bacteria, *Biol. J. Linn. Soc.* **6**:337–347.

Levine, N. D., Corliss, J. O., Cox, F. E. G., Deroux, G., Grain, J., Honigberg, B. M., Leedale, G. F., Loeblich, A. R., III, Lom, J., Lynn, D., Merinfeld, E. G., Page, F. C., Poljansky, G., Sprague, V., Vávra, J., and Wallace, F. G., 1980, A newly revised classification of the protozoa, *J. Protozool.* **27**:37–58.

Lighthart, B., 1969, Planktonic and benthic bacterivorous protozoa at eleven stations in Puget Sound and adjacent Pacific Ocean, *J. Fish. Res. Board Can.* **26**:299–304.

Lighthill, J., 1976, Flagellar hydrodynamics, *SIAM Rev.* **18**:161–230.

Linley, E. A. S., Newell, R. C., and Bosma, S. A., 1981, Heterotrophic utilization of mucilage released during fragmentation of kelp (*Ecklonia maxima* and *Laminaria pallida*). I. Development of microbial communities associated with the degradation of kelp mucilage, *Mar. Ecol. Prog. Ser.* **4**:31–41.

Lohmann, H., 1911, Über das Nannoplankton und die Zentrifugierung Kleinster Wasserproben zur Gewinnung desselben in lebendem Zustande, *Int. Rev. Gesamten Hydrobiol. Hydrogr.* **4**:1–38.

Lohmann, H., 1920, Die Bevölkerung des Ozeans mit Plankton, *Arch. Biontol. (Ges. Naturforsch. Freunde)* **4**:1–617.

Lucas, I. A. N., 1968, A new member of the chrysophyceae bearing polymorphic scales, *J. Mar. Biol. Assoc. U. K.* **48**:437–441.

May, R. M., 1973, *Stability and Complexity in Model Ecosystems,* Princeton University Press, Princeton, New Jersey.

Mignot, J. P., 1965, Étude ultrastructurale de *Cyathomonas truncata* From (flaggellé cryptomonadine), *J. Microsc. (Paris)* **4**:239–252.

Mignot, J. P., 1974, Étude ultrastructurale des *Bicoeca,* protistes flagellés, *Protistologica* **10**:543–565.

Moestrup, Ø., 1982, Flagellar structure in algae: A review, with new observations particularly on the Chrysophyceae, Phaeophyceae (Fucophyceae), Euglenophyceae, and *Reckertia, Phycologia* **21**:427–528.

Moestrup, Ø., and Thomsen, H. A., 1976, Fine structural studies on the flagellate genus *Bicoeca, Protistologica* **12**:101–120.

Moiseev, E. V., 1980, The zooflagellates in the open parts of the Black Sea, in: *Pelagic Ecosystem of the Black Sea* (P. P. Chirchowa, ed.), pp. 174–179, Nauka, Moscow (in Russian).

Newell, R. C., Lucas, M. I., and Linley, E. A. S., 1981, Rate of degradation and efficiency of conversion of phytoplankton debris by marine micro-organisms, *Mar. Ecol. Prog. Ser.* **6**:123–136.

Page, F. C., 1976, *An Illustrated Key to Soil Amoebae,* Freshwater Biological Association, Scientific Publication no. 34, Ambleside, Cumbria, England.

Parke, M., and Leadbeater, B. S. C., 1977, Check-list of British marine choanoflagellida—Second revision, *J. Mar. Biol. Assoc. U. K.* **57**:1–6.

Patterson, P. J., and Fenchel, T., Insights into the evolution of heliozoa (Protozoa, Sarcodina) as provided by ultrastructural studies on a new species of flagellate from the genus *Pteridomones, Biol. J. Linn. Soc.* **34**:381–403.

Pennick, N. C., and Clarke, K. J., 1972, *Paraphysomonas butcheri* sp. nov. a marine, colourless, scale-bearing member of the chrysophyceae, *Br. Phycol. J.* **7**:45–48.

Pomeroy, L. R., and Johannes, R. E., 1968, Occurrence and respiration of ultraplankton in the upper 500 metres of the ocean, *Deep-Sea Res.* **15**:381–391.

Pratt, D. M., 1959, The phytoplankton of Narragansett Bay, *Limnol. Oceanogr.* **4**:425–440.

Preisig, H. R., and Hibberd, D. J., 1982, Ultrastructure and taxonomy of *Paraphysomonas* (Chrysophyceae) and related genera 1, *Nord. J. Bot.* **2**:397–420.

Pringsheim, E. G., 1952, On the nutrition of *Ochromonas, Q. J. Microsc. Sci.* **93**:71–96.

Robertson, M. L., Mills, A. L., and Zieman, J. C., 1982, Microbial synthesis of detritus-like particles from dissolved organic carbon released by tropical seagrasses, *Mar. Ecol. Prog. Ser.* **7**:279–285.

Ruinen, J., 1938, Notizen über Salzflagellaten. II. Über die Verbreitung der Salzflagellaten, *Arch. Protistenkd.* **90**:210–258.

Sherr, E. B., and Sherr, B. F., 1983, A double-staining epifluorescence technique to assess frequency of dividing cells and bacteriovory in natural populations of heterotrophic microprotozoa, *Appl. Environ. Microbiol.* **46**:1388–1393.

Sherr, B. F., Sherr, E. B., and Berman, T., 1982, Decomposition of organic detritus: A selective role for microflagellate protozoa, *Limnol. Oceanogr.* **27**:765–769.

Sherr, B. F., Sherr, E. B., and Berman, T., 1983, Grazing, growth and ammonium excretion rates of a heterotrophic microflagellate fed with four species of bacteria, *Appl. Environ. Microbiol.* **45**:1196–1201.

Sherr, B. F., Sherr, E. B., and Newell, S. Y., 1984, Abundance and productivity of heterotrophic nanoplankton, *J. Plankt. Res.* **6**:195–203.

Sieburth, J. McN., 1979, *Sea Microbes,* Oxford University Press, New York.

Sieburth, J. McN., and Davis, P. G., 1982, The role of heterotrophic nanoplankton in the grazing and nurturing of planktonic bacteria in the Sargasso and Carribean Seas, *Ann. Inst. Oceanogr. Paris* **58**:285–296.

Silver, M. W., J. G. Mitchell, and Ringo, D. C., 1980, Silicious nanoplankton. II. Newly discovered cysts and abundant choanoflagellates from the Weddell Sea, Antarctica, *Mar. Biol.* **58**:211–217.

Skuja, H., 1939, Beitrag zur Algenflora Lettlands II, *Acta Horti Bot. Univ. Latv.* **11/12**:41–169.

Skuja, H., 1948, Taxonomie des Phytoplanktons einiger Seen in Uppland, Schweden, *Sym. Bot. Ups.* **9**:1–399.

Skuja, H., 1956, Taxonomische und biologische Studien über das Phytoplankton schwedischer Binnengewässer, *Nova Acta Reg. Soc. Sientarum Ups. Ser. 4,* **16**:1–404.

Sleigh, M. A., 1964, Flagellar movement of the sessile flagellates *Actinomonas, Codonosiga, Monas* and *Poteriodendron, Q. J. Microsc. Sci.* **105**:405–415.

Sorokin, Y. I., 1977, The heterotrophic phase of plankton succession in the Japan Sea, *Mar. Biol.* **41**:107–117.

Sorokin, Y. I., 1979, Zooflagellates as a component of the community of eutrophic and oligotrophic waters in the Pacific Ocean, *Oceanology* **19**:316–319.

Sorokin, Y. I., 1981, Microheterotrophic organisms in marine ecosystems, in: *Analysis of Marine Ecosyststems* (A. R. Longhurst, ed.), pp. 293–342, Academic Press, New York.

Sorokin, Y. I., and Mikheev, V. N., 1979, On characteristics of the Peruvian upwelling ecosystem, *Hydrobiologica* **62**:165–198.

Sorokin, Y. I., and Paveljeva, E. B., 1972, On the quantitative characteristics of the pelagic ecosystem of Dalmee Lake (Kamchatka), *Hydrobiologica* **40**:519–552.

Stewart, V., Lucas, M. I., and Newell, R. C., 1981, Heterotrophic utilization of particulate matter from the kelp *Laminaria pallida, Mar. Ecol. Prog. Ser.* **4**:337–348.

Stout, J. D., 1980, The role of protozoa in nutrient cycling and energy flow, in: *Advances in Microbial Ecology,* Vol. 4 (M. Alexander, ed.), pp. 1–50, Plenum Press, New York.

Stout, J. D., Bamforth, S. S., and Lousier, J. D., 1982, Protozoa, in: *Methods of Soil Analysis, 2. Chemical and Microbiological Properties,* pp. 1103–1120, Agronomy Monograph 9, American Society of Agronomy–Soil Science Society of America, Madison, Wisconsin.

Swale, E. M. F., 1969, A study of the Nanoplankton flagellate *Pedinella hexacostata* Vysotskii by light and electron microscopy, *Br. Phycol. J.* **4**:65–86.

Swale, E. M. F., 1973, A study of the colourless flagellate *Rhynchomonas nasuta* (Stokes) Kent, *Biol. J. Linn. Soc.* **5**:255–264.

Takahashi, E., 1981, Loricate and scale-bearing protists from Lützow-Holm Bay, Antarctica, *Antarct. Rec.* **73**:1–22.

Thomsen, H. A., 1973, Studies on marine choanoflagellates I. Silicified choanoflagellates of the Isefjord (Denmark), *Ophelia* **12**:1–26.

Thomsen, H. A., 1975, An ultrastructural survey of the chrysophycean genus *Paraphysomonas* under natural conditions, *Br. Phycol. J.* **10**:113–127.

Thomsen, H. A., 1976, Studies on marine choanoflagellates II. Fine-structural observations on some silicified choanoflagellates from the Isefjord (Denmark) including the description of two new species, *Norw. J. Bot.* **23**:33–51.

Thomsen, H. A., 1978, Nanoplankton from the Gulf of Elat (Gulf of Aqaba) with particular emphasis on choanoflagellates, *Isr. J. Zool.* **27**:34–44.

Thomsen, H. A., 1979, Electron microscopical observations on brackish-water nanoplankton from the Tvärminne area, SW coast of Finland, *Acta. Bot. Fenn.* **110**:11–37.

Thomsen, H. A., 1982, Planktonic choanoflagellates from Disko Bugt, West Greenland, with a survey of the marine nanoplankton of the area, *Meddl. Grønl. Biosci.* **8**:3–35.

Thomsen, H. A., and Boonruang, P., 1983a, A microscopical study of marine collared flagellates (Choanoflagellida) from the Andaman Sea, SW Thailand: Species of *Stephanacantha* gen. nov., and *Platypleura* gen. nov., *Protistologica* **14**:193–214.

Thomsen, H. A., and Boonruang, P., 1983b, Ultrastructural observations on marine choanoflagellates (Choanoflagellida, Acanthoecidae) from the coast of Thailand: Species of *Apheloecion* gen. nov., *J. Plankt. Res.* **5**:739–753.

Thomsen, H. A., and Moestrup, Ø., 1983, Electron microscopical investigations on two loricate choanoflagellates (Choanoflagellida), *Calotheca alata* gen. et sp. nov. and *Syndetophyllum pulchellum* gen. et comb. nov., from Indo-Pacific localities, *Proc. R. Soc. Lond. B* **214**:41–52.

Thomsen, H. A., Zimmermann, B., Moestrup, Ø., and Kristiansen, J., 1981, Some new freshwater species of *Paraphysomonas* (Chrysophyceae), *Nord. J. Bot.* **1**:559–581.

Throndsen, J., 1969, Flagellates of Norwegian coastal waters, *Nytt Mag. Bot. (Oslo)* **16**:161–216.

Throndsen, J., 1974, Planktonic choanoflagellates from North Atlantic waters, *Sarsia* **56**:95–122.

Van Es, F. B., and Meyer-Reil, L. A., 1982, Biomass and metabolic activity of heterotrophic marine bacteria, in: *Advances in Microbial Ecology,* Vol. 6 (K. C. Marshall, ed.), pp. 111–170, Plenum press, New York.

Vickerman, K., 1976, The diversity of the kinetoplastid flagellates, in: *Biology of Kinetoplastida,* Vol. 1 (W. H. R. Lumsden and D. A. Evans, eds.), pp. 1–34, Academic Press, New York.

Williams, P. J. leB., 1981, Incorporation of microheterotrophic processes into the classical paradigm of the planktonic food web, *Kiel. Meeresforsch. Sonderh.* **5**:1–28.

3

r- and *K*-Selection and Microbial Ecology

JOHN H. ANDREWS and ROBIN F. HARRIS

1. Introduction

The essence of the concept of *r*- and *K*-selection is that organisms strive to maximize their fitness for survival in either uncrowded (*r*-selection) or crowded (*K*-selection) environments. Fitness is defined following ecological convention as the proportion of genes left in the population gene pool (Pianka, 1983, p. 10). The terms *r* and *K* refer, respectively, to the maximum specific rate of increase (maximum specific growth rate minus minimum specific death rate) of an organism and to the density of individuals that a given environment can support at the population equilibrium. Since both *r* and *K* can vary within a species and are subject to modification, the division of natural selection into *r*- and *K*-selection is of considerable basic interest in evolutionary ecology.

In this chapter we review and attempt to interpret in a microbial context the basis for *r*- and *K*-selection. We stress key contributions to the concept, rather than present an exhaustive historical chronology. Details regarding the terminology and mathematical derivations are given in an Appendix (see Section 7), so that they are accessible without disrupting the text. (A list of symbols and abbreviations is also appended in Section 7.4.) A conspicuous shortcoming of ecological concepts is that *mechanistic* explanations are typically lacking. A mechanistic interpretation is

JOHN H. ANDREWS • Department of Plant Pathology, University of Wisconsin, Madison, Wisconsin 53706. ROBIN F. HARRIS • Departments of Soil Science and Bacteriology, University of Wisconsin, Madison, Wisconsin 53706.

emphasized to clarify the idea and to provide focal points for testing hypotheses. An overview of the topic, together with our analysis and others that have been or might be used, appears in the flowsheet in Fig. 1.

2. The Concept of r- and K-Selection

2.1. The Postulate

We explore the notion of r- and K-selection as a postulate (Fig. 1); i.e., a hypothesis posed as an essential premise for a chain of reasoning. r- and K-traits and their associated environments seem to be significant factors to consider in population biology; however, we do not know how valid the postulate is—our approach is not an attempt to establish primacy, but rather to present the postulate and evaluate the evidence for it. The premise advanced concerns the tradeoff between fitness in crowded versus uncrowded environments. Thus, according to the postulate, organisms face a suite of "either–or" r/K-related choices such as: high rate of acquisition of nutrients or high affinity for nutrients; high stress resistance of spores or high sensitivity of spores to stimulation; doing one thing well ("specialists") or many things indifferently ("generalists").

A critical analysis of the concept requires that something be known about what constitutes "crowding" and about the nature of the environments concerned. The major differences affecting fitness between uncrowded and crowded environments are summarized in Table I. These criteria set the stage for comparisons of life history strategies among organisms in the two environments. However, such comparisons can best be made only for species occupying the same trophic level, because whether a species is limited by resources or some other mechanism (e.g., predation) is related to its trophic position (Wilbur *et al.,* 1974). Put briefly, uncrowded environments provide for "r-conditions": population densities are low, and density-dependent growth factors are negligible. Population regulation is typically mainly through density-independent mechanisms (e.g., sporadic storms, desiccation, or temperature extremes), which affect essentially the same proportion of individuals at all densities. Conversely, crowded environments provide for "K-conditions": population densities are high and populations are limited by density-dependent controls, such as food supply, toxic metabolites, or predation, which cause proportionate (per capita) changes in birth or death as population density changes. Hence, crowding implies more than just competition for food or space. This broader interpretation is noteworthy and is often overlooked (e.g., Parry, 1981).

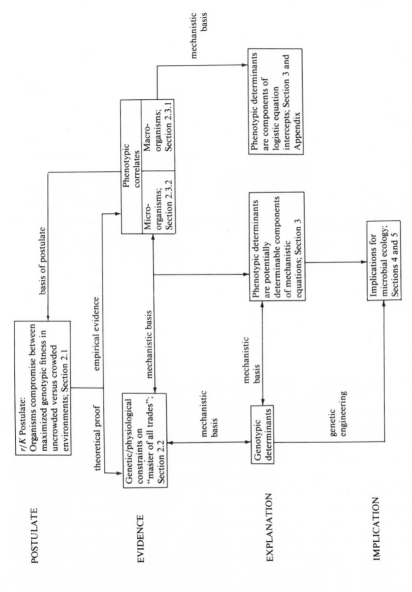

Figure 1. *r*- and *K*-selection: An overview of the organization of this chapter; a synopsis of evidence for and explanations of the concept; and its implications for microbial ecology.

Table I. Some Key Attributes of Crowded versus Uncrowded Environments
Which Affect Fitness

Environment	Population density	Per capita food supply	Effect on potential growth inhibition or death acceleration		Potential death acceleration due to predation or parasitism
			Inadequate food	Toxic metabolites	
Uncrowded	Low	High	Low	Low	Low
Crowded	High	Low	High	High	High

Uncrowded environments are typically unpredictable and transitory. Species colonizing such habitats will do best if they have a high r (in effect, in microbiological terms, a high μ_{max}; Table II), which promotes discovery of habitat → reproduction → dissemination before a saturated condition (K-selection) or loss of the habitat occurs. On the other hand, organisms that compete well in terms of seizing and retaining a particular habitat and efficiently extracting energy from it will be favored under crowded conditions. Genotypes with a high r will be selectively favored in the former situation; those with a high K, in the latter.

The ecological dogma is that a high r occurs at the expense of a low K, or a relatively low r is the sacrifice for a high K (e.g., Wilson and Bossert, 1971, pp. 111–112; Roughgarden, 1971). Apparently, organisms cannot thrive under both conditions; i.e., they cannot maximize both parameters. This does not mean that the two forms of selection are mutually exclusive. Both r and K are subject to evolutionary adjustment upward or downward; the relative degree of the corresponding selective pressure is determined largely by environmental stability (MacArthur and Wilson, 1967).

Put another way, with the focus on the environment rather than the organism, environments can be characterized on an r/K basis, determined by the extent and periodicity of transient r- and K-conditions prevailing over time. Any *environment* is thus an integrator of the instantaneous *conditions* of which it is comprised. We define an r-condition to be when the environment/population relationship (defined quantitatively later) is such that members of the population in question grow at their maximum specific rate of increase r. An r-condition may arise temporarily because of an increase in the food supply and/or a catastrophic decline in the population density. A K-environmental condition exists when the environment/population relationship is such that the specific rate of increase of the population is close to or at zero and the population density is correspondingly close to or at the carrying capacity K. In oligotrophic waters, for example, this state commonly exists because the inherently low rate of food supply favors fixation of the per capita rate of food supply in the maintenance rather than the growth-supporting range. Hence, in theory, an r/K environmental continuum can be visualized, delimited at the r end by an exclusively r-environment in which r-conditions exist continuously, and at the K end by an exclusively K-environment in which extreme K-conditions exist continuously. In practice, most natural microbial environments will show characteristic oscillations between extreme r- and extreme K-conditions, the relative frequency and extent of which will dictate the location of the environment along the r/K continuum and the related pressure for microbial fitness. Although the role of environmental stability in r- and K-selection is a recurrent theme implicit in much of the literature, we explicitly emphasize its importance and focus on it in this review.

In predictably extreme (harsh) environments, such as hydrothermal vents, the arctic, hot springs, deserts, or the highly saline conditions of the Salton Sea, adaptations to the physical environment could be prominent. Greenslade (1983) proposed a habitat template partitioned into areas dominated by three selection processes: r-selection, where habitat predictability is low; K-selection, where the habitat is favorable and its predictability is high; and A-selection in unfavorable habitats, particularly those that are predictably unfavorable. However, we believe and explain later that, in addition to any other considerations, it is informative to consider all environments from an r/K perspective.

In overview, comparisons of organisms based on r- and K-selection are made least ambiguously (1) among species on the same trophic level; (2) in nonextreme environments; and (3) when the descriptors "crowded" and "uncrowded" can be given a basis for quantitation.

2.2. Theoretical Proof

As indicated in Fig. 1, we neither develop here a theoretical proof, nor do we know of an existing derivation confirming that the tradeoff inherent in the r/K postulate is inevitable. Obviously, no organism is "master of all trades." That this is the case reflects the inherent genetic and physiological constraints common to all living things, together with the temporal and spatial variability in habitats. Instead, a divison of labor is necessitated and is the ultimate reason that there are so many species (MacArthur, 1972, p. 61; Tilman, 1982, pp. 244–259). Tempest *et al.* (1983) rationalized from ecophysiological criteria that, for microorganisms in particular, the size of the genome must necessarily be kept to a minimum. This would support the microbial corollary of the r/K postulate that the genetic capability to produce a suite of high-geared and an alternate suite of low-geared enzymes is unlikely to occur in a single microorganism. From thermodynamic considerations (Harris, 1982) the energy cost of operating a suite of enzymes capable of both fast and slow catalysis would be very high and this, combined with recognition that energy tends to be the major growth-limiting factor for microorganisms, supports another microbial corollary of the r/K postulate, that a single microorganism is unlikely to be capable of production and operation of a suite of enzymes having both fast and slow catalytic properties.

2.3. Some of the Evidence

The concept of r- and K-selection is intuitively reasonable and indeed there is much circumstantial evidence from both macroecology and microbial ecology that it exists. The seminal ideas were contributed largely by Dobzhansky (1950), who compared evolution in the tropics and temperate latitudes. Actually, it is usually overlooked that the great naturalist Wallace (1878), in his remarks on tropical plant and animal life (pp. 65–68, 121–123), anticipated many of Dobzhansky's conclusions. Dobzhansky surmised that adaptation in the species-rich tropics is primarily to a harsh biological environment, while the fewer species in colder realms have to contend mainly with the physical environment. Put simplistically, the outcome of different evolutionary pressures between the two regions is competitiveness (high K) or productivity (high r), respectively (Dobzhansky, 1950).

2.3.1. Macroecology

Dobzhansky's rather abstract theme was echoed in work by numerous macroecologists. Cody (1966), for example, used the principle of time and energy allocation to formulate a general theory on clutch size in birds. In temperate zones, he viewed energy as being used to increase r, whereas in the tropics clutches are smaller, which he said reflected the tradeoff between reproduction, predator avoidance, and competition. Different phenotypes were considered more fit in different environments. Predictions were made that all stable environments would favor reduced clutches. [However, there are other hypotheses regarding the evolution of clutch size and Cody's ideas have been challenged (Stearns, 1976)].

The theory of r- and K-selection originated formally with the publication of MacArthur and Wilson's (1967) classic text, *The Theory of Island Biogeography*. From their quantitative analysis of island colonization and species interactions, MacArthur and Wilson made predictions about the ecology of colonizing species at different stages of the colonizing sequence. The most obvious of the predictions from island biogeographic theory that pertain to *r*- and *K*-selection are as follows. (1) *r*-Selection is expected in rigorously seasonal environments subject to flushes of resources (or in habitats subject to other kinds of predictable or unpredictable disturbance) where survivors recolonize during favorable periods. Conversely, *K*-selection is anticipated in stable, hospitable environments. (2) Newly colonizing species are subject to *r*-selection but, once entrenched, will be influenced more by *K*-selection. Species long established should evolve for efficiency (unit output of biomass or numbers per unit of energy input) of collecting and converting resources into progeny. Species repeatedly going extinct and recolonizing small islands should evolve for productivity (unit output of biomass or numbers per unit time); they should reproduce in large numbers to consume the resources before competitors arrive. These predictions likely have some value in efforts to identify the selection pressures responsible for the evolution of interrelated characteristics in an organism.

Not surprisingly, the idea was hit upon almost immediately (e.g., Pianka, 1970) that possession of particular traits could be associated with life under putative *r*- or *K*-conditions. This development had two significant implications. First, whereas particular features of an organism (see later this section) might be correlated with certain environmental parameters, there may not be a cause-and-effect relationship (Wilbur *et al.,* 1974). Second, the notion was broadened and became focused on the *presumed consequence* of *r*- and *K*-selection rather than the *cause*. The circularity or feedback in the *r/K* concept is apparent if it is stated in terms of traits: *r/K*-selection results in a distinct suite of phenotypic features; possession of the appropriate features implies an *r*- or *K*-selection regime (Fig. 1). Put this way, the postulate becomes a tautology (Peters, 1976) and, although not by any means worthless, it can be phrased differently (Andrews, 1984b) or developed mechanistically as in Section 3, so that it is amenable to experimental tests (Section 4).

Pianka (1970) proposed that an *r*-selection regime was associated with such correlates as early reproduction, high *r* values, variable environments, and small body size, whereas *K*-selection was associated with delayed reproduction, lower *r* values, more stable environments, and larger body size. Further, he suggested that because of such correlates, organisms could be positioned along a hypothetical *r/K* continuum. These ideas provided the basis for subsequent elaborations [for reviews see Stearns (1976, 1977, 1980) and Parry (1981)], including the enlarged scope of *r/K*-selection, and use of the terms to broadly classify organisms on a comparative basis as so-called "*r*-strategists" or "*K*-strategists." The idea of ranking organisms along a spectrum achieved considerable popularity; Southwood (1977), for example, related it specifically to types of insect pests ("*r*- or *K*-pests") and the prospective control strategies for each group. Although this system of classifying organisms seems convenient and superficially informative, if used in the extreme it can be quite misleading. One could compare different taxa within the microbial world, for example, but it would not be meaningful to align bacteria, butterflies, and elephants along an r/K gradient. As Begon and Mortimer (1981, p. 153) indicate, this is because each species is subject to unique design constraints and frame of reference on the world so that, beyond closely defined limits (e.g., the same trophic level noted earlier), their environments are effectively incomparable.

The foregoing exemplify the dozens of studies with animals, including salamanders, lizards, fish, birds, and mammals, on r- and K-selection and the larger issue of life history evolution [reviewed by Stearns (1977)]. As summarized later, many, but not all, "fit" the r/K scheme. Analyses are complicated by several factors, such as behavioral patterns and the long periods of time over which life history efforts (e.g., reproduction) are made (Begon and Mortimer, 1981).

The concept of r- and K-selection has been used also to interpret plant life history strategies. Plant studies have typically centered on the differential allocation of resources between reproductive and vegetative tissue as a measure of r- or K-selection (e.g., Gadgil and Solbrig, 1972; Gaines et al., 1974; McNaughton, 1975; Wilbur, 1976). For example, Gadgil and Solbrig (1972) studied four biotypes of dandelions differing in reproductive/vegetative allocation according to the degree of disturbance of their respective habitats within a local area. The common biotype in a highly disturbed site channeled a higher proportion of its biomass to reproduction at the expense of competitiveness (leaf biomass) compared to the biotype from the undisturbed site. Although such studies avoid some of the shortcomings of animal experiments noted above and may be appealing conceptually, they suffer from a number of problems. Among these are the difficulties in assessing "disturbance" and in isolating the experimental variable, crowding, as the causative factor. Parry (1981) criticized the use of reproductive effort as an inappropriate measure of r- and K-selection and indicated that it could reflect other differences, such as demographic structure, in the populations.

The most influential modification of r- and K-selection for plant life histories was developed by Grime (1974, 1979), who proposed that there are three major determinants influencing vegetation: competition, stress, and disturbance. According to Grime, each stimulates a distinct response by the plant. Competition involves mutual demand on "the same units of light, water, mineral nutrients or space" (Grime, 1974, p. 27). Both stress and disturbance inhibit development of a large standing crop by "restricting primary production" or by "damage to the vegetation," respectively. These three determinants were modeled as an equilateral triangle, the corners of which represented the extremes. Thus, the gradient in influence of each factor and the predicted location of habitat types together with their associated plant species could be viewed on a triangular template. Grime (1979) expanded this scheme and related it to *r*- and *K*-selection, which he modified to *R-C-S* selection. Plants that he called ruderals develop under *r*-conditions ("*R*-selection") of low stress/ high disturbance. The converse permutation, high stress/low disturbance, which he designated "*S*-selection," promoted a stress-tolerant strategy, comparable to Pianka's *K*-selection regime. Plants termed competitors, subjected to "*C*-selection" (low stress/low disturbance), occupied a position about midway on the *r/K* continuum. For each selection category, Grime (1979) developed correlates analogous to those of Pianka (1970). Significantly, although Pianka's (1970) list of correlates noted particular environmental attributes associated with *r*- or *K*-selection, Grime was the first to really develop the idea of *r*- or *K*-environments as determinants in selection.

In overview, there is a compelling but not unequivocal observational basis for *r*- and *K*-selection in both animal and plant ecology, particularly the latter. The postulate (Fig. 1) predicts the occurrence of *r*- and *K*-"traits" or correlates, and the fact that these exist strengthens the premise. Begon and Mortimer (1981, pp. 150–158) feel there is mutual reinforcement between strategy and environment: an *r*- (or *K*-) environment is viewed as promoting a corresponding strategy, which, by positive feedback, leads to the environment's appearing, *from the organism's standpoint,* as being even more *r*- (or *K*-) selective.

That the *r/K* concept is unable to account for *all* life history strategies (Begon and Mortimer, 1981, p. 155) or that its predictions are inconsistent with some evidence (Wilbur *et al.,* 1974; Stearns, 1977) is not unexpected and is certainly no reason for discarding the postulate. What this situation suggests is that more refined testing of the idea and a better understanding of its explanatory powers are needed.

2.3.2. Microbial Ecology

Although microbial ecologists may have been thinking conceptually like macroecologists, it was not until the mid 1970s that the two disciplines were bridged with respect to terminology. Among the more substantial syntheses was that by Esch *et al.* (1977), who used Pianka's (1970) correlates to group life history features of animal parasites within the *r/ K* framework. Subsequently, Andrews and Rouse (1982) proposed that plant parasites could be positioned along an *r/K* continuum relative to each other based on (1) the relative allocation of resources to maintenance, growth, and reproduction (cf. Gadgil and Solbrig, 1972), and (2) the nature of the parasitic association, especially regarding acquisition of nutrients and effect on the host. These and other papers (e.g., Jennings and Calow, 1975; Swift, 1976) established two things. First, the r/K concept of macroecology had a direct, although generally unrecognized parallel in microbiology. Second, phenotypic correlates of microbes, analogous to those for plants and animals, could be associated with a putative selection regime.

Pugh (1980) made the connection between microbial ecology and Grime's (1974, 1979) stress–competition–disturbance template for plants, which he broadened to include the fourth permutation, high stress/high disturbance. Pugh (1980) modified Grime's definition of disturbance to "such dramatic alteration of the normal that (i) the biomass is reduced (e.g. after fungicide application); or (ii) a new environment is superimposed on the existing one, and results in a relative reduction in living fungal biomass" (pp. 2–3). For example, the massive input of leaves to the soil ecosystem in the autumn was seen as a type (ii) disturbance in that it results in a smaller amount of fungal biomass per unit volume of soil. Fungal species acting as ruderals or r-strategists, comprising the preponderance of fungi isolated from soil, were placed in the low-stress/high-disturbance category. Where disturbance is low, selection for competitive ability was viewed as increasing along the stress gradient, at the end of which (high stess/low disturbance) were placed fungi decomposing recalcitrant substrates (e.g., cellulose, lignin, chitin). Their subsistence was attributed to simple nutrient requirements or to efficiency or adaptability in using the nutrients available. Finally, the high-stress/high-disturbance habitat, which Pugh represented by a leaf surface, supposedly confronted microbe populations with both extreme environmental fluctuations and high competition for nutrients.

Pugh (1980) and others (e.g., Gerson and Chet, 1981) noted the similarity between r/K concepts and terms such as zymogenous (allochthonous) and autochthonous, long familiar to microbial ecologists. Also, at least with respect to nutrients, the properties of Poindexter's (1981) oligotrophic bacteria resemble those of a K-strategist, whereas her copiotrophs (eutrophs) are analogous to r-strategists (Hirsch et al., 1979). However, as discussed later, the terms are not synonomous.

The strongest experimental evidence for r- and K-selection in macroecology or microbial ecology derives from many studies over the years on competition between bacteria (e.g., Veldkamp et al., 1984). This body of work relates bacterial phenotypic properties (particularly specific growth rate versus nutrient concentration relationships) to competitive success and thus is mechanistic in approach. The tradeoff identified is that the organism that does well at low nutrient concentrations has a low

μ_{max}. Unfortunately, because of disciplinary isolation, these kinds of experiments have not been incorporated into the r/K context.

To summarize, it is interesting that there is persuasive, albeit largely indirect, evidence for *r*- and *K*-selection in both disciplines. Historically, the main difference in approach has been that the microbiological r/K nomenclature is used mainly descriptively, without a quantitative basis, whereas macroecologists attempted to develop equations quantitatively defining r/K differences between organisms. Ironically, the conceptual conclusions of these equations are defendable, but the equations themselves are suspect and mechanistically a dead end, whereas microbial systems, which probably offer the best models for a mechanistic analysis of *r*- and *K*-selection, await exploitation.

3. Mechanistic Basis of *r*- and *K*-Selection

We consider *r*- and *K*-selection with respect to the entire life cycle, including both the active and resting phases. The inactive phase is especially important for microbes because, when stressed, they are uniquely able to adopt a distinct mode of existence. The response of elephants to severe crowding would be emigration and eventual death; fungi and bacteria, on the other hand, isolate themselves morphologically and physiologically from the environment, e.g., by shutting off active transport.

Understanding the mechanistic basis of *r*- and *K*-selection inevitably involves reference to models of population dynamics. As noted by Roels and Kossen (1978) and Esener *et al.* (1983), a major problem in modeling is to balance sufficient realism with complexity: on the one hand, if the model does not contain at least the most important mechanisms, it does not describe reality; on the other hand, if the model is too complex, it can be almost impossible to test experimentally. The basic difference in complexity between macro- and microbiological systems and the consequent limitations with respect to modeling approaches based on *r*- and *K*-seleciton concepts must be recognized. The more complex the biological system, the more difficult it is to delve deeper than the most gross phenotypic expressions (birth/growth and death) of population dynamics. The relative simplicity of microbial systems, particularly bacteria, offers the highest potential for mechanistic interpretation of the effect of crowding on population dynamics, past the gross growth/death relationships, to the causative level (food limitation, toxicity, stress, predation/parasitism) and, for crowding with respect to food, to a level recognizing phenotypic properties of rate and efficiency of food acquisition and use (Pirt, 1975;

Harris, 1981; Button, 1983). The contrast between macro- and microsystems is analogous to attempts at explaining biological phenomena from the standpoint of ultimate versus proximate causation.

3.1. Population Dynamics of the Active Phase

The conventions we use for the specific (per capita and per unit biomass) determinants of population kinetics are summarized in Table II. In general, these conventions are consistent with those commonly used in general ecology (per capita) and microbiology (per unit biomass). Since the conventions used in the literature for the specific rate of increase and the specific rate of growth are highly ambiguous, we draw attention in particular to the following. There is no generally accepted, simple term for the unspecified (potentially variable) specific rate of increase (more properly, rate of change); we use the cumbersome $(1/X)\ dX/dt$ and the less descriptive y (our convention) terms interchangeably. The specific rate of increase refers only to the increase in living, and not living plus dead, organisms (i.e., it is in effect the net or observed specific growth rate) and is equal to the specific rate of biomass production (gross or real specific growth rate, abbreviated to specific growth rate μ) minus the specific death rate (λ): $(1/X)\ dX/dt = \mu - \lambda$. It is important to recognize [e.g., expansion of Eq. (3) in Section 3.1.1] that the specific rate of food consumption q is related directly to the specific rate of biomass production μ and not to the specific rate of increase: $q = A\mu = A[(1/X)\ dX/dt + \lambda]$, where A is the proportionality coefficient (specific growth requirement).

In this section we review and expand currently available models of microbial population dynamics in r and K terms. The logistic equations are included for historical reasons only, since they have no mechanistic basis. Certain aspects of the mechanistic equations should be recognized. The equations have been simplified as much as possible to focus on quantifiable phenotypic properties of primary importance as determinants of ecological competence. Currently available experimental techniques for estimating certain equation components (e.g., nutrient concentrations) require modification and improvement to overcome complexity and detection limit shortcomings. The equations as written are most applicable to bacterial systems, and assume that all cells in a population behave identically. Emphasis has been placed on population dynamics with respect to food resources, largely because food limitation is in practice the major controlling factor, but also because of the current lack of appropriate mechanistic equations for modeling toxic and predation/parasitism effects on microbial populations.

Table II. Conventions Used for Determinants of Population Kinetics[a]

Physical and biotic conditions	Specific rate of change (increase)	Specific (per capita)						Specific (per unit biomass)			
		Population density	Food density	Birth rate	Death rate	Generation time	Fecundity	Production (growth rate)	Food consumption rate	Death rate	Doubling time
Unspecified	$(1/X)\, dX/dt,$	X	S	b	d	t_g	D	μ	q	γ	t_d
All optimized	r^{max}	—	—	b_0^{max}	d_0^{min}	—	—	μ_{MAX}	q^{MAX}	γ_{MIN}	—
Uncrowded	r	—	—	b_0	d_0	—	—	μ_{max}	q^{max}	γ_{min}	—
Saturated	0	K	S_K	—	—	—	—	μ_K	q_K	γ_K	—
Oversaturated	—	—	—	—	—	—	—	λ	—	—	—

[a] The population rate of increase dX/dt and population density X may be expressed in terms of numbers of organisms ($dX/dt = dN/dt$; $X = N$) or total biomass ($dX/dt = dM/dt$; $X = M$); for interconversion, $M = \alpha N$, where α is the per capita biomass. t_g is the generation time between birth and reproduction; D is the number of daughter cells produced per microorganism; $b = (\ln D)/t_g$; $\mu = (\ln 2)/t_d$ ($\mu = b$ for binary fission at biomass doubling, e.g. bacterial). The usual subscript rather than superscript notation for the max, min, and other qualifiers is generally used in the equations, and with the following exceptions in the text: for competitive comparison (e.g., Fig. 2, μ_I^{max} and μ_{II}^{max} are used rather than μ_{max}; K_I^n and K_{II}^n rather than K_S; S_I^m and S_{II}^m rather than S_m; etc.); for internal consistency (e.g., Fig. 1A, $r^{max} = b_0^{max} - d_0^{min}$). Unless specified otherwise, q and q_s are used interchangeably.

3.1.1. Logistic Equations

The following is a review of the basic concepts of the most commonly used logistic equations; expansion of these equations is detailed in the Appendix. The logistic equations arose from the desire by macro-ecologists to express in quantitative terms the commonly observed phenomenon that under uncrowded conditions the per capita rate of increase r (growth rate corrected for death) is maximized and that with increasing crowding, problems of inadequate food supply for all population members, toxic metabolic product accumulation, and predation/parasitism-related death become increasingly severe, resulting in a decline in the per capita rate of increase (due to a decline in the per capita growth and/or an increase in the per capita death rate) until the population density equilibrates at its carrying capacity K. The simplest mathematical expression for the change from an r- to a K-condition is that the specific rate of increase (y axis) declines linearly with increasing population density (x axis) from a maximum (r, at mathematical zero population density) to zero (at a maximized population density K). The resulting straight-line relationship (MacArthur, 1972; Pielou, 1969) as shown in Fig. 2 is given by

$$\frac{1}{X}\frac{dX}{dt} = y = r - kX \tag{1}$$

where k is the proportionality coefficient defining the density dependence of the specific rate of increase y; in terms of the y- and x-axis intercepts, Eq. (1) becomes the Verhulst–Pearl logistic equation,

$$\frac{1}{X}\frac{dX}{dt} = y = r - \frac{r}{K}X \tag{2}$$

From Eq. (2) it can be seen that at low X, y will be dominated by r, whereas when $X \to K$, y will be dominated by K. For two organisms, organism I with a high r and low K, and organism II with a low r and high K (Fig. 2), the r-organism (I) will outcompete the K-organism (II) at low population densities (X less than the competitive crossover density, C) and vice versa at high densities ($X > C$). Expansion of the r and K terms into density-dependent and density-independent birth and death components is detailed in the Appendix (Section 7.1).

As illustrated in Fig. 2, increasing population density is accompanied by decreasing food density, implying that the reduction in the specific rate of increase is mechanistically associated with food limitation. The inset

Figure 2. Comparative specific rate of increase versus population density [after MacArthur (1972) and Wilson (1975)] and population feeding rate [after Pielou (1969)] for two organisms. Two organisms showing Verhulst–Pearl logistic growth are envisaged as competing in natural selection. At low population densities, the specific rate of increase (net growth rate) is dominated by the intrinsic rate of increase r: organism I outcompetes organism II because $r_I > r_{II}$. At high population densities, the specific rate of increase is dominated by the carrying capacity K: organism II outcompetes organism I because

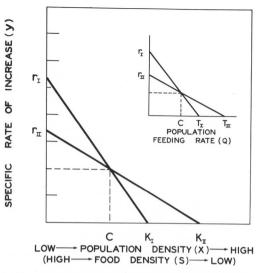

$K_{II} > K_I$. At equilibrium, when the specific rates of increase are zero, each population is by definition at its carrying capacity. Organism I will prevail when the environment fluctuates enough to keep populations constantly growing at a density less than the competitive cross-over density C (*r*-selection); organism II will win in environments stable enough to permit the populations to remain at a density greater than C (*K*- selection). Since high food density is often correlated with low population density, this coordinate is also shown. The inset is the population feeding rate analogue (Smith equation) of logistic growth. T is the rate of food consumption at equilibrium.

in Fig. 2 illustrates a commonly quoted logistic relationship between the specific rate of increase in the feeding rate (and, hence, the specific rate of population increase) and the rate of food consumption Q:

$$\frac{1}{Q}\frac{dQ}{dt} = \frac{1}{X}\frac{dX}{dt} = y = r - kQ$$

$$\frac{1}{X}\frac{dX}{dt} = y = r - \frac{r}{T}Q \tag{3}$$

where T is the rate of food consumption when the population reaches equilibrium. As identified in the Appendix (Section 7.3), expansion of Eq. (3) leads to a concave relationship between the specific rate of increase and population density (Smith, 1963; Pielou, 1969).

The insurmountable defect of the logistic equations is that the mechanistic basis of the decline in the specific rate of increase with increasing population density is not explicitly recognized (Williams, 1972; Roels and Kossen, 1978). Use of logistic equations in macroecology is understandable in the sense that development of mechanistic alternatives is severely complicated (perhaps insurmountably for the immediate future) by the complexity of higher life forms. This is not the case for microorganisms. For bacteria, in particular, use of logistic equations for defining population dynamics within empirically determined r and K boundary conditions is *not* to be recommended. The lack of mechanistic basis for such an assumed logistic relationship means it is unlikely that the logistic approach would be experimentally verifiable or of any predictive value.

3.1.2. Equations Based on the Nature of Crowding

In natural ecosystems, reductions in the specific rate of population increase with increasing population density probably reflect responses to a combination of food and space limitation, toxic stress, and predation/parasitism crowding phenomena. However, for many ecosystems, crowding with respect to food resources is generally considered to be the major growth-limiting determinant. Mechanistic equations for growth phase metabolism and the transition from growth phase to vegetative resting phase metabolism are presented in the Appendix (Section 7.2); these equations are the basis of the competitive food acquisition and growth curves in Fig. 2 and the relationship between rate of food supply and population density in Fig. 3. The components of the equations appear as correlates in Table III for r- and K-selected microorganisms. Equations describing the effect of toxic metabolites on growth are given in Pirt (1975); quantitative models for predator/prey population dynamics are reviewed by Williams (1980).

3.1.2a. Dependence of Population Dynamics on Food Density. Figure 3 illustrates competitive food acquisition and growth by two organisms: organism I (r-organism) has a high q^{max} (maximum specific rate of food acquisition/use) and a high (low-affinity) K_s (half-saturation coefficient); organism II (K-organism) has a low q^{max} and a low K_s (high affinity). For simplicity, both organisms in Fig. 3 are assumed to be food density (S)-limited at high S ($S_r = \infty$, where S_r is the food density below which the rate of food uptake is controlled by the rate of food acquisition), show similar q^m (specific maintenance energy rate) and S_n (food density at which the rate of food uptake is exactly counterbalanced by the rate of food efflux; i.e., $q = 0$) properties, and show a prolonged transition from growth to resting metabolism (i.e., Fig. 3 does not recognize any shift from active to resting metabolism).

Table III. Some Life History Features and Fitness-Determining Phenotypic Traits of *r*- or *K*-Selected Microorganisms under Various Environmental Conditions

	Organism	
Trait	*r*-Strategist	*K*-Strategist

I. Life history

Longevity of growth phase	Short	Long
Rate of growth under uncrowded conditions	High	Low
Relative food allocation during transition from uncrowded to crowded conditions	Shift from growth and maintenance to reproduction (spores)	Growth and maintenance
Population density dynamics under crowded conditions	High population density of resting biomass; high initial density (and/or high resistance to stress) compensates for (and/or ameliorates) death loss	High equilibrium population density of highly competitive, efficient, growing biomass; growth replacement compensates for death loss
Response to enrichment	Fast growth after variable lag	Slow growth after variable lag
Mortality	Often catastrophic; density-independent	Variable
Migratory tendency	High	Variable

II. Fitness-determining phenotypic traits of growth phase in:

(a) Uncrowded environments

Maximum specific rate of increase (r^{max}, r)	High	Low
Minimum specific death rate (d_0^{min}, d_0)	Variable	Low
Maximum specific growth or birth rate (μ^{max}, b_0^{max}, b_0, D)	High	Low
Maximum specific rate of food acquisition (q_s^{max})	High	Low
Nutritional range supporting μ_{max}	Variable	Variable
Minimum food density supporting μ_{max} (S_r)	High	Low
Efficiency of food conversion to biomass (Y_s, E_{ATP}, Y_{ATP})	Low	High
Maintenance requirement (q_s^m, q_{ATP}^m)	High	Low

(continued)

Table III. (*continued*)

Trait	Organism	
	r-Strategist	*K*-Strategist
Resistance to density-independent population controls (catastrophes: desiccation, temperature extremes, washout, flooding, compaction)	Low	Variable–high
(b) Resource-limited (crowded) environments		
Ability to acquire food at very low food densities (K_s)	Low (high K_s)	High (low K_s)
Nutritional range	Variable	Variable
Efficiency of food conversion to biomass (Y_s, E_{ATP}, Y_{ATP})	Moderate	High
Maintenance requirement (q_s^m, q_{ATP}^m)	High	Low
Minimum specific growth rate (μ_{min})	Low	Very low
Food density needed to support μ_{min} (S_c)	Low	Very low
Food density needed to support maintenance requirement (S_m)	Low	Very low
Inhibitory metabolite production	Variable–low	Variable–high
Resistance to density-dependent mortality (starvation, toxins, predation/parasitism)	Low	High
Resistance to density-independent mortality (catastrophes)	Low	Variable–high
(c) Inhibitory metabolite-limited (crowded) environments		
Tolerance to inhibitory chemicals (antibiotics, staling products, pH)	Variable–low	Variable–high
Production of inhibitory chemicals	Variable–low	Variable–high
(d) Predation/parasitism-limited (crowded) environments		
Tolerance to predation/parasitism	Low	Variable–high
Palatability	High	Low
Maximum specific growth rate (μ_{max}, b_0, D)	High	Low

Table III (*continued*)

Trait	Organism	
	r-Strategist	*K*-Strategist
III. Fitness-determining phenotypic traits of resting phase		
Rapidity of shift from active to resting metabolism upon starvation	Variable; relatively rapid	Variable; relatively slow
Resistance of resting vegetative phase to density-dependent and density-independent mortality	Very low	High
Food density below which leakage occurs for resting vegetative organisms (S_n)	Low	Very low
Food density triggering dormancy break for resting vegetative organisms	High, and/or highly selective	Low, not highly selective
Tendency to form spores	High	Low
Spore size	Small	Large
Resistance of spores to density-dependent and density-independent stress	High–very high	High
Food density triggering dormancy break for spores	Very high, and/or very highly selective	High, not highly selective

Under high food density conditions, $q \approx q^{max}$ and $\mu = \mu_{max}$, where [from Eq. (A18), Section 7.2]

$$\mu_{max} = \frac{q^{max}}{A^b} - \frac{q^m}{A^b} \qquad (S_r = \infty; S \gg K_s) \qquad (4)$$

$$\approx \frac{q^{max}}{A^b} = q^{max} Y^b \qquad (q^{max} \gg q^m) \qquad (5)$$

In accordance with Eq. (4), $q_I^{max} > q_{II}^{max}$ results in $\mu_I^{max} > \mu_{II}^{max}$ (Fig. 3), identifying the importance of q^{max} as a major determinant of μ_{max}. Figure 3 is concerned solely with the dependence of the specific rate of food acquisition q and the specific rate of biomass production μ on food density S. Experimentally, μ versus S relationships are commonly derived with the implicit assumption that microbial death is negligible. This may well be the case for cultures showing an exponential rate of biomass increase ($r = \mu_{max} - \gamma_{min} \approx \mu_{max}$ for $\gamma_{min} \approx 0$, where γ_{min} is the minimum specific

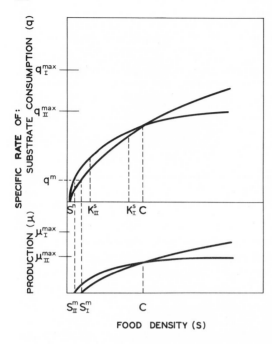

Figure 3. Comparative specific rates of food consumption q and specific rates of growth μ versus food density S for two organisms. Two organisms, showing saturation kinetics for q with respect to S, no minimum specific growth rate, and having similar maintenance requirements q^m are envisaged as competing in natural selection. At high food density, q and μ are dominated by q^{max}: for $s \gg K_s$, organism I outcompetes organism II because $q_I^{max} > q_{II}^{max}$ and thus $\mu_I^{max} > \mu_{II}^{max}$. With decreasing S, the food affinity property K_s becomes an increasingly more important determinant of q: for S less than the competitive crossover food density C, organism II outcompetes organism I because $K_{II}^s < K_I^s$ (low K_s means high food affinity) and thus $\mu_{II} > \mu_I$. Equilibrium ($\mu = 0$) occurs for organism I at $S = S_I^m$ (where $q_I = q_I^m$); at S_I^m, organism II is still capable of growth since at S_I^m, $q_{II} > q_{II}^m$ and $\mu_{II} > 0$. Equilibrium occurs for organism II at $S = S_{II}^m$ (where $q_{II} = q_{II}^m$); at S_{II}^m, $q_I < q_I^m$ and thus organism I must ultimately shift to resting metabolism or die since endogenous metabolism (decay) cannot support the maintenance energy requirement for active metabolism indefinitely.

death rate). However, mechanistic interpretation of biomass kinetics, particularly under productive (high population density) conditons, requires consideration of the death component and related r/K concepts. As long as population density remains below X_r (the level at which crowding with respect to toxic stress and predation/parasitism affects the specific growth and/or death rates), the specific rate of increase will remain at r [from Eqs. (A24) and (A18)]:

$$\frac{1}{X}\frac{dX}{dt} = \mu_{max} - \gamma_{min} \tag{6}$$

$$= \frac{q^{max} - (q^m + \gamma_{min}A^b)}{A^b} = r \qquad (S_r = \infty; S \gg K_s; X < X_r) \tag{7}$$

where A^b is the specific growth requirement for biosynthesis. Equation (7) identifies the importance of q^{max}/μ_{max} as the major determinant of r. Since the death rate term has a much higher proportionate effect on *r* the lower the μ_{max}, minimizing γ_{min} and the effect of crowding on γ are potentially more important for *K*- than *r*-organisms (Table III). With decreasing food density, the K_s rather than the q^{max} property becomes an increasingly more important determinant of q and thus μ, resulting in a crossover for the competitive q and μ curves at a critical food density (Fig. 3) [from Eq. (A15)–(A17)]:

$$\mu = \frac{q^{max}}{A^b} \frac{S}{K_s + S} - \frac{q^m}{A^b} \qquad (S_m \leq S \leq S_r) \tag{8}$$

$$= \frac{q^{max}}{K_s} \frac{S}{A^b} - \frac{q^m}{A^b} = \frac{\alpha_{max}S}{A^b} - \frac{q^m}{A^b} = \alpha_{max}SY \qquad (S_m \leq S \ll K_s) \tag{9}$$

where α_{max} is the maximum specific affinity for the nutrient and Y is the specific growth yield. The high food affinity (low K_s) tradeoff by the *K*-organism against a high q^{max} is thus a major determinant of the high competitive ability of the *K*-organism at very low food densities (Table III). The energy requirement and energy efficiency properties, and death sensitivity to suboptimal food availability, also become increasingly more important competitive properties with decreasing food density [from Eqs. (A11)–(A13) and (A23)]:

$$\frac{1}{X} \frac{dX}{dt} = \mu - \gamma = \frac{q - q^m}{A^b} - \gamma \tag{10}$$

$$= \frac{q^{max}S/(K_s + S) - q^m_{ATP}/E_{ATP}}{A_a + A^b_{ATP}/E_{ATP}} - \gamma \qquad (S_m \leq S \leq S_r)$$

where q^m_{ATP} is the specific rate of ATP generation/use for maintenance, E_{ATP} is the efficiency of dissimilatory ATP generation; A^b_{ATP} ($= 1/Y^b_{ATP}$) is the specific amount of ATP generated/used for biosynthesis. Accordingly, a low q^m_{ATP}, low A^b_{ATP} (high Y^b_{ATP}), and high E_{ATP} and a low γ would be expected to be characteristic of a *K*-organism (Table III). The food density S_m at which $q = q^m$ and thus $\mu = 0$ is given by [rearrangement of Eqs. (A13) and (A11), recognizing that at $S = S_m$, $\mu = 0$ by definition]

$$S_m = \frac{qK_s}{q^{max} - q} = \frac{(\mu A^b + q^m)K_s}{q^{max} - (\mu A^b + q^m)} = \frac{q^m K_s}{q^{max} - q^m} \tag{11}$$

From Eq. (11) and as illustrated in Fig. 3, the lower K_s property of the *K*-organism than the *r*-organism results in a much lower S_m for the *K*-organism. From a competitive standpoint, this means that at the food density supporting the maintenance energy requirements of the K-organism, the r-organism would not be able to support active metabolism by exogenous food supply means, and would tend to burn itself up by endogenous metabolism (decay) during the transition from growth metabolism to passive (much lower q_{ATP}^m) resting phase metabolism. At the food density S_n at which food uptake is counterbalanced by intracellular metabolite leakage, the specific decay rate γ and thus the specific rate of decrease will be attenuated by a low maintenance energy requirement q_{ATP}^m and a high energy efficiency of endogenous metabolism $E_{ATP/x}$ [from Eqs. (A23) and (A20)]:

$$\frac{1}{X}\frac{dX}{dt} = -(\lambda + \gamma) = -\left(\frac{q_{ATP}^m}{E_{ATP/x}} + \gamma\right) \quad (S = S_n) \quad (12)$$

and accordingly a K-organism might be expected to show a high $E_{ATP/x}$, a low q_{ATP}^m, and a low γ during the growth/resting phase transition. If the S_n for the r-organism is higher than the S_n for the K-organism, then in competitive culture the K-organism will reduce the food density to $<S_n$ for the r-organism, potentially resulting in concentration gradient-induced leakage and a resultant higher specific decay rate and a higher specific death rate (and thus a more negative specific rate of change) for the r-organism in mixed as compared to pure culture [from Eqs. (A23) and (A21)]:

$$\frac{1}{X}\frac{dX}{dt} = -(\lambda + \gamma) = -\left(\frac{q_{ATP}^m}{E_{ATP/x}} + l + \gamma\right) \quad (S < S_n) \quad (13)$$

For many microorganisms, it is likely that S_r will be relatively low (i.e., that q and thus μ_{max} are not limited by food acquisition at high S). In addition, the food density signalling cell cycle initiation for "shut down" or "growth precursor" cells (Dow *et al.*, 1983) is probably higher than S_m, resulting in the existence of a critical S value S_c supporting a minimum specific growth rate μ_{min} [rearrangement of Eqs. (A13) and (A11)]:

$$S_c = \frac{(\mu_{min}A^b + q^m)K_s}{q^{max} - (\mu_{min}A^b + q^m)} \quad (14)$$

For $\mu_{min} \approx 0.05\ \mu_{max}$ (Pirt, 1975; Konings and Veldkamp, 1980), an r-organism would have a higher μ_{min}, triggered at a higher S_c, than a K-organism (Table III). It is possible that with decreasing μ, actively growing microorganisms become more energy efficient (e.g., show a higher E_{ATP}); however, this will merely change the values for μ_{min}/S_c or S_m.

Figure 4. The instantaneous relationship between population density X and the rate of resource supply R to the population, and how these parameters pertain to *r*- and *K*-selection. The upper line described by q_s^{max} is the rate needed to sustain growth at the intrinsic rate of increase *r*. The lower line q_s^m is the minimal rate for maintenance only. Population densities such that $R/X \geq q_s^{max}$ are resource-unlimited, i.e., subjected to *r*-selection. The zone between q_s^{max} and q_s^m is resource-limited. As the population density approaches q_s^m, the selective advantage shifts from productivity to competitiveness, i.e., populations are subjected increasingly to *K*-selection. q_s^m represents an equilibrium condition where supply balances the demand to maintain the population of active biomass at the carrying capacity K with respect to resources. Over population occurs below q_s^m. Populations adjust back to equilibrium by death or conversion from active to inactive (resting) biomass. Note that K_{max} can be set by constraints such as predation or toxic product accumulation below the K determined by resources alone.

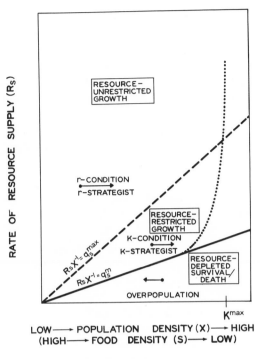

3.1.2b. Dependence of Population Dynamics on Rate of Food Supply. Mechanistic interpretation of the relationship between food supply rate R and population dynamics (Fig. 4) requires recognition that the food density pool S controlling the rate of food acquisition by each member of the population q is a function of the rate of food demand by the population Q (which increases as population density X increases, $Q = qX$), as well as of R:

$$\frac{dS}{dt} = R - Q = R - qX \tag{15}$$

where dS/dt is the rate of change (increase or decrease) in food density. From Eq. (15), as long as Q is $<R$, the food density pool will increase; but with increasing X, Q will eventually become equal to, or transitionally exceed, R, resulting in a decrease in the food density pool to growth-limiting levels. From a practical standpoint, under uncrowded conditions, *r*-phase growth can be maintained for a given food supply rate as long as the food density remains above that needed to support μ_{max} and

the population density remains below that causing an increase in the specific death rate. With increasing population density, the specific rate of increase will ultimately decrease because of the decrease in q-controlled μ caused by increasing X ($q = Q/X = R/X$) and because the specific death rate γ will tend to increase with increasing X [from Eq. (A22), recognizing that Q and thus q are controlled by R under these conditions]:

$$
\begin{aligned}
\frac{1}{X}\frac{dX}{dt} &= \mu - \gamma = \frac{q - q^{m}}{A^{b}} - \gamma \\
&= \frac{R/X - q^{m}}{A^{b}} - \gamma \qquad \left(\frac{Q}{X} = \frac{R}{X} < q^{max}\right)
\end{aligned}
\tag{16}
$$

At the carrying capacity, the specific rate of increase will become zero and the boundary condition for maintaining the maximized population density constant at K (i.e., that the specific rate of increase is zero rather than negative) (Fig. 4) is given by [from Eq. (A22), recognizing that Q is controlled by R]

$$
\begin{aligned}
\frac{1}{X}\frac{dX}{dt} &= \frac{R/K - (q^{m} + \gamma_{K}A^{b})}{A^{b}} = \frac{R/K - (q^{m} + \mu_{K}A^{b})}{A^{b}} \\
&= \frac{R/K - q_{K}}{A^{b}} = 0 \qquad \left(\frac{Q}{X} = \frac{R}{X} = \frac{R}{K} = q_{K}\right)
\end{aligned}
\tag{17}
$$

where γ_{K}, μ_{K}, and q_{K} are the specific death and growth rates and specific rate of food consumption/use, respectively. A transient increase in X past the $K = R/q_{K}$ boundary (either because of population overshoot or reduction in R) will result in X adjustment back (death and/or shift to resting phase) to the R/q_{K} boundary (Fig. 4). Under relatively low food supply rate (low R) conditions it is likely that the carrying capacity will be controlled by food limitation rather than by toxic metabolite and/or predation/parasitism mechanisms (Fig. 4). Accordingly, the specific death rate for relatively low-R systems will tend to be relatively low and will consequently be counterbalanced by a relatively low specific growth rate (in Fig. 4, $\gamma_{K} = \mu_{K} \approx 0$ over a relatively high X range). The carrying capacity with respect to resources will be highly dependent on energy efficiency under these conditions [from Eq. (17) rearranged and Eqs. (A11) and (A12)]:

$$
\begin{aligned}
K &= \frac{R}{q_{K}} = \frac{R}{q^{m} + \gamma_{K}A^{b}} = \frac{R}{q^{m} + \mu_{K}A^{b}} \\
&= \frac{RE_{ATP}}{q^{m}_{ATP} + \gamma_{K}(A_{a} + A^{b}_{ATP})} = \frac{RE_{ATP}}{q^{m}_{ATP} + \mu_{K}(A_{a} + A^{b}_{ATP})}
\end{aligned}
\tag{18}
$$

Because the primary determinant of competitive fitness under food-limiting conditions is the K_s property, for *K*-organisms of similar K_s properties, the organism with the highest energy efficiency will achieve the highest carrying capacity [Eq. (18)]. With increasing population density in rich environments (high *R*), the carrying capacity will ultimately be restricted by something other than food limitation (Fig. 4) and the food density will consequently increase, since *Q* is no longer determined by *R*. Under such circumstances the boundary maximum for the specific death rate will be the maximum specific growth rate [from Eq. (18) with $\gamma_K = \mu_K = \mu_{max}$]:

$$K_{max} = \frac{R_{max}}{q^m + \gamma_K A^b} = \frac{R_{max}}{q^m + \mu_{max} A^b} \qquad (19)$$

where K_{max} is the carrying capacity with respect to unlimited food supply, and R_{max} is the minimum rate of food supply needed to support $Q_{max} = q^{max} K_{max}$.

Because of their generally very high specific growth rates, microorganisms show a high potential for an explosive increase in population density under nutrient-sufficient conditions; accordingly, even under high nutrient supply conditions *R*, population density *X* soon reaches a level such that the food available to each member of the population ($q = R/X$) rapidly approaches or transiently declines below q^m. For certain ecosystems, periodic or sustained reductions in biomass density by catastrophic events (e.g., periodic desiccation or dilution) or biotic mechanisms (e.g., cropping by predators) ensure growth conditions for much of the time, but even in these systems microorganisms frequently must be prepared to survive periodic starvation for extended times. To counter the inevitable reduction in per capita rate of food supply with time, microorganisms have developed mechanisms for short- and long-term survival in the absence of food. For short-term survival, as relevant to the feast-and-famine environment of the gut, microorganisms like *Escherichia coli* are best served to maintain, under famine conditions, a metabolically alert state characterized by a full complement of growth metabolism enzyme and active transport systems (Koch, 1979). This allows the *E. coli* cells remaining [e.g., attached to the gut wall (Freter, 1984)] after the catastrophic event of fecal elimination to respond immediately to the transitional feast condition resulting from food ingestion. *Escherichia coli*, with its characteristic properties of a high μ_{max} and broad range of high q^{max} heterotrophic nutritional options, thus acts as a classical (non-spore-forming) *r*-organism when considered within the context of its natural habitat.

3.2. Population Dynamics of the Resting Phase

While maintenance of an active metabolic state is feasible for short-term starvation survival, long-term survival necessitates a shift to resting phase metabolism (Morita, 1982). In addition to a general reduction in intracellular metabolic activity, a major difference between resting- and non-resting-phase metabolism is that resting cells turn off their energy-draining, proton motive force-driven transport systems (Konings and Veldkamp, 1983) rather than spend energy in a losing battle to take up sufficient food against a substantial concentration gradient to at least pay for the cost of transport. A major concern that resting microorganisms potentially have is to prevent leakage of intracellular organic metabolites by passive transport in response to the very low extracellular metabolite concentration maintained in natural ecosystems by highly effective, scavenging K-organisms. The decay equations [Eqs. (12) and (13)] are relevant in principle to resting phase organisms, with the proviso that the q^m term reflects minimized ATP expenditure for intracellular and osmoregulatory control oriented almost exclusively toward retention of cell integrity rather than exploitation of the environment. It is evident that long-term survival cannot be accomplished in the face of a high rate of endogenous metabolism and intracellular metabolite leakage.

Although mechanistic equations are not available for quantitative description of biomass kinetics and metabolism during the transition from the growth to the resting phase and during the transition back from the resting to the growth phase, certain qualitative generalizations can be made. The phenotypic traits of a successful r-organism under the unpredictable condition of transient enrichment and long-term starvation characteristic of many ecosystems may be summarized as follows. During periods of food limitation the organism should be in a metabolically inert spore state protected from physical and biotic stress by a tough, well-insulated outer layer; the related relative insensitivity of such a spore to minor perturbations in external nutrient conditions would be advantageous, since competitive growth by the germinated spore would be short-lived at best under such conditions. Dormancy break by the spore should be triggered only by a major nutrient perturbation (Table III) capable of supporting sufficient vegetative growth to more than compensate for any death during the transition from spore to vegetative back to spore state (thereby resulting in a net increase rather than decrease in spore density). From a sociobiological standpoint, the growth phase may be considered to be merely a spore's way of making another spore (cf., Wilson, 1975, p. 3). Within this context, the signal for the vegetative phase to initiate spore germination, whether in positive (e.g., density-dependent production of trigger biochemicals) or negative (e.g., nutrient deficiency) form, must be given while there are still sufficient nutrients available to support subsequent spore formation (Table III).

Fluctuations in food availability (caused by periodic reductions in the rate of food supply to the ecosystem or by competitive depletion of food density to an inaccessible level) require all non-spore-forming microorganisms to be able to survive long-term starvation in the vegetative state. Vegetative resting cells are inevitably more susceptible to physical and biotic stress than inert spores. Decay by endogenous metabolism [Eq. (12)] is, by definition, substantially less for resting cells than for cells in transition from growth to resting phase metabolism. A carry-over of the high-geared, relatively inefficient metabolism shown by growth-phase *r*-organisms to the resting phase would result in a higher decay rate for *r*- than *K*-organisms. Conversely, a carryover of the higher resistance to concentration gradient-induced leakage [Eq. (13)] of growth-phase *K*- as compared to *r*-cells would result in a lower decay and death rate of resting-phase *K*- as compared to *r*-cells. Resting vegetative cells are potentially more sensitive than spores to changes in environmental nutrient conditions. For *r*-organisms, a requirement for a nutrient concentration sufficiently high to drive nutrients by passive transport mechanisms into resting cells would ensure dormancy break only under conditions conducive to subsequent competitive growth; resting vegetative cells of *K*-organisms might be expected to be more sensitive to perturbations by lower concentration of nutrients than *r*-organisms (Table III), but it is likely that the nutrient concentration required to break dormancy is still substantially higher than that triggering the shift from active to resting metabolism.

4. *r*- and *K*-Selection as an Experimental Paradigm

The r/K postulate and the "Everything is everywhere—the environment selects" concept can be consolidated into a general hypothesis that the phenotypic fitness traits of microbial populations are determined at the first-order level by the r/K nature of their environment. Since experimental evaluation of the validity and scope of this hypothesis requires a mechanistic understanding of what constitutes an r/K-environment, we first summarize in mechanistic terms the concept of the r/K environmental continuum, with emphasis on crowding with respect to resources as a major criterion. In accordance with Eq. (15) and Fig. 4, extreme r/K-environmental conditions may be defined as follows: extreme *r*-environmental condition,

$$R \geq \Sigma(q^{\max}X) \qquad (X < X_r) \qquad\qquad (20)$$

extreme K-environmental condition,

$$R \leq \Sigma(q^m X) \qquad\qquad (21)$$

Equation (20), defining the extreme r-environmental condition, specifies that the rate of food supply R to the microbial population X is more than adequate to allow all members of the population to consume food as fast as necessary ($q^{\max}$) to support unrestricted growth ($\mu_{\max}$ condition with respect to resources) and that the population density is sufficiently low ($X < X_r$) that toxicity and predation/parasitism phenomena are negligible (r-condition). Equation (21), defining the extreme K-environmental condition, specifies that the rate of food supply to the microbial population is inadequate to meet even the maintenance needs q^m of all members of the population. As mentioned previously, the time course of the r- and K-conditions prevailing in a natural environment locates the environment along the r/K continuum and identifies the related selection pressures for ecological fitness. It is particularly important from a phenotypic fitness standpoint to recognize the implications of Eq. (20) with respect to the diversity of r-environments: an r-environmental condition may arise from an increase in the rate of food supply R and/or a decrease in the population density X. For example, an r-condition may arise transitionally in soil because of an organic matter perturbation (increase in R); on the other hand, an extreme r-condition may arise transitionally on a leaf surface because of rainfall washout of the microbial population (decrease in X). Finally, it should be reemphasized that in the absence of catastrophic reductions in biomass, microbial growth [increase in X, Eq. (20)] inevitably results in a shift (commonly rapid because of the high growth rates of microorganisms) from the inherently unstable r-condition to the equilibrium K-condition. Interestingly, this dynamic situation is directly analogous to environmental disturbances affecting macroorganisms (Weins, 1977; Grime, 1979) and fungi (Pugh, 1980). The microbial system offers the prospect of a mechanistic approach to quantifying and evaluating the specific factors involved.

The general r/K postulate gives rise to several experimentally testable specific hypotheses. We first briefly review experimental methods for establishing controlled r- and K-conditions and characterizing r/K phenotypic traits (Table III) and then evaluate the conceptual basis and experimental implications of some of the more broad-based, specific hypotheses relevant to general microbial ecology.

4.1. Experimental Characterization of r/K Phenotypic Traits

Emphasis here is placed on identification of experimental approaches rather than description and critical evaluation of specific experimental techniques and data interpretation procedures.

The forms, amounts, and physiological state (metabolically active or inactive, dividing or not dividing) of microbial populations may be characterized, with varying success, depending on the nature of the ecosystem, by a variety of nondestructive and selective staining microscopic techniques (Atlas, 1982; van Es and Myer-Reil, 1982; Fry and Humphrey, 1978). Mass balance and viability analyses in batch culture allow characterization of nutrient-unrestricted specific rate of increase r, specific growth rate μ_{max}, specific death rate γ, specific rate of food consumption q, and growth efficiency A ($= 1/Y \approx A^b = 1/Y^b$) during the exponential phase, and endogenous decay λ and starvation-related death γ during the stationary/death phase (Pirt, 1975; Harris, 1981). In principle, fed batch cultures (van Versefeld *et al.*, 1984) allow characterization of carrying capacity K and specific rates of food consumption, growth, and death, phenotypic properties shown at equilibrium, for a controlled rate of food supply R. Biomass and food density kinetics under crowded oligotrophic or copiotrophic (depending on R) conditions provide information on (1) physiological strategy (e.g., carrying capacity maximized by use of food solely for maintenance; or use of food to support a smaller K of biomass growing at μ_{min}, with inevitable development of a secondary population of resting cells), and (2) competitive ability (prevailing food density S) under K-conditions. Continuous culture allows (Pirt, 1975; Harris, 1981; Kuenen and Harder, 1982) (1) characterization of μ and q versus S relationships, including the food affinity coefficient K_s, and (2) by use of Eq. (A11) and appropriate mass balance analyses, characterization of the maintenance energy requirement q^m and the assimilatory A_a, dissimilatory A_d^b, and total A^b ($= 1/Y^b$) food requirements for growth. Continuous cultures may also be used for bioassay determination of relative competitive ability (Kuenen and Robertson, 1984; Veldkamp *et al.*, 1984) between a given organism and a well-characterized control as a function of steady-state dilution rate–nutrient concentration condition. The efficiency of dissimilatory ATP generation may be estimated from experimentally determined dissimilatory pathway (key metabolic enzymes and functional proton translocating sites) considerations.

In summary, although techniques are available in principle for measuring the components of the mechanistic equations, the limitations of the prevailing state of the art in experimental microbial ecology should be recognized. A major current research challenge is development and validation of techniques for routine determination of limiting nutrient concentration in the competitive ecological range (nano- to micromolar) and quantitative discrimination among growing, metabolically active, resting, and dormant biomass. A long neglected issue brought into focus by the r/K concept is the need for research approaches allowing measurement and interpretation of specific metabolic rates and growth yields in microbial cultures where death, lysis, and cryptic growth cannot be assumed to be negligible. Modeling and methodology advances in this area in particular will require continual interplay between theory and experimentation.

4.2. Specific Hypotheses

A central specific hypothesis is, "The phenotypic fitness traits of natural microbial populations reflect the r/K nature of their environment." Experimental evaluation of this hypothesis necessitates characterization of natural microbial environments in terms of their location on the r/K environmental continuum (as determined by the nature and periodicity of r- and K-conditions prevailing over time in the environments). The potential for predictive success is highest for environments located on the extremes of the continuum. Choice of sampling times must involve consideration of preceding as well as prevailing r/K-conditions. For a postulated K-dominated environment, such as the surface sediment of an oligotrophic lake, the predominant population would be hypothesized to consist of resting and active vegetative cells of K-selected microorganisms characterized by relatively low q^{max}/μ_{max} and K_s properties and other life history traits consistent with K-strategists (Table III). Recognizing the relatively long periods between transient r-conditions, caused, for example, by settled algal blooms, r-strategists would tend to be present largely as inert spores. For a hypothesized r-dominated environment, such as the gut, the predominant population would be expected to consist of active cells of r-selected microorganisms such as $E.\ coli$ characterized by relatively high q^{max}/μ_{max} properties traded off against a relatively low competitive ability under low-food-density conditions (high K_s). As discussed previously, because of the regularity of the feast/famine and microbial biomass reduction regime, spores are an inappropriate resting mechanism; and the rapid regrowth capability needed to rebuild the population following fecal elimination effectively selects against characteristically low-μ_{max} K-organisms.

We are aware that many factors control to varying degrees the competitive ecological relationships in natural ecosystems, and that, consequently, the phenotypic traits of major fitness-determining importance may not be directly r- and K-related (e.g., salt tolerance in saline environments). However, it is still relevant to characterize such environments at a first-order level in terms of their location on the r/K continuum since this establishes the nature of the dynamic abiotic and biotic selective pressures exerted by the environments on the microbial community.

A second specific hypothesis is, "Microorganisms subjected to r/K environmental conditions develop corresponding r/K phenotypic fitness traits." Evaluation of this hypothesis involves subjecting pure cultures to controlled r and K experimental conditions, followed by characterization (Section 4.1) and comparison of their phenotypic traits in accordance with predictions (Table III). Pure culture sources should include stock microorganisms, well characterized in terms of metabolic pathways, nutritional requirements, growth rate and efficiency, q and μ versus S relationships, and physiological strategy for growth and survival under K-conditions. Additionally, microorganisms dominating in mixed cultures subjected specifically to well-defined r and K cultural conditions should be isolated as a source of unequivocally r- and K-selected test organisms. Continuous r-conditions may be achieved by maintaining batch cultures under such conditions that unrestricted exponential growth occurs at all times. This can be accomplished by a variety of serial transfer (Luckinbill, 1978), removal/replacement, and/or catastrophic death-dealing shock treatments. Continuous K conditions develop inevitably in fed batch cultures. Continuous cultures may be established to provide a range of steady-state, food-limited conditions supporting varying growth rates and corresponding food densities. In addition, non-food-limited continuous cultures may be established to evaluate adaptive response to crowding with respect to phenomena other than food limitation, such as the presence of toxic metabolites. The nonrecombinant DNA mutation approach (Lein, 1983) may be used to accelerate the adaptive time scale. A basic question relevant to the potential uniqueness of r and K (contrasted with, for example, alternate energy sources) as selective pressures is how broad the mutational boundaries are for an r-organism to develop K properties and vice versa. What is the relative magnitude of the price that must be paid for such shifts in r and K-properties (e.g., what does it cost for an increase in specific growth rate for a K-organism in terms of life cycle complexity, nutritional versatility, or starvation survival capability)? It is known that, under selective pressure, microorganisms may adaptively produce enzymes with increased specific affinities V_{max}/K_m, reflecting both an increase in the maximum velocity V_{max} and a decrease in the half-saturation coefficient K_m of the enzyme (Hall, 1984). This raises the question of whether it is possible for an organism to develop selectively a suite of such enzymes so that it would be genotypically capable of competing in both an uncrowded and a crowded environment: the r/K postulate specifies that this is not possible.

The last hypothesis we examine is, "The competitive ability of specific microorganisms for short- and long-term success in a given environment is a function of the compatibility of their r/K phenotypic fitness traits with the r/K nature of the environment." The underlying premise of this test is again that there are two opposing types of selection, which result in a tradeoff; i.e., a K-strategist will be less fit under conditions of r-selection and vice versa. Experimentally, populations adapted to r- and K-selection should, when reintroduced to nature, become established best in the environments that present the corresponding selective pressure (Andrews, 1984b). Experimental techniques for evaluating competitive ability under controlled r- and K-conditions and in natural environments are analogous in principle to those described for batch, fed batch, and continuous culture (e.g., Kuenen and Robertson, 1984) supplemented by appropriate microscopic and viability methodology (Section 4.1). Evaluation of competitive ability in mixed cultures is greatly facilitated by the availability of marked (e.g., auxotrophic or antibiotic-resistant) strains (Luckinbill, 1978; Andrews, 1984b) identical physiologically to the parent except for the marker property. If this hypothesis proves to be valid, then a major practical implication is genetic engineering of suites of r/K-traits that are compatible/incompatible with specific kinds of r/K-environmental conditions, thereby allowing for short- and/or long-term establishment of target microbial strains in natural ecosystems in accordance with maximized biological control options (e.g., development of a rhizobium legume inoculant that is highly competitive for root infection but "self-destructs" overwinter and thus can be replaced at will as improved strains are developed).

5. Implications of *r*- and *K*-Selection

The r- and K-selection theory originated as an incisive attempt to describe opposing selection forces influencing the evolution of colonizing populations. Subsequently, the concept has been broadened considerably. It can and has been used to group or *classify* organisms independently of individual species characteristics; to *explain* relationships between various adaptive traits; and to *predict* how an organism might change its life history traits to maximize r or K; i.e., how it might survive in particular environments (Gould, 1977; Stearns, 1977). The concept can be instructive conceptually when used in the first manner, provided that it is remembered that the traits are merely *correlates:* the distinction between correlation and causation must not be confused. r- and K-type environments may select for organisms with r- and K-type traits, but the observation that such features exist does not in itself mean that they *resulted* from the purported environmental selection. To invoke r- or K-selection in an explanatory or predictive manner requires that crowding be implicated in the evolution of the feature or, more appropriately, the suite of features in question. This has been difficult or impossible to demonstrate convincingly for several reasons. First, the organism as a unit evolves and not specific traits *per se;* second, the nebulous semantics and various, often conflicting, usages of the concept have confused the central message; third, limitations in macroecology on experimental design and manipulation have generally precluded rigorously testing a single variable (crowding). As is evident from the foregoing mechanistic analysis, microbial ecologists have the systems to circumvent the latter two of these obstacles.

We feel that, particularly for microorganisms, it is useful conceptually to recognize that r/K selection pressures exerted by an environment are a function of the nature and periodicity of the *r*- and *K*-conditions prevailing over time in the environment. This gives rise to the concept of an r/K environmental continuum described earlier. The power of this approach lies in the fact that *r*- and *K*-conditions are definable in terms of mechanistically based relationships between the organism and its environment: an *r*-condition is defined as existing when the environment/organism relationship R/X (where R is the rate of food supply and X is the population density) is such ($R/X \geq q^{max}$) that all individual members of the population are capable of growing at their maximum rate of increase *r*; a *K*-condition is defined as existing when the environment/organism relationship is such ($q^m \leq R/X < q^{max}$) that the rate of increase is approaching zero and population density is approaching the carrying capacity K. These relationships, with mechanistic emphasis on crowding with respect to resources, can be expanded into equations identifying the phenotypic properties under selective pressure for maximized ecological competence under specific *r*- and *K*-conditions. Experimentally, the equations provide a basis for evaluating and characterizing specific phenotypic fitness traits, and ultimately testing the r/K postulate that there is a fundamental and major tradeoff between microorganisms that do well in uncrowded versus crowded environments. Within this context we believe that the r/K concept may be usefully applied to all environments, including extremely adverse physiochemical environments, where an overriding determinant of growth and survival is the organism's ability to handle the adversity. Such adverse environments are potentially subject to variable oscillations in *r*- and *K*-conditions, and the location of the environment on the r/K environmental continuum is relevant to whether the environment will be essentially sterile (*r* end) or will tend to support a resource-limited carrying capacity population (*K* end).

It is also instructive to redefine the ecological oligotrophic/copiotrophic and autochthonous/zymogenous terms within the context of the r/K-environmental condition concept. An oligotrophic condition is currently defined as existing in an aquatic environment showing a very low rate of food supply [approaching zero or a fraction of a mg C/liter per day (Poindexter, 1981; Kjelleberg, 1984)]. In principle, a low rate of food supply (low R) may be associated with a high or low food density and may support an r- or K-condition, depending on the prevailing population density X. The implicit assumption that the food density under oligotrophic conditions is very low, combined with the designated low R, is consistent with classifying an oligotrophic condition as a food-limited, low-carrying-capacity, K-condition. Accordingly, an oligotroph may be characterized as a food-limited, K-selected organism. A copiotrophic (eutrophic) condition is currently defined as existing in an aquatic environment showing a relatively high rate of food supply [$>$50-fold higher than for an oligotrophic condition (Poindexter, 1981)]. Except for recognition that a copiotrophic condition is not an oligotrophic condition, this definition provides no information on the nature of the selective pressures exerted on the microbial inhabitants. For an environment characterized by continuous perturbation with a high rate of food supply, the selective pressures will oscillate from r-selection, immediately following the perturbation, to K-selection, controlled by various combinations of crowding with respect to food limitation, toxic byproduct, and/or predation/parasitism. The common practice of isolating copiotrophs by batch liquid or solid medium enrichment techniques operationally defines copiotrophs as r-selected organisms. A similar rationale may be applied to redefinition of autochthonous and zymogenous terminology, although ambiguities in ecological definition of these terms is a complicating factor. Within the context of soil microbial ecology semantics, autochthonous and zymogenous mean the same operationally as the aquatic oligotrophic and copiotrophic terms, respectively. In keeping with historical and current generally accepted meanings, we suggest the following mechanistically explicit definitions: (1) an autochthonous and an oligotrophic condition should be defined as a low-nutrient-flux (low-R) K-condition; (2) a zymogenous condition should be defined as a high-nutrient-concentration (high-S and thus in effect a high instantaneous R) r-condition; and (3) a copiotrophic (eutrophic) condition should be defined either as a high-nutrient-concentration r-condition (zymogenous or r-copiotrophic) or a high-nutrient-flux K-condition (K-copiostrophic). Since a K-condition, for example, can be either oligotrophic or copiotrophic, the two systems of nomenclature are not synonomous; i.e., terms such as autochthonous or eutrophic describe particular types of r- and K-conditions. Finally, to avoid ambiguities, we recommend against continued use of the term autochthonous as a descriptor for resident or indigenous (as distinct from invader or pathogenic) microorganisms.

　　　The limitations imposed by r- and K-selection necessitate "choices" by the organism, each with inherent advantages and disadvantages. The lifestyle alternatives are essentially three. A microbial species might be an *r*- or a *K*-strategist, thriving in the corresponding environment, but at the expense of barely surviving or becoming locally extinct elsewhere. For example, a *K*-organism could not recover fast enough after a catastrophic event to become prominent in an *r*-environment. Since all habitats have the potential for periodic catastrophes, *r*-organisms could survive in *K*-environments, but in a cryptic fashion as spores or similar forms. The third option is life as a "diplomat," where the extremes are negotiated in favor of a middle course where the microbe does neither very well nor very badly under either condition.

The *r*/*K*-selection concept carries important implications for the manipulation of microorganisms, including genetic engineering. Although the specific microbial characteristics determining successful establishment on the phylloplane, on the rhizoplane, or in the colon may be unique, broadly speaking, the traits are essentially those of a colonist or *r*-strategist as outlined in Table III. These and similar criteria may serve both as guides to suitable habitats for isolating candidate microbes and for selecting or engineering specific isolates. A good biocontrol agent, for instance, might be an *r*-organism engineered without the resistant stage so that it would colonize rapidly, preclude the pathogen during the vulnerable stage of the host, and then gradually decline.

Our list of correlates and those of others (e.g., Pianka, 1970; Stearns, 1976) suggest that many important life history traits *may* have evolved in response to *r*- and *K*-selection. This could apply also to the timing of expression of these features; e.g., the onset of particular stages in the life cycle (Andrews, 1984a,b). There are well-established precedents in the entomological literature, such as the transition from wingless to winged forms of aphids, or from the solitary to the gregarious (migratory) state in locusts. In microbes, crowding may trigger alterations in enzyme properties or repression or derepression of parts of the genome. This, in turn, could be reflected in various phenotypic changes, such as encapsulation, production of secondary metabolites, alteration in adhesive properties, or host specificity. Dobzhansky (1950) drew attention to adaptive polymorphism in *Drosophila*, whereby a species may comprise several biotypes, each well adapted to a particular environment. High genotypic and phenotypic plasticity in microbes is another manifestation of the same thing. In fact, the cells of a so-called "homogeneous" bacterial population may be shown upon scrutiny to be quite different in their behavior. Among other possiblities, they could adopt different strategies in response to the same cue. Interestingly, such diversification within a "clonal" population may be the precursor to tissue differentiation in higher organisms.

Although the basis of the DNA content in eucaryotes is poorly understood, one could speculate that the size of the DNA "brain" required to direct the life cycle in a microbe approximates a sum of the subgenomes for each phase of that life cycle. This presents an interesting question: either organisms have a finite coding capacity, in which case as life cycle complexity increases presumably physiological scope (and selective advantage relative to competitors?) for each component decreases, or more complex microbes have a greater capacity to enlarge their genomes by such means as acquiring extranuclear DNA, heterokaryosis, and parasexuality. In less abstract terms, does a heteroecious rust fungus requiring multiple hosts to complete its life cycle have in aggregate a larger genome, or less physiological dexterity, than its autoecious counterpart? What do these restrictions imply for life history strategies?

The problem an organism faces from the standpoint of life cycle tactics is deciding what suite of compatible properties is needed for each part of the life cycle. What are the physiological restrictions associated with the choice of a specific grouping? Put another way, what clusters of phenotypic traits are compatible and incompatible with one another? How do these differ in a high-geared versus a low-geared organism?

If the hypothesis is sustained that phenotypic traits of microbial populations are determined at the first-order level by the r/K nature of their environment, then there are important ecological implications. These include the ability to predict at least the general properties of organisms isolated from a particular locale, the likely response of different populations to catastrophes, and the probability that microbes will colonize when introduced into a given environment. The r/K analysis would supplement traditional habitat descriptors such as substrate type, electron donor, etc. Microbial ecologists would gain a new perspective on habitat characteristics, and knowledge of obvious practical value for genetic engineering and biocontrol discussed above. Despite hazards in determining successional patterns in litter and other substrates (Swift, 1982), the r/K postulate may prove to be a useful model for interpreting resource/competitive relationships.

6. Concluding Remarks

The major contribution of r- and K-selection as a postulate is that, in a concise statement, it synthesizes, simplifies, and provides a means for interpreting disparate facts pertaining to much of natural selection. At worst, its proponents can be accused of embracing a vague concept that may ultimately explain little of the variation among organisms. Even if this accusation is eventually substantiated, the concept has stimulated new research and new perspectives on the evolution of life history traits; as such, it would generate subsequent, more accurate postulates. Perhaps most importantly, the concept has provided an as yet largely unexploited bridge between macro- and microbiology. It is ironic indeed that at a time when the *r*- and *K*-selection concept is derided as passé by many plant and animal ecologists on the well-founded grounds that rigorous experimental proof is virtually unattainable in their discipline, microbiologists, who have the systems to subject the idea and its predictions to a rigorous test, are just beginning to broach the subject.

Shortly after he and Wilson formally proposed the idea, MacArthur (1972, p. 230) commented that *r*- and *K*-selection provided a convenient but not the only means for subdividing all natural selection. Undoubtedly, the issue over what specific pressures are embodied in *r*- and *K*-selection will remain controversial. The matter will hinge largely on whether crowding is broadly or narrowly defined. Regardless, it is only one of many interacting constraints or evolutionary hurdles confronting organisms; others include design, seasonality, and the trophic and demographic structure of populations. Determining the actual role *r*- and *K*-selection plays, among other forces, in molding life histories will continue to present an important and fascinating challenge to ecologists for the forseeable future.

7. Appendix

7.1. Interpretation of the Verhulst–Pearl Logistic Equation

This section interprets the Verhulst–Pearl equation in terms of the specific birth and death rate semantics and concepts commonly used in general ecology (Wilson and Bossert, 1971; Wilson, 1975; Pianka, 1983). Following the approach of Wilson and Bossert (1971), we derive the logistic equation starting with the basic concept that the per capita rate of population change (increase) $(1/X)\, dX/dt$ represents the difference between the per capita birth rate b and the per capita death rate d:

$$\frac{1}{X}\frac{dX}{dt} = b - d \tag{A1}$$

It is then assumed that under optimal physical conditions b is maximized at a constant value b_0^{max} and d is minimized at a constant value d_0^{min} at low population densities, but beyond a certain threshold density K', b decreases linearly and d increases linearly with increasing X:

$$b = b_0^{max} - k_b^{min}X \quad (K' = 0) \tag{A2}$$
$$d = d_0^{min} - k_d^{min}X \quad (K' = 0) \tag{A3}$$

where b_0^{max} and d_0^{min} represents the optimized population density-independent (zero subscripts) components of b and d, respectively; and k_b^{min} and k_d^{min} represent properties defining the contributions of the density-dependent components ($k_b X$ and $k_d X$) to b and d, respectively. In effect, the k_b and k_d terms are determinants of the magnitude of density-dependent decreases and increases in b_0^{max} and d_0^{min}, respectively. Substitution of Eq. (A2) and (A3) into (A1) gives

$$\frac{1}{X}\frac{dX}{dt} = (b_0^{max} - k_b^{min}X) - (d_0^{min} + k_d^{min}X) \qquad (K' = 0) \qquad (A4)$$

and this can be rearranged to give the logistic equation:

$$\frac{1}{X}\frac{dX}{dt} = (b_0^{max} - d_0^{min}) - (k_b^{min} + k_d^{min})X \qquad (K' = 0) \qquad (A5)$$

$$= r^{max} - k^{min}X \qquad (K' = 0) \qquad (A6)$$

where r^{max} is the (density-independent) maximum specific rate of increase (shown under uncrowded, optimized physical conditions) and k^{min} represents the minimum impact of increasing population density on attenuation of r^{max}. As identified in Eqs. (A1) and (A5) and illustrated in Fig. 5, as the per capita birth rate decreases and the per capita death rate increases with increasing population density, the specific rate of increase declines proportionately until, when $b = d$, it becomes zero; at this point $X = K^{max}$, the maximum carrying capacity (achieved under optimal physical environmental conditions). The components of the K^{max} constant are seen to be [from Eq. (A5) with $(1/X)\,dX/dt = 0$ and $X = K^{max}$]

$$K^{max} = \frac{b_0^{max} - d_0^{min}}{k_b^{min} + k_d^{min}} \qquad (K' = 0) \qquad (A7)$$

Similarly, the k^{min} constant is given by [from Eq. (A6) with $(1/X)\,dX/dt = 0$ and $X = K^{max}$]

$$k^{min} = r^{max}/K^{max} \qquad (K' = 0) \qquad (A8)$$

Substitutions of Eq. (A8) into (A6) gives the Verhulst–Pearl logistic equation:

$$\frac{1}{X}\frac{dX}{dt} = r^{max} - \frac{r^{max}}{K^{max}}X \qquad (K' = 0) \qquad (A9)$$

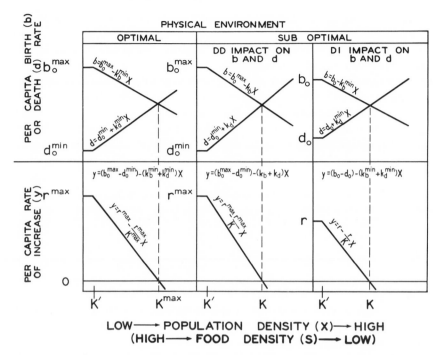

Figure 5. Dependence of per capita rate of increase y $[=(1/X)\,dX/dt]$ on per capita birth rate b and per capita death rate d [after Wilson (1975), Wilson and Bossert (1971), and Pianka (1983)]. At low population densities X, below a critical value K' [following the semantics of Wilson (1975) and identified, for simplicity, to be zero in this example], b and d and thus $y = b - d$ are assumed to be constant; with increasing X, b decreases and d increases linearly, resulting in a linear decrease in y; at equilibrium, $b = d$, $y = 0$, and the population is at the carrying capacity ($X = K$). Under optimal physical conditions (left side of the figure) the density-independent (b_0 and d_0) and density-dependent ($k_b X$ and $k_d X$) components of y are optimized to give $r = b_0^{max} - d_0^{min} = r^{max}$ and $K = (b_0^{max} - d_0^{min})/k_b^{min} + k_d^{min}) = K^{max}$. Under suboptimal physical conditions, as manifested by a deleterious change in the density-dependent (DD) components ($k_b X$ and $k_d X$) of y (middle of the figure), $r = b_0^{max} - d_0^{min} = r_{max}$ is unchanged, but y at any given $X > K$ is decreased, including $K < K^{max}$. Under suboptimal physical conditions, as manifested by a deleterious change in the density-independent (DI) components (b_0 and d_0) of y (right side of the figure), $r < r^{max}$ and y at any given X is decreased, and $K < K^{max}$.

Suboptimal environmental conditions may cause a change in the k determinants of the density-dependent components of b and d and/or may cause a change in the density-independent (b_0 and d_0) components (Fig. 5). It should be recognized that a change in the density-independent (b_0 and d_0) properties will cause a change in the relationship between specific rate of increase and population density (and consequently also a change in K), thereby creating potential ambiguities in the use of density-dependent and density-independent terminology (Murray, 1982).

In summary, we emphasize that the above equations are presented more from the standpoint of acquainting the reader with the b and d semantics commonly used in the general ecology literature than as a mechanistic approach to understanding r- and K-selection. The basic mechanistic limitation of the logistic equation still holds: none of the equations explicitly recognizes the mechanistic basis of the effect of increased population density (crowding) on birth rate/death rate and the resultant rate of increase.

7.2. Primary Equations Based on Specific Rate and Efficiency of Food Consumption

The basic relationships between the specific rate of food acquisition/consumption q and its components, the population rate of food consumption Q and food supply R, food density S, specific growth rate μ, and specific rate of increase $(1/X)dX/dt$ [largely following the approach and semantics of Pirt (1975) as modified and expanded in Harris (1981)], are summarized as follows:

$$q = \frac{Q}{X} \tag{A10}$$

$$= q^m + q^b$$

$$= q^m + \mu A^b = q^m + \mu(A_a + A_d^b) = q^m + \frac{\mu}{Y^b} \tag{A11}$$

$$= \frac{q_{ATP}^m}{E_{ATP}} + \mu\left(A_a + \frac{A_{ATP}^b}{E_{ATP}}\right) = \frac{q_{ATP}^m}{E_{ATP}} + \mu\left(A_a + \frac{1}{Y_{ATP}^b}\right) \tag{A12}$$

$$= q^{max}\frac{S}{K_s + S} \quad (S_n \leq S \leq S_r)$$

$$= \alpha S \quad (S_n \leq S \leq S_r) \tag{A13}$$

$$= q^{max} \quad (K_s \ll S \ll S_r) \tag{A14}$$

$$= \frac{q^{max}}{K_s}S \quad (S_n < S \ll K_s)$$

$$= \alpha_{max}S \quad (S_n < S \ll K_s) \tag{A15}$$

$$\mu = \frac{q}{A} = qY$$

$$= \frac{q^b}{A_b} = \frac{q - q^m}{A_b} = (q - q^m)Y^b \tag{A16}$$

$$= \frac{q^{max}S}{A^b(K_s + S)} - \frac{q^m}{A^b} \quad (S_m \leq S \leq S_r)$$

$$= \alpha SY = \frac{\alpha S}{A} = \frac{\alpha S - q^m}{A^b} \quad (S_m \leq S \leq S_r) \tag{A17}$$

$$= \mu_{max} = \frac{q^{max} - q^m}{A^b} = \frac{q^{max}}{A^b} \quad (q^{max} \gg q^m; S_r = \infty; S \gg K_s) \tag{A18}$$

$$= \mu_{max} \frac{S}{K_s + S} \quad (q^{max} \gg q^m; S_r = \infty; S \gg S_m) \tag{A19}$$

$$\lambda = \frac{q^m_{ATP}}{E_{ATP/x}} \quad (S = S_n) \tag{A20}$$

$$= \frac{q^m_{ATP}}{E_{ATP/x}} + l \quad (S < S_n) \tag{A21}$$

$$\frac{1}{X}\frac{dX}{dt} = \mu - \gamma \quad (S \geq S_m)$$

$$= \frac{q - q^m}{A^b} - \gamma = \frac{q - (q^m + \gamma A^b)}{A^b} \quad (S \geq S_m) \tag{A22}$$

$$= -(\lambda + \gamma) \quad (S < S_m) \tag{A23}$$

$$= \mu_{max} - \gamma_{min} = r \quad (S \geq S_r; X \leq X_r) \tag{A24}$$

$$K = \frac{R}{q_K} = \frac{R}{q^m + \gamma_K A^b} = \frac{R}{q^m + \mu_K A^b} \quad (R \leq R_{max}) \tag{A25}$$

where the components of the equations are defined in the list of General Symbols and Abbreviations (Section 7.4). The equations are listed in self-deductive sequence. Detailed development and discussion of the equations are beyond the scope of this review; certain key points, and explanations of equations not used directly in the text but included for comparison with the microbial modeling literature, are as follows. The basic concepts are (1) recognition that the specific rate of food consumption/use q partitions additively into various assimilatory and dissimilatory components [Eq. (A11)], and (2) assumption that the specific rate of food consumption/use follows saturation kinetics below the food density level at which food acquisition limits growth [Eq. (A13)]. Implicit in Eq. (A17) is that exogenous substrate metabolism will take precedence over endogenous metabolism for meeting maintenance energy requirements (i.e., endogenous metabolism/decay does not occur as long as $q \geq q^m$). For comparison with the mechanistic equations of Button (1983), the relationship between his and our symbolism is: $-dA/dt = -dS/dt = Q$; $a_A^0 = \alpha_{max}$; $X = X$; $A = S$; $a_A = \alpha$; $V_{max} = q^{max}X$; $K_A = K_s$; $v = Q$; $\mu_A = \mu$; $\mu_{max} = \mu_{max}$; $K_\mu = K_{s(\mu)}$. Accordingly, for example, Button's Eq. (1) is identical to Eq. (A15) (recognizing that $Q = qX$), Eq. (2) is identical to Eq. (A13) (recognizing that $\alpha = q/s$), Eq. (3) is identical to Eq. (A1) (recognizing that $Q = qX$), and Eq. (5) is identical to Eq. (A17) (recognizing that $q = \alpha S$, $\mu = 0$ at $S = S_m$, and $\mu = \mu_{max}$ at $S \geq S_r$). Our α (rather than a_A) terminology for the specific affinity of the organism for the nutrient follows that used traditionally for describing nutrient uptake kinetics by roots (Nye and Tinker, 1977). As emphasized by Button (1983) and Veldkamp *et al.* (1984), a major advantage of using α_{max} (the slope of the q versus S relationship in the S region, where q is first order with respect to S) is that α_{max} can be determined experimentally, independent of whether or not uptake is controlled by simple Michaelis–Menten kinetics. Only if the latter case holds is $\alpha_{max} = q^{max}/K_s$ [Eq. (A15)], sometimes called the "specificity" of an enzyme-catalyzed reaction (Hall, 1984). The empirical rather than mechanistic basis of the Monod μ versus S saturation equation is well recognized (Roels and Kossen, 1978; Button, 1983; Veldkamp *et al.,* 1984): the highly restrictive conditions under which the Monod equation is applicable are identified in Eq. (A19).

7.3. Expansion of the Smith Equation

Equation (3), rearranged into the form

$$\frac{1}{X}\frac{dX}{dt} = r\frac{T - Q}{T} \tag{A26}$$

states that the specific rate of biomass increase is proportional to $(T - Q)/T$, the proportion of the rate of food supply not yet utilized (Pielou, 1969; Smith, 1963). The Q term consists of a maintenance and a (growth rate-dependent) biosynthesis component [from Eqs. (A10), (A11), and (A22)]:

$$Q = qX = q^m X + q^b X \tag{A27}$$

$$= q_X^m + A^b\mu X = q^m X + A^b\left(\frac{1}{X}\frac{dX}{dt} + \gamma\right)X \tag{A28}$$

When saturation is reached, $dX/dt = 0$, and by definition $X = K$ and $Q = R$, so that [from Eq. (A25)]

$$T = (q^m + \gamma_K A^b)K \tag{A29}$$

Substitution of Eqs. (A28) and (A29) into Eq. (A26) gives

$$\frac{1}{X}\frac{dX}{dt} = r\frac{(q^m + \gamma_K A^b)K - (q^m + \gamma A^b)X - A^b\,dX/dt}{(q^m + \gamma_K A^b)K} \tag{A30}$$

For $\gamma = \gamma_K$, and putting $q^m + \gamma_K A^b = c_1$ and $A^b = c_2$, we find that Eq. (A30) becomes

$$\frac{1}{X}\frac{dX}{dt} = \frac{r(c_1 K - c_1 X - c_2\, dX/dt)}{c_1 K} \tag{A31}$$

Rearrangement of Eq. (A31) and simplification by putting $c_1/c_2 = c$ gives (Pielou, 1969; Smith, 1963)

$$\frac{1}{X}\frac{dX}{dt} = \frac{r(K - X)}{K + (r/c)X} \tag{A32}$$

where $c = q^m/A^b + \gamma_K$ is the specific rate of biomass replacement at saturation (Smith, 1963). It should be recognized that a basic assumption of Eq. (A32) is that the specific death rate is constant over the entire range of population density and food availability conditions existing between uncrowded growth and saturation. If Eq. (A32) holds (Pielou, 1969; Smith, 1963), the specific rate of increase (net specific growth rate) will decrease in a concave fashion as population density increases (i.e., it will decrease rapidly with increasing X while X is small and less rapidly as X becomes larger). The major mechanistic limitation of the Smith equation is that, although the Q and R terms are expanded into mechanistically meaningful subcomponents, the starting equation [Eq. (A27)] has no explicit mechanistic basis.

7.4. General Symbols and Abbreviations

A Specific amount of substrate required/consumed for biosynthesis and maintenance ($dS/dX = 1/Y$, for negligible death), abbreviated from A_s.

A_a Specific amount of substrate assimilated for biosynthesis, abbreviated from A_{a-s}

A^b Specific amount of substrate assimilated and dissimilated for biosynthesis, abbreviated from A_s^b

A_d^b Specific amount of substrate dissimilated for biosynthesis, abbreviated from A_{d-s}^b

A_{ATP}^b Specific amount of ATP required/consumed for biosynthesis ($= 1/Y_{ATP}^b = 1/Y_{ATP}^{max}$)

b Specific (per capita) birth rate

b_0 Maximum b for a given, not necessarily optimal, physical environment under uncrowded ($X < X_r$) conditions; a population density-independent property

b_0^{max} Maximum b realizable by an organism under uncrowded ($X <$ X_r) optimal physical environmental conditions; a population density-independent property

D Fecundity: number of daughter cells produced per reproductive event by a microorganism

d Specific (per capita) death rate

d_0 Minimum d for a given, not necessarily optimal, physical environment under uncrowded ($X < X_r$) conditions; a population density-independent property

d_0^{min} Minimum d realizable by an organism, shown under uncrowded ($X < X_r$) optimal physical environmental conditions; a population density-independent property

E_{ATP} Efficiency of dissimilatory ATP generation, abbreviated from $E_{ATP/s}$

$E_{ATP/x}$ Efficiency of endogenous metabolism

K Carrying capacity of an environment (with respect to resources, unless specified otherwise): equilibrium population density of food-consuming (nondormant) cells; in terms of b and d, $K = (b_0 - d_0)/(k_b + k_d)$; in terms of R, $K = R/q_K$

K_{max} Maximum K under optimal physical environmental conditions; in terms of b and d, $K_{max} = (b_0^{max} - d_0^{min})/(k_b^{min} + k_d^{min})$; in terms of R, $K_{max} = R_{max}/q_K$

K_s Half-saturation coefficient for q: food density at which $q = 0.5q^{max}$

$K_{s(\mu)}$ Half-saturation coefficient for μ: food density at which $\mu = 0.5\mu_{max}$

k Proportionality coefficient determining the density dependence of the specific rate of population density change (increase) in the Verhulst–Pearl logistic equation ($k = k_b + k_d$)

k_b, k_b^{min} Proportionality coefficient determining the density dependence of the birth rate component of the specific rate of population density increase in the Verhulst–Pearl logistic equation ($b = b_0 - k_bX$); the superscript min identifies optimal physical environmental conditions

k_d, k_d^{min} Proportionality coefficient determining the density dependence of the death rate component of the specific rate of population density increase in the Verhulst–Pearl logistic equation ($d = d_0 + k_dX$); the superscript min identifies optimal physical environmental conditions

l Specific leakage rate of intracellular metabolites

q Specific rate of food acquisition/consumption for biosynthesis and/or maintenance [$(1/X) dS/dt$], abbreviated from q_s

q^b Specific rate of food consumption for biosynthesis, abbreviated from q_s^b

q^m Specific rate of food consumption for maintenance, abbreviated from q_s^m

q_{ATP}^m Specific rate of ATP production/use for maintenance

q^{max} Maximum potential specific rate of food acquisition/consumption with respect to food density

q_K Specific rate of food acquisition/consumption at equilibrium ($= q^m + \lambda_k A^b = q^m + \mu_k A^b$)

R Rate of food supply to a population, abbreviated from R_s

R_{max} Maximum rate of food supply to a population beyond which there is no increase in K

r Maximum specific (per capita) rate of population density change (increase) for a given, not necessarily optimal, physical environment and uncrowded ($X < X_r$; $S > S_r$) conditions ($= b_0 - d_0 = \mu_{max} - \gamma_{min}$); Sometimes called the intrinsic rate of increase

r_{max} Maximum r realizable by an organism, shown under uncrowded ($X < X_r$; $S > S_r$) optimal physical environmental conditions ($= b_0^{max} - d_0^{min} = \mu_{max} - \gamma_{min}$)

S Food density (concentration)

S_c Minimum food density required for cell cycle (growth) initiation: at S_c, $\mu = \mu_{min}$

S_m Food density supporting $q = q^m$ (below S_m, decay by endogenous metabolism must occur)

S_n Food density at which $q = 0$ (below S_n, in addition to decay by endogenous metabolism, decay by intracellular metabolite leakage potentially occurs)

S_r Minimum food density supporting μ_{max} and r (for negligible density-dependent death); For $S_r = \infty$, $\mu = \mu_{max}$ is limited by the rate of food acquisition (rather than, for example, the rate of intracellular metabolism)

t_d Biomass doubling time [$= (\ln 2)/\mu$]

t_g Population density generation time [$= (\ln D)/b$]

X Population density of living (and unless specified otherwise, physiologically active) organisms expressed in terms of numbers of individuals N or biomass M.

X_r Maximum population density above which the specific rate of increase $< r$

Y Specific growth yield with respect to food ($= dX/dS$), abbreviated from Y_s

Y_{ATP}^b Specific growth yield with respect to ATP, excluding the effect of maintenance energy ($= 1/A_{ATP}^b$); commonly called Y_{ATP}^{max}

y Specific rate of population density change (increase) under unspecified conditions [$(1/X)\, dX/dt = b - d = \mu - \gamma$]; sometimes called the Malthusian parameter

α_{max} Maximum specific affinity for food uptake ($= q_{max}/K_s$ for simple saturation kinetics)

γ Specific (per unit biomass) death rate

γ_K Specific (per unit biomass) death rate at equilibrium ($= \mu_K$)

γ_{min} Minimum specific (per unit biomass) death rate for a given, not necessarily optimal, physical environment, shown under uncrowded conditions

γ_{MIN} Minimum specific (per unit biomass) death rate realizable by an organism, shown under uncrowded, optimal physical environmental conditions

λ Specific (per unit biomass) decay rate $[-\lambda = (q_{ATP}^m/E_{ATP/x}) + l]$

μ Specific (per unit biomass) gross rate of biomass production (growth) under unspecified conditions $[=(1/X)\,dX/dt + \gamma]$

μ_K μ at equilibrium ($= \gamma_K$)

μ_{max} Maximum μ for a given, not necessarily optimal, physical environment and uncrowded ($X < X_r$; $S > S_r$) conditions

μ_{MAX} Maximum μ realizable by an organism, shown under uncrowded ($X < X_r$; $S > S_r$), optimal physical environmental conditions

μ_{min} Minimum μ, shown at S_c; may characteristically be a function of μ_{max}

ACKNOWLEDGMENTS. This is a contribution from the College of Agricultural and Life Sciences, University of Wisconsin-Madison. Support by the National Science Foundation to John H. Andrews (grant DEB-8110199) and by the U.S. Department of Agriculture to Robin F. Harris (Hatch grant 2495) are gratefully acknowledged.

References

Andrews, J. H., 1984a, Life history strategies of plant parasites, in: *Advances in Plant Pathology*, Vol. 2 (D. S. Ingram and P. H. Williams, eds.), pp. 105–130, Academic Press, London.

Andrews, J. H., 1984b, Relevance of r- and K-theory to the ecology of plant pathogens, in: *Current Perspectives in Microbial Ecology* (M. J. Klug and C. A. Reddy, eds.), pp. 1–7, American Society for Microbiology, Washington, D.C.

Andrews, J. H., and Rouse, D. I., 1982, Plant pathogens and the theory of r- and K-selection, *Am. Nat.* **120**:283–296.

Atlas, R. M., 1982, Enumeration and estimation of microbial biomass, in: *Experimental Microbial Ecology* (R. G. Burns and H. J. Slater, eds.), pp. 84–102, Blackwell, Oxford.

Begon, M., and Mortimer, M., 1981, *Population Ecology*, Sinauer, Sunderland, Massachusetts.

Button, D. K., 1983, Differences between the kinetics of nutrient uptake by microorganisms, growth and enzyme kinetics, *Trends Biochem. Sci.* **8**:121–124.

Cody, M. L., 1966, A general theory of clutch size, *Evolution* **20**:174–184.

Dobzhansky, T., 1950, Evolution in the tropics. *Am. Sci.* **38**:209–221.

Dow, C. S., Whittenbury, R., and Carr, N. G., 1983, The "shut down" or "growth precursor" cell—An adaptation for survival in a potentially hostile environment, in: *Microbes in Their Natural Environment* (J. H. Slater, R. Whittenbury, and J. W. T. Wimpenny, eds.), pp. 187–247, Cambridge University Press, Cambridge.

Esch, G. W., Hazen, T. C., and Aho, J. M., 1977, Parasitism and *r*- and *K*-selection, in: *Regulation of Parasite Populations* (G. W. Esch, ed.), pp. 9–62, Academic Press, New York.

Esener, A. A., Roels, J. A., and Kossen, N. W. F., 1983, Theory and application of unstructured growth models: Kinetic and energetic aspects, *Biotechnol. Bioeng.* **25**:2803–2841.

Freter, R., 1984, Factors affecting conjugal plasmid transfer in natural bacterial communities, in: *Current Perspectives in Microbial Ecology* (M. J. Klug and C. A. Reddy, eds.), pp. 105–114, American Society for Microbiology, Washington, D.C.

Fry, J. C., and Humphrey, N. C. B., 1978, Techniques for the study of bacteria epiphytic on aquatic macrophytes, in: *Techniques for the Study of Mixed Populations* (D. W. Lovelock and R. Davies, eds.), pp. 1–29, Academic Press, New York.

Gadgil, M., and Solbrig, O. T., 1972, The concept of *r*- and *K*-selection: Evidence from wild flowers and some theoretical considerations, *Am. Nat.* **106**:14–31.

Gaines, M. S., Vogt, K. J., Hamrick, J. L., and Caldwell, J., 1974, Reproductive strategies and growth patterns in sunflowers *(Helianthus), Am. Nat.* **108**:889–894.

Gerson, U., and Chet, I., 1981, Are allochthonous and autochthonous soil microorganisms *r*- and *K*-selected? *Rev. Ecol. Biol. Sol.* **18**:285–289.

Gould, S. J., 1977, *Ontogeny and Phylogeny,* Harvard University Press, Cambridge, Massachusetts.

Greenslade, P. J. M., 1983, Adversity selection and the habitat templet, *Am. Nat.* **122**:352–365.

Grime, J. P., 1974, Vegetation classification by reference to strategies, *Nature* **250**:26–31.

Grime, J. P., 1979, *Plant Strategies and Vegetation Processes,* Wiley, New York.

Hall, B. G., 1984, Adaptation by acquisition of novel enzyme activities in the laboratory, in: *Current Perspectives in Microbial Ecology* (M. J. Klug and C. A. Reddy, eds.), pp. 79–86, American Society for Microbiology, Washington, D.C.

Harris, R. F., 1981, Effect of water potential on microbial growth and activity, in: *Water Potential Relations in Soil Microbiology* (J. F. Parr, W. R. Gardner, and L. F. Elliot, eds.), pp. 23–95, Soil Science Soceity of America, Madison, Wisconsin.

Harris, R. F., 1982, Energetics of nitrogen transformation, in: *Nitrogen in Agricultural Soils* (F. J. Stevenson, ed.), pp. 833–899, American Society of Agronomy, Madison, Wisconsin.

Hirsch, P., Bernhard, M., Cohen, S. S., Ensign, J. C., Jannasch, H. W., Koch, A. L., Marshall, K. C., Matin, A., Poindexter, J. S., Rittenberg, S. C., Smith, C. D., and Veldkamp, H., 1979, Life under conditions of low nutrient, in: *Strategies of Microbial Life in Extreme Environments* (M. Shilo, ed.), pp. 357–372, Verlag Chemie, Weinheim.

Jennings, J. B., and Calow, P., 1975, The relationship between high fecundity and the evolution of entoparasitism, *Oecologia (Berl.)* **21**:109–115.

Kjelleberg, S., 1984, Effects of interfaces on survival mechanisms of copiotrophic bacteria in low-nutrient habitats, in: *Current Perspectives in Microbial Ecology* (M. J. Klug and C. A. Reddy, eds.), pp. 151–159, American Society for Microbiology, Washington, D.C.

Koch, A. L., 1979, Microbial growth in low concentrations of nutrients, in: *Strategies of Microbial Life in Extreme Environments* (M. Shilo, ed.), pp. 261–279, Verlag Chemie, Weinheim.

Konings, W. N., and Veldkamp, H., 1980, Phenotypic responses to environmental change, in: *Contemporary Microbial Ecology* (D. C. Ellwood, J. N. Hedger, M. J. Latham, J. M. Lynch, and J. H. Slater, eds.), pp. 159–191, Academic Press, New York.

Konings, W. N., and Veldkamp, H., 1983, Energy transduction and solute transport mechanisms in relation to environments occupied by microorganisms, in: *Microbes in Their Natural Environments* (J. H. Slater, R. Whittenbury, and J. W. T. Wimpenny, eds.), Cambridge University Press, Cambridge.

Kuenen, J. G., and Harder, W., 1982, Microbial competition in continuous culture, in: *Experimental Microbial Ecology* (R. G. Burns and H. J. Slater, eds.), pp. 342–367, Blackwell, Oxford.

Kuenen, J. G., and Robertson, L. A., 1984, Competition among chemolithotrophic bacteria under aerobic and anaerobic conditions, in: *Contemporary Microbial Ecology* (M. J. Klug and C. A. Reddy, eds.), pp. 306–313, American Society for Microbiology, Washington, D.C.

Lein, J., 1983, Strain development with non-recombinant DNA techniques, *Am. Soc. Microbiol. News* 49:576–579.

Luckinbill, L. S., 1978, r- and K-selection in experimental populations of *Escherichia coli*, *Science* 202:1201–1203.

MacArthur, R. H., 1972, *Geographical Ecology,* Harper and Row, New York.

MacArthur, R. H., and Wilson, E. O., 1967, *The Theory of Island Biogeography,* Princeton University Press, Princeton, New Jersey.

McNaughton, S. J., 1975, r- and K-selection in *Typha, Am. Nat.* 109:251–261.

Morita, R. Y., 1982, Starvation-survival of heterotrophs in the marine environment, in: *Advances in Microbial Ecology,* Vol. 6 (K. C. Marshall, ed.), pp. 171–198, Plenum Press, New York.

Murray, B. G., 1982, On the meaning of density dependence, *Oecologia (Berl.)* 53:370–373.

Nye, P. H., and Tinker, P. B., 1977, *Solute Movement in the Soil–Root System,* University of California Press, Berkeley.

Parry, G. D., 1981, The meanings of r- and K-selection, *Oecologia (Berl.)* 48:260–264.

Peters, R. H., 1976, Tautology in evolution and ecology, *Am. Nat.* 110:1–12.

Pianka, E. R., 1970, On r- and K-selection, *Am. Nat.* 104:592–597.

Pianka, E. R., 1983, *Evolutionary Ecology,* 3rd ed., Harper and Row, New York.

Pielou, E. C., 1969, *An Introduction to Mathematical Ecology,* Wiley, New York.

Pirt, S. J., 1975, *Principles of Microbe and Cell Cultivation,* Blackwell, London.

Poindexter, J. S., 1981, Oligotrophy: Fast and famine existence, in: *Advances in Microbial Ecology,* Vol. 5 (M. Alexander, ed.), pp. 63–89, Plenum Press, New York.

Pugh, G. J. F., 1980, Strategies in fungal ecology, *Trans. Br. Mycol. Soc.* 75:1–14.

Roels, J. A., and Kossen, N. W. F., 1978, On the modelling of microbial metabolism, *Prog. Industrial Microbiol.* 14:95–203.

Roughgarden, J., 1971, Density-dependent natural selection, *Ecology* 52:453–468.

Smith, F. E., 1963, Population dynamics in *Daphnia magna* and a new model for population growth, *Ecology* 44:651–663.

Southwood, T. R. E., 1977, The relevance of population dynamic theory to pest status, in: *Origins of Pest, Parasite, Disease and Weed Problems* (J. M. Cherrett and G. R. Sagar, eds.), pp. 35–54, Blackwell, London.

Stearns, S. C., 1976, Life-history tactics: A review of the ideas, *Q. Rev. Biol.* 51:3–47.

Stearns, S. C., 1977, The evolution of life history traits: A critique of the theory and a review of the data, *Annu. Rev. Ecol. Syst.* 8:145–171.

Stearns, S. C., 1980, A new view of life-history evolution, *Oikos* 35:266–281.

Swift, M. J., 1976, Species diversity and the structure of microbial communities in terrestrial habitats, in: *The Role of Terrestrial and Aquatic Organisms in Decomposition Processes* (J. M. Anderson and A. Macfadyen, eds.), pp. 185–222, Blackwell, London.

Swift, M. J., 1982, Microbial succession during the decomposition of organic matter, in: *Experimental Microbial Ecology* (R. G. Burns and J. H. Slater, eds.), pp. 164–177, Blackwell, London.

Tempest, D. W., Neijssel, O. M., and Zevenboom, W., 1983, Properties and performance of microorganisms in laboratory culture; their relevance to growth in natural ecosystems, in: *Microbes in Their Natural Environment* (J. H. Slater, R. Whittenbury, and J. W. T. Wimpenny, eds.), pp. 119–149, Cambridge University Press, Cambridge.

Tilman, D., 1982, *Resource Competition and Community Structure,* Princeton University Press, Princeton, New Jersey.

Van Es, F. B., and Myer-Reil, L. A., 1982, Biomass and metabolic activity of heterotrophic marine bacteria, in: *Advances in Microbial Ecology,* Vol. 6 (K. C. Marshall, ed.), pp. 111–170, Plenum Press, New York.

Van Verseveld, H. W., Chesbro, W. R., Braster, M., and A. H. Stouthauser, 1984, Eubacteria have 3 modes of growth keyed to nutrient flow, *Arch. Microbiol.* **137:**176–184.

Veldkamp, H., van Gemerden, H., Harder, W., and Laanbroek, H. J., 1984, Competition among bacteria: An overview, in: *Current Perspectives in Microbial Ecology* (M. J. Klug and C. A. Reddy, eds.), pp. 279–290, American Society for Microbiology, Washington, D.C.

Wallace, A. R., 1878, *Tropical Nature and Other Essays,* MacMillan, London.

Wiens, J. A., 1977, On competition and variable environments, *Am. Sci.* **65:**590–597.

Wilbur, H. M., 1976, Life history evolution in seven milkweeds of the genus *Asclepias, J. Ecol.* **64:**223–240.

Wilbur, H. M., Tinkle, D. W., and Collins, J. P., 1974, Environmental certainty, trophic level, and resource availability in life history evolution, *Am. Nat.* **108:**805–817.

Williams, F. M., 1972, Mathematics of microbial populations, with emphasis on open systems, in: *Ecological Essays in Honor of G. Evelyn Hutchinson* (E. S. Deevey, ed.), pp. 387–426, Archon, Hamden, Connecticut.

Williams, F. M., 1980, On understanding predator–prey interactions, in: *Contemporary Microbial Ecology* (D. C. Ellwood, J. N. Hedger, M. J. Latham, J. M. Lynch, and J. H. Slater, eds.), pp. 349–375, Academic Press, New York.

Wilson, E. O., 1975, *Sociobiology—The New Synthesis,* Belknap Press of Harvard University Press, Cambridge, Massachusetts.

Wilson, E. O., and Bossert, W. H., 1971, *A Primer of Population Biology,* Sinauer, Sunderland, Massachusetts.

4

Iron Transformations by Freshwater Bacteria

J. GWYNFRYN JONES

1. Introduction

Although the bacteria involved in the iron cycle have been recognized since the last century, they have received scant attention compared with those responsible for the cycling of carbon, nitrogen, and sulfur. This is hardly surprising; although iron forms 5% by weight of the earth's crust and is of considerable economic importance, the involvement of bacteria in the global iron cycle is of little quantitative significance (Nealson, 1983). In the presence of oxygen and at near neutral pH, conditions which prevail over much of this planet, the oxidation of iron and its precipitation and deposition as the ferric form, Fe(III), is essentially a chemical process. The reaction is, however, dependent on pH, ferrous iron [Fe(II)] concentration, temperature, and ionic strength of the solution. In a freshwater system where the last two components were relatively stable, Davison and Seed (1983) found no evidence for biological mediation of the reaction. Given a solubility product of 10^{-38} M for Fe(OH)$_3$ and therefore a probable maximum concentration of free Fe(III) at neutrality of 10^{-17} M what, then, is the likely involvement of bacteria in the iron cycle of freshwater systems?

This chapter is concerned with three aspects of iron transformation by bacteria, and most of the information presented will come from stratified lakes, largely because they are such convenient systems for the study

J. GWYNFRYN JONES • Freshwater Biological Association, Ambleside, Cumbria LA22 OLP, England.

of iron cycling. The three processes, Fe(III) reduction, Fe(III) deposition, and Fe acquisition, are treated very differently, largely because the involvement of bacteria is very different in each case, as, also, is the level of our understanding.

Bacteria are important agents in Fe(III) reduction in marine (Sørensen, 1982) and freshwater (Jones et al., 1983) sediments, but, although the ability to reduce certain metal species is widespread among bacteria, the mechanisms involved are little understood. Reports on the involvement of bacteria in iron deposition probably have a longer history in the scientific literature than any other aspect of the iron cycle. This was undoubtedly due, in part, to the macroscopic appearance of ferric oxyhydroxide flocs in streams, bogs, and other waterlogged situations, and the ease with which the bacteria, often of distinctive morphology, could be recognized. Our understanding of the process of deposition in bacteria other than prosthecate forms has, however, advanced remarkably little, and the success rate in isolating and culturing the organisms involved has been poor. Although it is now possible to make some general statements about the involvement of extracellular polymers in metal deposition, both the taxonomy and physiology of those bacteria would be advanced significantly if more isolates were obtained. As mentioned above, the concentration of soluble iron in oxygenated systems is likely to be extremely low, yet, with the exception of certain lactobacilli (Archibald, 1983), bacteria have an absolute requirement for iron, the mechanism for the capture of which must have evolved in parallel with the appearance of oxygen in the Earth's atmosphere. The role of iron scavenging in the success of animal pathogens has provided a basic understanding of the mecha-

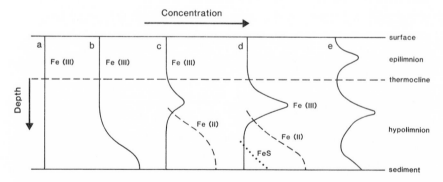

Figure 1. The distribution of iron in a stratified eutrophic lake during (a) Winter, (b) early summer, (c) midsummer, (d) late summer, (e) turbidity profile of the water column in late summer; the peak in the epilimnion is caused by phytoplankton.

nisms involved and this information is now being applied in the study of bacterial iron uptake in both terrestrial and aquatic environments.

It will be clear from these opening paragraphs that this chapter does not attempt to provide a comprehensive review of the iron cycle. The acidophilic iron-oxidizing bacteria, such as *Thiobacillus* and *Leptospirillum*, which conserve energy by Fe(II) oxidation, are not considered. This is partly because their role is of less significance in most (neutral pH) lakes, and partly because their importance in acid mine drainage and metal recovery from low-grade ores has ensured adequate coverage elsewhere (Kelly *et al.*, 1979; Lundgren and Silver, 1980). Similarly, the broader aspects of geochemical cycling of iron and its role on a global scale are covered in an excellent review by Nealson (1983). The present chapter will seek to examine in greater depth the microbiology of iron reduction, deposition, and acquisition in freshwater lakes.

2. The Iron Cycle in Fresh Water

The pioneering work of Mortimer (1941, 1942) on the geochemical cycling processes in stratified eutrophic lakes has received scant attention in microbiological circles. His research on Esthwaite Water (English Lake District) laid the foundation on which our understanding of the iron cycle in lakes has been constructed. It is most convenient to consider this cycle on a seasonal basis, as illustrated in Fig. 1. During winter the iron is present as particulate or colloidal Fe(III) and its concentration is constant with depth of the isothermal water column (Fig. 1a). Reported concentrations in natural waters range from 0.002 to 12.5 μM, with those in the English Lake District ranging from 0.35 μM in the oligotrophic lakes to 2.7 μM in the more eutrophic waters, such as those studied by Mortimer. With the onset of thermal stratification the water of the hypolimnion is, in essence, isolated by a temperature and therefore a density gradient. However, the biological consumption of oxygen in the hypolimnion proceeds particularly in the surface sediment, where the microbial biomass is three to four orders of magnitude greater than in the water column. As oxygen is consumed and the redox potential of the surface sediments drops to about $+200$ mV, ferrous iron is released into the water column but is immediately reoxidized and precipitates as ferric oxyhydroxide, producing the characteristic curve illustrated in Fig. 1b. As summer progresses, the Fe(III) concentration in the hypolimnion builds up and eventually reductive processes cause its redissolution as Fe(II). Reduction is most intense at the sediment–water interface and therefore a gradual upward dissolution of the Fe(III) is observed, resulting in a peak of Fe(III) immediately below the thermocline (Fig. 1c). In the meantime,

Fe(III) reduction continues in the sediment and Fe(II) is released into the now anoxic water column. If significant sulfate reduction occurs in the sediment, then FeS may be formed in the deeper water (Fig. 1d). The processes that establish aerobic/anaerobic boundaries and the interactions of the iron and sulfur cycles are complex, however, with biological and chemical components both contributing to the depth of the aerobic zone (Novitsky et al., 1981). The progress of the events described above may be followed by measuring the turbidity of the water in situ (Fig. 1e) and this technique has contributed to an understanding of both the dynamics of the iron cycle (Davison et al., 1981) and the association of certain "iron bacteria" with zones of particulate iron in the water column (Jones, 1981).

Of the few detailed studies of iron cycling in lakes, the most complete are those of Davison et al. (1981) and Verdouw and Dekkers (1980). Redox events in Esthwaite Water, the lake studied by Mortimer, Davison, Tipping, and co-workers, are dominated by iron, most of which enters the lake during the winter months. The bulk (70–90%) of this iron accumulates in the sediment, which has a mean iron content of approximately 50 mg/g (dry weight) and a C:Fe ratio of about 3:1. Only a small proportion of this annual iron load (3–18%) becomes involved in the internal iron cycle via reduction and dissolution. A similar small amount (less than 1% of the total iron content of the 0- to 12-cm layer of sediment) was reduced in Lake Vechten (Verdouw and Dekkers, 1980). Although of considerable importance to the redox processes within the lake, Fe(III) dissolution in Lake Vechten was equivalent to $<1\%$ of the potential reducing equivalents generated by primary production. Similar low values were obtained when the potential reducing power available to pure cultures of Fe(III)-reducing bacteria was calculated (Jones et al., 1984a).

The reductive component of the lake iron cycle could not be explained in chemical terms and a biological mechanism was demonstrated (Jones et al., 1983). The processes of oxidation and hydration, on the other hand, were chemical, and the product, amorphous spherical or ellipsoidal particles with diameters ranging from 0.05 to 0.5 μm (Tipping et al., 1981, 1982), bore a close resemblance to the "iron bacteria" referred to under the genus name Siderococcus (Kutuzova, 1974). Similarly, oxidized manganese particles resembled the "bacterium" Metallogenium, but the biological status of such particles requires further investigation (see Section 4). Another feature of Fe(III) particles in the lakes studied is that they are not comprised purely of iron, which accounts for only 30–40% of their weight. Humic acids play a significant part in their formation and surface chemistry (Tipping and Woof, 1983; Tipping and Cooke, 1982). The maintenance of Fe(III) oxide in an amorphous state also contributes to its availability for reduction in the water column.

A more detailed account of the chemistry of the processes described above is given by Davison and Tipping (1984) and the remainder of this section will be devoted to the bacteria associated with iron cycling in lakes. The distribution of these bacteria is shown, in general terms, in Fig. 2. The greatest number of Fe(III)-reducing bacteria is found in the lake sediments and their activity is reflected in the release of Fe(II) into the water column, usually after the removal of nitrate from overlying water. Iron(III)-reducers may also be found in the water column, where their viable counts may be as much as an order of magnitude higher than those of Fe(III)-depositing bacteria (Jones *et al.*, 1983). Magnetotactic bacteria are also encountered in the benthic zone and the ability to accumulate intracellular magnetite appears to be a fairly widespread characteristic (Blakemore *et al.*, 1979). It is in the zone of Fe(II) oxidation and precipitation, immediately below the metalimnion, that most of the "iron bacteria" are found. The various groups are often separated into discrete horizontal bands, possibly reflecting tolerance of, or requirement for, oxygen. Many Fe(III)-depositing bacteria isolated from the water column are conventional heterotrophs possessing little to distinguish them from other aquatic bacteria. There exist, however, morphologically distinct groups, known to microbiologists for decades, which still remain to be isolated and characterized. Among these are the genera *Siderococcus, Naumaniella, Ochrobium,* and *Siderocapsa.* Section 4, devoted to iron depositors, will attempt to draw together, admittedly in a subjective manner, what little is known about such bacteria.

The epilimnion, as illustrated in Fig. 2, is designated a zone of iron acquisition, but this is qualified by a question mark. Nealson (1983) con-

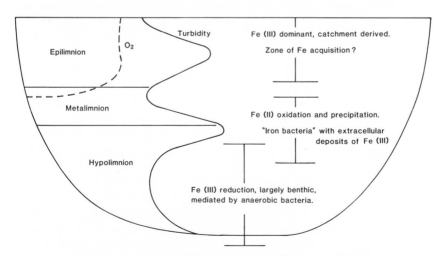

Figure 2. Zonation of the iron cycle in a stratified eutrophic lake.

cluded that Fe(II) was rare in aerobic environments, except in acidic lakes, such as those described by Colliene (1983), where photoreduction occurred. Evidence now suggests that photoreduction of iron–humic complexes is more widespread, and that the Fe(II) formed is sufficient to support algal (and presumably bacterial) growth in the euphotic zone (Finden et al., 1984). The most efficient iron-scavenging bacteria are therefore likely to be found in humic-poor (arid-zone?) water bodies or in the hypolimnia of deep, oligotrophic lakes where conditions are aerobic but there is insufficient light to photoreduce the Fe(III).

All the above comments have referred to water bodies during summer stratification. Similar events occur in rivers, but the iron cycle rarely dominates redox events. The exception to this is the drainage of catchments in which water-saturated soils and/or iron-rich minerals occur. In such areas, stream and river beds may be covered by growths of "iron bacteria." The iron cycle also continues during winter, but becomes spatially compressed and often confined to the sediment. Reduction of Fe(III) continues below the top few millimeters of sediment, but is limited by temperature and the supply of reducing power. However, many of the bacteria associated with Fe(III) deposition are found, and apparently grow, in the surface sediment layer, only migrating into the water column when conditions become more reduced during the summer (Jones, 1981).

3. Iron(III) Reduction

Iron(III) reduction might be considered as the step that initiates the internal iron cycle of lakes. Iron enters the lake from the catchment in an insoluble ferric form, and it is not until reduction and dissolution occur that the redox cycle described earlier can take place. Although these events had been described in detail, there was no chemical mechanism to account for the bulk of the Fe(II) formed. Sørensen (1982) showed that pasteurization stopped iron reduction in marine muds. Jones et al. (1983) achieved the same effect by the addition of $HgCl_2$ to freshwater sediments, and also demonstrated that the process had a temperature optimum of approximately 30°C. Both these results suggested a biological mechanism and the likely involvement of bacteria in Fe(III) reduction.

3.1. Iron(III)-Reducing Bacteria

The review by Nealson (1983) lists members of the genera *Bacillus, Aerobacter, Pseudomonas, Escherichia, Proteus, Clostridium, Achromobacter,* and *Staphylococcus* as Fe(III)-reducers, to which may be added

Arthrobacter (Shakhobova, 1981), *Micrococcus* (Woolfolk and Whiteley, 1962), *Corynebacterium* (Pfanneberg and Fischer, 1984), *Vibrio, Paracoccus, Bacteriodes, Desulfovibrio,* and *Desulfotomaculum* (Jones *et al.,* 1984b). Clearly no generalization can be made from such a list, since it contains both aerobes and anaerobes. Similarly, the tests for Fe(III) reduction have been conducted under many different conditions, some of which, as will be shown later, can have a marked effect on the results obtained. Iron(III) reduction may also be mediated by *Thiobacillus* and *Sulfolobus* when grown on sulfur as an energy source (Brock and Gustafson, 1976). This was considered to be a bacterial catalytic process, although work with cell-free extracts of *T. ferrooxidans* and *T. thiooxidans* suggested that it was not an enzymic process, but possibly dependent on reducing substrates formed during growth of the cells (Kino and Usami, 1982). The ability to reduce Fe(III) is therefore widespread among bacteria, and has been observed in organisms from many different environments.

3.2. The Mechanism of Iron(III) Reduction

In spite of its widespread occurrence, there is little information on the biochemical mechanism(s) of Fe(III) reduction. The proposals made to date, based often on indirect evidence, include the following.

3.2.1. Bacteria Alter the Environment to Permit Chemical Reduction of Fe(III)

There is little doubt that this occurs in nature and many laboratory experiments, and careful controls are required to distinguish the chemical and biological components of reduction. With mixed systems, such as sediment, the presence of a distinct temperature optimum and absence of increased rates of reduction at higher temperatures are good measures of biological activity (Brock, 1978; Jones *et al.,* 1983). In laboratory cultures, on the other hand, such results could be due entirely to an indirect effect of bacterial metabolism. The controls should therefore include, as a minimum, incubation of uninoculated medium, partially metabolized medium (during the experiment), and spent medium (at the end of the experiment) in the presence of Fe(III), to determine reduction due to medium components or bacterial metabolites. Mild chelating agents, such as citrate, can result in so much chemical reduction that it masks biological activity, and a lower level and less consistent reduction is observed with metabolites such as pyruvate. The extent of such activity will depend on the redox potential of the medium, the presence of reducing agents, and the gas phase used. Results obtained with freshwater bacteria and those from culture collections (Jones *et al.,* 1984a,b) were corrected for chemical reduction, but there is insufficient information to determine whether this was done in other published work. The production of reducing substances also varies with the organism and the substrate involved. Two *Vibrio* strains were isolated from profundal anoxic lake sediment; one fermented glucose and the other malate. The spent medium of the former (the cells were removed by centrifugation and filtration) reduced significant quantities of Fe(III) (Jones *et al.,* 1983), whereas no reduction was detected with the latter (Jones *et al.,* 1984a).

3.2.2. Bacteria Use Fe(III) As a Hydrogen Sink

It is assumed in this instance that the bacterium diverts hydrogen by a mechanism not clearly defined, but in some way associated with the regeneration of NAD and, on occasions, substrate level phosphorylation (SLP):

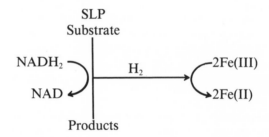

Such a process may or may not result in net energy conservation, depending on the organism and the scale of Fe(III) reduction. Of the many soil bacteria isolated by Ottow and co-workers, some did not reduce nitrate (see Section 3.2.3) and were assumed to possess what was designated as a ferrireductase system (Munch and Ottow, 1977). There was no evidence to indicate whether the enzyme functioned via SLP or an electron transport chain, but Fe(III) reduction was inhibited by Mn(IV) and not by nitrate. On thermodynamic grounds alone it might be predicted that Mn(IV) would be reduced before Fe(III) and this occurs even when the manganese is provided as a macrocrystalline form of MnO_2 (Jones *et al.,* 1984a). The specificity of the reductase system appears to vary with the organism involved. Woolfolk and Whiteley (1962) reported the reduction of a wide range of inorganic compounds by molecular hydrogen using cell extracts of *Micrococcus lactilyticus.* Similarly, Jones *et al.* (1984b) showed that, while a glucose-fermenting *Vibrio* reduced a wider range of metals, a lithotrophic hydrogen-utilizer was more specific. This suggested that the fermenter possessed a very general mechanism which could be used as a hydrogen sink, whereas that in the lithotroph may have been more closely linked to energy conservation (see Section 3.2.3).

Energy conservation associated with Fe(III) reduction by a heterotrophic anaerobe was demonstrated by Jones et al. (1984a). They isolated a malate-fermenting *Vibrio*, the molar growth yield of which increased by approximately 30% in the presence of Fe(III). The reduction of Fe(III) was accompanied by a decrease in the quantity of ethanol formed and a concomitant increase in acetate concentration. Thus, there was a slight diversion of metabolism to more energetically favorable end products, the sort of result that might be expected if Fe(III) was linked to SLP reactions. Although there was good agreement, in terms of molar concentrations, between the quantity of ethanol diverted to acetate and the amount of Fe(II) produced, this was only sufficient to explain a small part of the apparent increase in molar growth yield. Therefore, only a small part of the potential reducing power of the substrate was diverted to Fe(III) reduction; this might explain why the majority of iron-reducing bacteria isolated do not appear to conserve energy by the process (Jones et al., 1983) and has interesting parallels with field observations (see Section 3.4).

3.2.3. Bacteria Reduce Fe(III) via an Electron Transport System

This aspect of iron reduction has received most attention, largely due to the studies of Ottow and coworkers on bacteria isolated from soil. The assumption made is that the reduction occurs via an electron transport chain with or without the involvement of a nitrate reductase:

$$NADH_2 \rightarrow Flavoprotein \rightarrow FeS \rightarrow Quinone \rightarrow Cyt\ b \rightarrow Cyt \rightarrow \begin{array}{c} O_2 \\ H_2O \end{array}$$

$$NAD$$

$$? \qquad ?$$

$$2Fe(III) \quad 2Fe(II) \quad 2Fe(III) \quad 2Fe(II)$$

Many of the Fe(III)-reducers examined were facultative anaerobes and possessed nitrate reductases (Ottow and Glathe, 1971). These enzymes were implicated in the process (Ottow and Munch, 1978; Munch and Ottow, 1983) when nitrate reductase-less mutants failed to reduce Fe(III) (Ottow, 1970). The sites of the mutations were not, however, established and some of the nit⁻ mutants continued to reduce Fe(III). The latter were considered to possess the ferrireductase systems referred to earlier (Munch and Ottow, 1977). Much of the evidence for the involvement of nitrate reductase in iron reduction relies on the inhibitory effect of nitrate on the process in both bacterial cultures and natural sediments (Sørensen, 1982; Jones et al., 1983). However, Jones et al. (1984b) also noted that Fe(III) may be reduced by bacteria that do not possess a nitrate reductase and, conversely, the possession of a nitrate reductase did not confer the ability to reduce iron. Lascelles and Burke (1978) reported that Fe(III) reduction by membrane preparations of Staphylococcus aureus was insensitive to 2-heptyl-4-hydroxyquinoline-N-oxide (HQNO) and azide. They concluded that Fe(III) was reduced by electron transport chain components that preceded cytochrome b, but that were preferentially oxidized by a nitrate reductase system, thus explaining the inhibitory effect of nitrate. On the other hand, Obuekwe et al. (1981) reported that under certain circumstances this effect may be due to the reoxidation of Fe(II) by the NO_2^- formed. This mechanism would appear to be of little significance in freshwater, since Fe(III) reduction does not begin until nitrate has been totally depleted and under such circumstances significant accumulations of nitrite are rare.

Given the redox potential of the Fe^{3+}/Fe^{2+} couple and the reaction

$$2Fe^{3+} + [H_2] \rightarrow 2Fe^{2+} + 2H^+, \qquad \Delta G^{\circ\prime} = 228.3 \text{ kJ/mole}$$

(Thauer et al., 1977) it would appear that electron transport phosphorylation linked to Fe(III) reduction is thermodynamically favorable. However, few attempts have been made to determine whether bacteria conserve energy by the process. Takai and Kamura (1966) reported significantly increased bacterial growth in the presence of Fe(III), proportional to the quantity of Fe(II) produced, but no evidence to support this statement was presented. A facultatively anaerobic and facultatively chemolithotrophic pseudomonad isolated by Balashova and Zavarzin (1979) reduced Fe(III) with H_2 as the electron donor. Plate counts were three orders of magnitude higher in the presence of Fe(III), but direct evidence of energy conservation was not obtained. Jones et al. (1983) described a similar H_2-utilizing organism, the ATP yield of which increased significantly as a result of Fe(III) reduction. The growth of such organisms is, however, extremely slow and it would appear that the major agents of Fe(III) reduction in lakes are more likely to be chemoorganotrophs. The mechanism(s) they employ in the reduction are, however, imperfectly understood.

3.3. Factors Affecting Iron(III) Reduction

Excluding those factors known to affect most biological processes (e.g., temperature, pH), two remain that are of particular relevance to Fe(III) reduction: the nature of the iron in the environment and whether bacterial attachment is a prerequisite of reduction.

Iron is found in many crystalline mineral forms (Nealson, 1983), and is also present as amorphous particles, in which the development of crystalline structure has been inhibited by humic substances, and as clay minerals, such as chlorite (Davison and Tipping, 1984). In general terms, and assuming the absence of significant quantities of chelated iron, the amorphous form of Fe(III) is the most labile and the clay mineral the least. Jones *et al.* (1983) observed that rates of reduction decreased along the series $FeCl_3 > FePO_4 > Fe_2O_3 >$ goethite [an orthorhombic crystalline form of α-FeO(OH)]. In such experiments the $FeCl_3$ would have hydrolyzed immediately on addition to the liquid medium, but its final form and degree of aggregation would depend on the organic matter content and ionic strength of that medium. Munch and Ottow (1982) also concluded that amorphous Fe(III) was reduced more rapidly than its crystalline forms, and that among the latter, those with high energy levels were preferentially reduced (e.g., lepidocrite > hematite > goethite). However, DeCastro and Ehrlich (1970) and Pfannenberg and Fischer (1984) found that goethite was reduced more readily than hematite. Since different organisms were involved in these experiments, and not enough is known about the mechanisms involved, it is not possible to draw general conclusions about bacterial reduction of mineral forms of Fe(III), except, perhaps, to add that even less is known about the reactivity of Fe(III) bound in clay mineral lattices.

The mineral forms of iron described above are, essentially, large, insoluble crystals. It is hardly surprising, therefore, that attachment of bacteria to the mineral enhances reduction (Munch and Ottow, 1982, 1983). This is also true of the native metal. Bacteria attached to mild steel caused removal of the crystalline coat followed by extensive colonization during which extensive production of fibrous extracellular polysaccharide occurred (Obuekwe *et al.*, 1981). Even when Fe(III) is provided as $FeCl_3$, separation of the iron and bacteria by a dialysis membrane decreases reduction by as much as 70% (Jones *et al.*, 1983). It would appear that contact between the highly insoluble forms of Fe(III) and bacteria is often an essential component of the reductive process. Extracellular reduction of Fe(III) by marine diatoms under aerobic conditions has been reported (Anderson and Morel, 1980, 1982) and this mechanism appears to be independent of photoreduction (Finden *et al.*, 1984). Such processes deserve further investigation.

3.4. The Role of Iron(III) Reduction in the Carbon Cycle

As has been mentioned, Fe(III) reduction may dominate redox events in eutrophic lakes, but it is equivalent to only a very small part of the potential reducing power of the system. Verdouw and Dekkers (1980) calculated that Fe(III) and Mn(IV) reduction accounted for <1% of the primary production of Lake Vechten. The total quantity of Fe(III) reduced in the anoxic hypolimnion of Esthwaite Water [~ 10 g Fe/m^2 (Davison et al., 1981)] was equivalent to only 0.8% of the potential reducing equivalents that enter as sedimentary organic carbon (an average of 67 g C/m^2 over 4 years). Similarly, the quantity of Fe(III) reduced by a facultative anaerobe, and the associated shift in fermentation end products, would explain only a small part of the observed increase in molar growth yield in the presence of iron (Jones et al., 1984a).

What, then, is the overall effect of Fe(III) reduction on the carbon cycle in the field? An observation by Jones et al. (1984a) may provide a small clue. They added Fe(III) to anoxic sediment slurry limited in nitrate and sulfate, which was then incubated under hydrogen and nitrogen atmospheres. In all cases the addition of the iron caused a significant decrease in the quantity of volatile fatty acids that accumulated (Table I). The higher concentrations of fatty acids observed under the hydrogen atmosphere were attributed to the effect of increased hydrogen partial pressure on their metabolism by the obligate proton-reducing component of a syntrophic partnership (Boone and Bryant, 1980; Mah, 1982). Added butyric acid was also removed more rapidly in the presence of Fe(III) and, although this did not alter the quantity of acetate present at the end of the experiment, significantly less methane was produced.

Clearly, the effect of Fe(III) on such a mixed community was quite different from the diversion of fermentation end products observed in a pure culture, and it is interesting to speculate on the mechanisms involved. These might include: (1) diversion of electron flow from nitrate reductase in the absence of nitrate as electron acceptor; (2) replacement of a syntrophic H$_2$ acceptor by a chemical electron acceptor [this has not been demonstrated in the laboratory (McInerney et al., 1979, 1981; Mah, 1982), but Fe(III) was not among the acceptors tested], and (3) alteration of the sediment redox potential by the Fe(III), thus allowing more complete substrate oxidation. The vast majority of Fe(III)-reducing bacteria isolated to date have been facultative anaerobes and therefore mechanism 2 appears to be the least likely.

Table I. The Decrease in Volatile Fatty Acid Accumulation in Anoxic Sediments Caused by the Addition of Fe(III)[a]

	Decrease in concentration (μmole/liter)			
	Acetate	Propionate	iso-Butyrate	Butyrate
Hydrogen atmosphere	3700	91	38	58
Nitrogen atmosphere	100	2	1	0.5

[a]Data taken from Jones et al. (1984a).

Regardless of how this increased anaerobic utilization of volatile fatty acids is achieved in the presence of Fe(III), it is of considerable interest to note that Fe(III) is now implicated in the bacterial suboxic diagenesis of organic matter in banded iron formations. It is suggested that isotopically light carbonate minerals resulted from early bacterial oxidation of isotopically light organic carbon. This occurred during a period when free oxygen was absent and Fe(III) (as the electron acceptor) was more abundant than in the postdiagenetic sedimentary rock (Walker, 1984) and certainly more abundant than in present-day lake sediments.

4. Iron(III) Deposition

The processes discussed in this section have been termed deposition rather than oxidation. This chapter is not concerned with acidophilic bacteria and, in the absence of conclusive evidence to the contrary, it is assumed that Fe(II) oxidation at neutral pH is essentially a chemical process. There is no doubt, however, that bacteria are associated with oxidized iron and that their cellular structures become encrusted with Fe(III). This often occurs on such a scale that the product is easily visible to the naked eye. It is for this reason that reports of "iron bacteria" go back to the last century. A better understanding of the process of Fe(III) deposition now exists, largely due to work on prosthecate and stalked bacteria, but progress in isolating other morphologically distinct "iron bacteria" has been remarkably slow. Their taxonomy has been based purely on morphological characteristics, and confusion in the nomenclature of some groups has arisen since their original description.

4.1. "Iron Bacteria"

"Iron bacteria" can be divided into two groups: those available in axenic culture and the taxonomy of which is reasonably well defined; and those, often morphologically distinct forms, that have been described for decades but remain to be characterized. The former have provided the greater part of the current understanding on iron deposition and therefore will be considered in greater depth in the next section. The morphologically distinct "iron bacteria," on the other hand, are the cause of some taxonomic confusion, not only because generic attribution has occasionally changed with time, but also because there has been some lack of acceptance of a common nomenclature. This section presents an unashamedly subjective view of these organisms. It attempts to identify areas of confusion, and proposes acceptance of a more limited range of "genera" until more axenic cultures are available.

Table II. "Iron Bacteria": Organisms Associated with the Deposition of Fe(III)

Morphological type	Genus or group	
	In axenic culture	Not yet isolated
Filamentous	*Sphaerotilus*	*Clonothrix*
	Leptothrix	*Crenothrix*
		Toxothrix
		Lieskeella
Prosthecate/appendaged	*Planctomyces* (*/Blastocaulis*)	*Gallionella*?
	Hyphomicrobium	*Metallogenium*
	Pedomicrobium	*Seliberia*?
Encapsulated/coccoid		*Ochrobium*
		Siderocapsa (*/Arthrobacter*)
		Naumanniella
		Siderococcus

The genera or groups to be considered are listed in Table II and are divided into those that are readily available in axenic culture and those that have either never been cultured or are not available for more general investigation (e.g., those organisms that may have been obtained in slide culture in the presence of other bacteria or those that have been cultured but whose cultures have since been lost). Not included in Table II are organisms previously described under the genera named *Ferribacterium, Siderobacter, Sideromonas, Siderosphaera, Hyalosoris,* and *Spirothrix,* either because the names have fallen into disuse or because some of these organisms may be considered as morphotypes of the *Siderocapsa/Arthrobacter* group (Gorlenko *et al.,* 1983). *Kusnezovia* and *Caulococcus* are rarely found and require further investigation before they can be distinguished from *Metallogenium* or *Siderococcus* (J. M. Schmidt and Zavarzin, 1981). Cyanobacteria (e.g., *Synechocystis*) and algae [several Chrysophytes, various members of the Volvocales, *Trachelomonas,* and the Chlorophytes *Pteromonas, Catena,* and *Dichotomosiphon* (Nicholls and Fung, 1982)] are also excluded, largely because the quantity of Fe(III) deposited by them is insignificant when compared with the bacterial component of the internal iron cycle of lakes.

The bacteria listed in Table II are illustrated in Fig. 3. The illustrations are not intended as a comprehensive guide, but more as an indication of the morphotypes reported in the literature. The following sections will include a brief annotated description of each genus/group, a few key references, and a subjective assessment of its status.

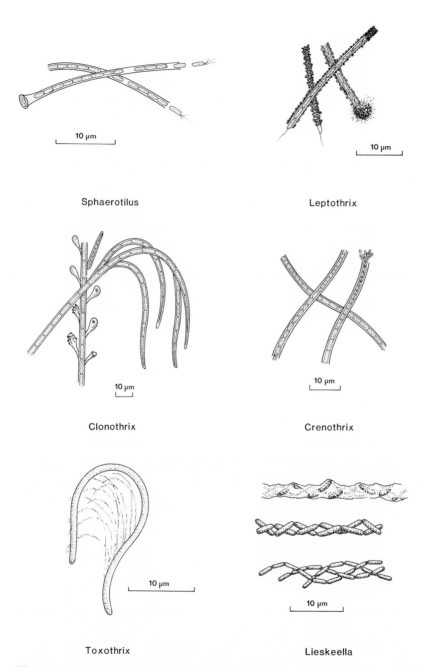

Sphaerotilus

Leptothrix

Clonothrix

Crenothrix

Toxothrix

Lieskeella

Figure 3. "Iron bacteria." A summary of features common to the morphotypes as described in the literature. (*continued*)

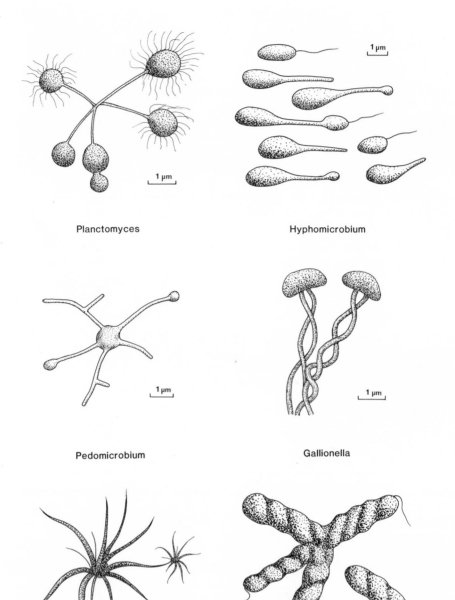

Planctomyces

Hyphomicrobium

Pedomicrobium

Gallionella

Metallogenium

Seliberia

Figure 3 (*continued*)

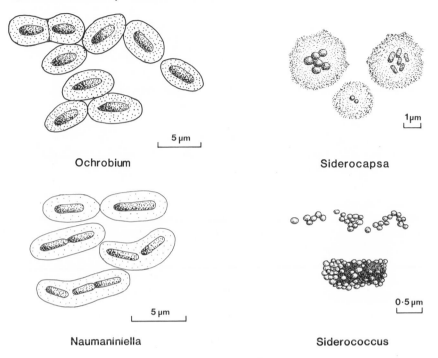

Ochrobium

Siderocapsa

Naumaniniella

Siderococcus

Figure 3 (*continued*)

4.1.1. The Filamentous Forms

Sphaerotilus/Leptothrix. These take the form of Gram-negative rods forming chains bound by a sheath. The single terminal cells are motile by bundles of polar flagella in *Sphaerotilus,* but more frequently by single flagella in *Leptothrix.* They possess a holdfast, which is the initiation point for sheath development. They require vitamin B12 or methionine for growth, and deposit PHB granules. Their G+C composition is 70 mole % (Mulder and Deinema, 1981). Pringsheim (1949) considered them to be sufficiently similar to be placed in the same genus. This view was shared by Wolfe (1958) and Dondero (1975), but not by Mulder (1974) and Mulder and van Veen (1974), who argued that Pringsheim failed to isolate *S. natans* from Fe(III) precipitates in which *L. ochracea* could be seen. Pringsheim also placed all *Leptothrix* morphotypes into a single species. The group is discussed in more detail by Mulder (1964) and van Veen *et al.* (1978). The major differences between the genera appear to be that *Sphaerotilus* responds rapidly to higher concentrations of organic carbon and is rarely encrusted with Fe(III) in the natural envi-

ronment, whereas *Leptothrix* is always found in such a state, with filaments of cells frequently leaving the iron-coated sheaths. Manganese
deposition/oxidation is found only in *Leptothrix* spp. The cells of *Leptothrix* morphotypes reported to date have a smaller diameter range (0.6–
1.5 μm) than those of *Sphaerotilus* (1.2–2.5 μm), although their length
ranges are similar.

Given similarities in the DNA composition and other features of the
isolates examined, it might be argued that axenic cultures of many more
strains are required before separation at genus level is justified. One
might even speculate that some of the diagnostic characteristics used
might be carried on extrachromasomal elements.

Clonothrix. This bacterium has the form of a multicellular filament
with false branching. Cells are large, 2–2.25 μm wide by 12–18 μm long,
forming filaments up to 1.5 cm in length. Cells taper at the tip of filaments. Some cells appear to form specialized (reproductive) bodies. Little
new has been reported and the most recent summary is given by Hirsch
(1974a). This organism is in need of rediscovery and more detailed investigation. Given its size, more recent reports might have been expected.

Crenothrix. The cells of this bacterium are enclosed in a thin sheath
with diameter ranging from 1–6 μm at the base to 6–9 μm at the tip of
the filament (Hirsch, 1974b). The filaments may be up to 1 cm long, containing cells which may be elongate or disc-shaped, depending on the
growth phase. Cells may divide in the horizontal or the vertical plane,
forming macro- or microgonidia. Studies of their ultrastructure indicate
that they are Gram-negative and contain membrane stacks similar to
nitrifiers or methylotrophs (Volker *et al.,* 1977). Consistent reports of this
morphologically distinctive bacterium provide hope that it may soon be
obtained in axenic culture.

Toxothrix. This is a gliding bacterium with cells 0.5–0.75 μm by 3–
6 μm, forming filaments up to 400 μm long (Hirsch and Zavarzin, 1974;
Hirsch, 1981). The highest reported growth temperature is 15.5°C and
cells explode when observed under the microscope. They are reported to
glide with the bend of a filament leading and may leave parallel tracks
and polymer filaments, which form bundles of twisted fibers. Repeated
deposits of twisted fibers often become encrusted with Fe(III). This
organism is morphologically similar to *Flexibacter/Herpetosiphon* and
the iron deposit may occur on the extracellular remains of a filament long
since departed. It must be shown to be different from other gliders to
retain its separate genus.

Lieskeella. This is a fragile bacterium, which bursts under the
microscope. Cells are rods with rounded ends 0.6 μm by 2–3 μm, forming
double chains, which wind to form a spiral enclosed in a slimy capsule.
The cells exhibit bipolar staining with methylene blue (Hirsch, 1974c).

This is another "genus" in need of rediscovery; no reports have emerged since Perfil'ev's early description of the bacterium.

4.1.2. Prosthecate and Appendaged Forms

Many members of this group undergo a complex life cycle, which may have contributed to the failure to isolate some morphotypes in the past.

Planctomyces (/Blastocaulis). Some morphotypes have been cultured. These Gram-negative cells are 0.3–1.7 μm in diameter, spherical, pear-shaped, or oblong on stalks 0.3–0.9 μm wide and up to 11 μm long (Hirsch and Skuja, 1974). Rosettes are commonly formed. Cells multiply by terminal or lateral buds. They appear to be oligotrophs, requiring only low concentrations of organic carbon. Marine strains exist (Staley and Bauld, 1981). Stalks may be simple or multifibrillar ribbons or ropes (J. M. Schmidt and Starr, 1981) and may be encrusted in iron. Iron(III) deposition depends on the site of origin and the morphotype involved (J. M. Schmidt *et al.,* 1982). Isolates are now available, but the taxonomy is unsatisfactory. Resistance to the temptation to create spurious "species" and the use of numbered morphotypes until more information is available (J. M. Schmidt and Starr, 1981) is an approach that could be used with profit elsewhere (see *Siderocapsa*).

Hyphomicrobium. These are Gram-negative cells, which multiply by budding at the tip of a hypha (or prostheca) (Hirsch, 1974d). Mature buds become motile with polar flagella. The cells are 0.5–1.0 μm by 1–3 μm. They are oligotrophs, as are virtually all the prosthecate/appendaged forms, and require CO_2. Many use C_1 compounds and often are successfully enriched with methylamine-containing media (Moore, 1981). It is assumed that the ability to produce a bud at a hyphal tip allows growth away from areas of heavy iron encrustation.

Pedomicrobium. This is similar to *Hyphomicrobium,* except that >1 hypha are produced per cell (Aristovskaya and Hirsch, 1974; Hirsch, 1968). They are heterotrophs, with cells that are oval, rod-, pear-, or bean-shaped, 0.4–2.0 μm long. Hyphae are usually of fairly constant width. They are more frequently found in soil, but freshwater forms are also found. Cultures have been deposited in collections (Ghiorse and Hirsch, 1978; Gebers, 1981).

Gallionella. These kidney-shaped or rounded cells, 2–3 μm long, produce a stalk (0.4–1.0 μm wide, but of variable length) at right angles to the long axis of the cell (Zavarzin and Hirsch, 1974a; Hanert, 1981a). There is heavy Fe(III) deposition on stalks, which may be equivalent to 90% by weight of the cell mass. Culture techniques include the use of FeS to provide Fe (Kucera and Wolfe, 1957) and formaldehyde to inhibit other bacteria (Nunley and Krieg, 1968). The stalks are fibrillar, and may branch and twist, although not in all cases (Heldal and Tumyr, 1983). Initial attachment and stalk development have been studied in depth (Hanert, 1970, 1974). They are microaerophilic and there is evidence of autotrophy (Hanert, 1968). Claims of lithotrophic metabolism remain to be substantiated. Some authorities place *Gallionella* with *Metallogenium* and *Siderococcus* as saprophytic mycoplasmas (Gorlenko *et al.,* 1983). On morphological grounds alone it is grouped here with appendaged bacteria, until further research with axenic cultures confirms its status.

Metallogenium. These are coccoid cells, 0.05–5 µm in diameter, which are claimed to develop into a spidery stellate form (Zavarzin and Hirsch, 1974b; Zavarzin, 1981). The filaments that develop (1–10 µm long) are primarily associated with Mn(IV) deposition, but Fe(III) is also found, depending on the water chemistry. Acidophilic forms, similar to *Gallionella,* were also observed (Walsh and Mitchell, 1972a,b, 1973), for which a G+C ratio of 72 mole % was reported. These organisms did not resemble those observed in neutral pH waters and no further research on them has been reported.

There is no evidence that the spidery stellate form is associated with living material (Gregory *et al.,* 1980), although mycoplasmas have been implicated (Gorlenko *et al.,* 1983). A fresh approach is required to throw some light on the controversy surrounding *Metallogenium.* In the meantime, the adoption of morphotypes rather than species might serve to emphasize uncertainties about its true nature.

Seliberia. These are gram-negative cells, spirally twisted 0.5–0.7 µm by 1–12 µm. They form star-shaped aggregates. The length of the spiral cells is related to the organic iron complex content of the medium (Aristovskaya, 1974; J. M. Schmidt and Swafford, 1981). They are generally considered to be soil organisms, but have also been isolated from small, standing water bodies. They are oligotrophic and do not grow on conventional nutrient-rich media. Further investigations are required to clarify the status of the genus and to assess the possibility of facultative autotrophy.

4.1.3. Encapsulated and Coccoid Forms

Many "genera" and "species" of encapsulated "iron bacteria" were established purely on morphological grounds. That the greater part of these are no longer recognized reflects an acceptance that such a classification was not acceptable until axenic cultures were obtained.

Ochrobium. These are ellipsoid to rod-shaped cells 0.5–3 µm by 1–5 µm, with single cells enclosed in a capsule (Zavarzin, 1974). The original description stated that some cells were partially surrounded by a thickening impregnated with iron (a torus) open at one end and through which flagella projected (see Fig. 3). The description of this phase closely matched that of *Pteromonas* (a eucaryotic alga of the same size), and more recent papers have restricted the genus name to totally closed encapsulated forms that may form clumps or chains (Dubinina, 1976; Dubinina and Kuznetsov, 1976; Jones, 1981; Gorlenko *et al.*, 1983). They are reported to be gas vacuolate. They are usually grouped with aerobic chemolithotrophs, but some morphotypes have been shown to be obligate anaerobes (Jones, 1981). Although one or two cell divisions are observed in mixed culture, the organism remains to be isolated. The genus name is retained because of consistency in morphotypes reported in North America, Europe, and the USSR, but its involvement in Fe(III) deposition may be doubtful (see later discussion).

Siderocapsa (/Arthrobacter). These cells are usually coccoid, 0.4–2.0 μm in diameter, and may be gas vacuolate (Zavarzin, 1974). One or more cells are enclosed in capsular material within and upon which Fe(III) is deposited; no clear capsular wall is visible. They are found on submerged plants, but planktonic forms have also been reported. A wide range of morphotypes has been reported (Drake, 1965; Svorcova, 1975, 1979; Hanert, 1981b), but the long "species" list requires supporting evidence in the form of axenic cultures. Dubinina and Zhdanov (1975) suggested that many of the morphotypes could be reproduced with a bacterium identified as a species of *Arthrobacter*. W.-D. Schmidt and Overbeck (1984) also claimed that *"Siderocapsa"* was merely the specialized growth stage of a pleomorphic bacterium. Whether *Arthrobacter* spp. account for all the *Siderocapsa* morphotypes or whether other bacterial genera are involved remains to be seen.

Naumanniella. These are rod-shaped cells, 0.7–3.0 μm by 1–10 μm, surrounded by a delicate regular capsule (Zavarzin, 1974). They are not included as an *Ochrobium* morphotype, because of consistent reports of budding within the capsule, and the absence of capsular material separating cells. For the same reason they should not be included in the genus *Siderocapsa* as proposed by Gorlenko *et al.* (1983). Axenic cultures are required.

Siderococcus. These are small (0.2–0.5 μm) spheres (Kutuzova, 1974; Dubinina and Kuznetzov, 1976; Gorlenko *et al.*, 1983) bearing such a close resemblance to amorphous iron particles found in lakes (Tipping *et al.*, 1981) that the onus lies on the microbial ecologist to prove that they are bacteria.

It will be clear from the foregoing that our understanding of the "iron bacteria" is far from complete. The available cultures of the *Sphaerotilus/Leprothrix* group and the prosthecate/appendaged forms are likely to provide new insights into iron deposition, but since so many "iron bacteria" remain to be isolated, it is not possible to propose a common mechanism for all forms. The taxonomy and physiology of certain genera are in particular need of clarification. Among the appendaged bacteria, the genus *Metallogenium* has long been the subject of controversy. Since it is primarily associated with Mn(IV) deposition (it is far more numerous and expresses many more morphotypes in lakes where the hypolimnion chemistry is dominated by the manganese rather than the iron cycle) and its status has been reviewed recently (Nealson, 1983), it is not considered in depth here. The organism responsible for the *Metallogenium* morphotype remains to be isolated and cultured with any consistency. An application of Koch's postulates to Mn(IV)-depositing bacteria in a manner similar to that proposed for the encapsulated forms may provide the answer.

Microbiologists have probably been least successful in their attempts to characterize the encapsulated "iron bacteria." Dubinina and Zhdanov (1975) isolated an *Arthrobacter* species that could, under appropriate conditions, reproduce the gross morphology of the genus *Siderocapsa*. This is not to say, however, that all forms of *Siderocapsa* are produced by arthrobacters, although a pleomorphic bacterium has been demonstrated in other planktonic morphotypes (W. D. Schmidt, 1984); other genera may also be involved. The microbial ecologist must apply the equivalent of Koch's postulates to the many and varied forms of "iron bacteria." Axenic cultures must be obtained consistently from the sample containing "iron bacteria," and these must then be reintroduced into natural samples to reproduce the relevant morphotype(s), from which, in turn, the original organism must be reisolated. Conventional culture techniques will continue to fail and a more imaginative approach is called for. It should be possible to reproduce bacteria-free hypolimnetic conditions in the laboratory. Similarly, some macrophytes (*Siderocapsa* morphotypes are often observed on aquatic plants) may now be grown in axenic culture. Construction of gradients of soluble metal and oxygen and the provision of suitable surfaces should provide conditions under which a given isolate may then express a wider range of morphotypes. The construction of "species" lists based on field morphotypes alone can no longer be acceptable. In certain circumstances (see the discussion of the genus *Ochrobium* in Section 4.3) this may even serve to perpetuate misconceptions about the bacteria concerned.

This is not to suggest that rapid advances may now be expected; were the "iron bacteria" that easy to characterize, this would have been done decades ago. It is of interest to note, however, that the problems they pose for the microbial ecologist have many parallels with those that have arisen from the adoption of the Bacteriological Code for the Cyanobacteria. Many ecologists are unable to accept this code unconditionally, not because it is the more logical classification for the "blue-greens," but because it makes no allowance for the validity of preserved material. Thus, many of the morphotypes considered to be relevant in the ecology of Cyanobacteria are not observed in axenic cultures. This may be because the axenic strains have lost the ability to reproduce these morphotypes or because it is not possible to reproduce the conditions in the laboratory under which they are expressed. If loss of ability to express a particular phenotype is to be avoided, then Koch's postulates must be applied to the "iron bacteria" as soon as axenic cultures become available.

4.2. The Mechanism of Iron(III) Deposition

Given the limited number of bacterial strains available for the study of Fe(III) deposition, few generalizations can be made, but these include the following. The sites at which Fe(III) is deposited on outer cell surfaces are usually anionic in charge. In Gram-positive bacteria teichoic acid phosphate groups and peptidoglycan carboxylate groups have been implicated, but secondary polymers may also be important. The more complex outer layers of Gram-negative bacteria may bind less Fe(III), largely due to the lower peptidoglycan content, but polar elements such as the phosphate groups of phospholipids may also be involved. The sheath of the *Sphaerotilus/Leptothrix* group, a protein–polysaccharide–lipid complex, is the main agent of Fe(III) deposition, the rate being two orders of magnitude slower in sheathless mutants (van Veen *et al.* 1978). Deposition is greatest when cells reach final stationary phase and appears to be independent of protein synthesis (Rogers and Anderson, 1976). The nature of the deposit, however, appears to depend on the species involved. Caldwell and Caldwell (1980) reported that in one species of *Leptothrix* the iron was amorphous, whereas in others it was deposited in the sheath as a hexagonal or fibrillar matrix. Fibrillar deposits of Fe(III) were also found within the cell and were therefore independent of sheath formation.

Iron deposition in budding bacteria was reported to be initiated at primary active sites (Hirsch, 1968; Ghiorse and Hirsch, 1978) and analysis of surface layers revealed the presence of polyanionic polymers, which were involved in iron deposition even in the presence of potent biological inhibitors or after heat treatments. Ghiorse and Hirsch (1979) proposed that positively charged iron hydroxides accumulated on the surface acidic polysaccharides (negatively charged), thus becoming negatively charged themselves. These would then attract more positive iron hydroxide by a mechanism independent of biological activity.

Examination of natural material, as opposed to laboratory cultures, has also shown that extracellular polymers are involved in Fe(III) deposition. This was shown to be true of both filamentous bacteria (Caldwell and Caldwell, 1980) and encapsulated forms. The proportion of Fe(III)-depositing encapsulated bacteria increased with depth in open oceanic samples (Cowen and Silver, 1984) and a major portion of the weakly bound iron was associated with these capsules. These observations are comparable to those made in stratified lakes (but on a much reduced vertical scale), where peaks of encapsulated iron bacteria are often observed in deeper water either within or below the metalimnion (Jones, 1981; Gorlenko et al., 1983). Observed Fe(III) deposition on the fibrillar stalks of naturally occurring appendaged bacteria cannot be attributed to a particular polymer and appears to depend on both the morphotype involved and the site from which it was isolated (J. M. Schmidt et al., 1982).

Given that mechanisms exist for both intracellular and extracellular deposition of iron, what does the bacterium gain from the process? There is insufficient evidence to show that bacteria conserve energy by Fe(III) deposition at neutral pH. The iron has been considered variously to confer stability on fragile cell wall structures, to protect the cell from oxygen, or to detoxify certain components (e.g., heavy metals) by their coprecipitation or binding in the ferric (oxy)hydroxide matrix. The advantages of intracellular deposition are more obvious. Iron(III) may be deposited to keep iron from reaching toxic concentrations. The iron content of the biota in general ranges from 50 to 200 μg/g dry weight, but values in excess of this are reported in certain bacteria. Intracellular fibrillar iron deposits have been observed in Leptothrix spp. (Caldwell and Caldwell, 1980) and the iron content of magnetotactic bacteria (Blakemore, 1975) may be as high as 20 mg/g dry weight. The deposition of magnetite in bacteria is a particularly interesting phenomenon. It may be observed in many benthic species and does not appear to be confined to any particular taxonomic group. The freshwater bacterium Aquaspirillum magnetotacticum deposits chains of cuboidal Fe_3O_4 particles which are approximately 50 nm wide. It is suggested that the magnetite is formed by the reduction of a hydrous ferric oxide precursor (Frankel and Blakemore, 1984), and that its presence ensures the downward migration of the bacteria to the sediments and away from toxic levels of oxygen. Apart from the advantages gained by this, it is of interest to note that, on a global scale, the prevalence of a particular magnetic orientation depends on the hemisphere in which the bacterium was found.

4.3. The Ecology of "Iron Bacteria"

Given the difficulties in isolating the vast majority of "iron bacteria," it follows that much of the information about their ecology is purely of a descriptive nature. It is possible to provide quantitative information about various morphotypes, but this cannot be used to determine whether the morphotype is a viable form of the organism or whether it is involved in particular physiological or geochemical processes. Detailed information on the distribution of "iron bacteria" in ground water and surface waters is given by Cullimore and McCann (1977), Kuznetsov (1970), and Gorlenko *et al.* (1983); this section will attempt to draw some general conclusions and, where possible, show how field data may be used to design more realistic isolation procedures.

The feature common to all sites where "iron bacteria" develop is the presence of a redox gradient, usually dominated by the Fe^{2+}/Fe^{3+} couple. The scale of the gradient in space or time controls the extent to which the "iron bacteria" grow and develop. Thus, in the water column of a stratified lake, such as that illustrated in Figs. 1 and 2, the "iron bacteria" may be found in a depth zone covering several meters and over which sharp gradients of oxygen and Fe(II) concentrations are observed. In very general terms the distribution of the individual groups of iron bacteria along these gradients is as illustrated in Fig. 4. The upper hypolimnion and the

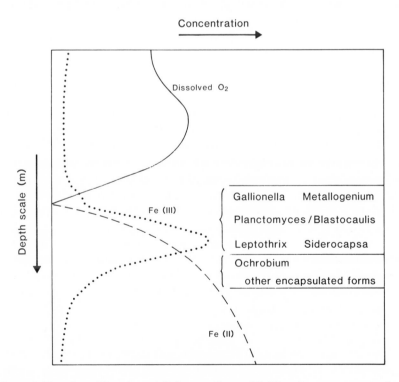

Figure 4. Zonation of "iron bacteria" along gradients of Fe(II) and oxygen concentration in a stratified eutrophic lake.

metalimnion are the zones where the bulk of the oxidized iron and the majority of "iron bacteria" are found. The majority are observed in water that contains traces of oxygen, but note that the peak in Fe(III) concentration may occur at depths where oxygen is not detectable. This does not imply that the peak was formed at those depths, nor that such water is always devoid of oxygen. Intermittent turbulence would be sufficient to introduce oxygen into the deeper water, thus causing the oxidation and precipitation of the iron. In addition to the occurrence of recognized "iron bacteria" such as *Leptothrix* and *Metallogenium* in the microaerophilic zone, *Ochrobium* has been observed consistently at depths below the peak of Fe(III) (Jones, 1981) and dividing and single cells peaked at this same depth. When sampled, taking care to exclude oxygen, further cell division in laboratory-incubated samples only occurred in the total absence of oxygen and was maximal in anoxic water from a depth 0.5 m greater than the population peak in the field. These observations were consistent with the suggestion that *Ochrobium* is an obligate anaerobe and that it may adjust its buoyancy in the field by the production of gas vacuoles. The results also served to emphasize the importance of precision in depth sampling (Cunningham and Davison, 1980) if conclusions about the metabolism of an organism are to be drawn from field data. If, as suggested, *Ochrobium* and possibly other encapsulated forms (which develop in the same zone) are obligate anerobes, what, then, is their role in the iron cycle? Such bacteria are usually grouped with the chemolithotrophs (Zavarzin, 1974; Hanert, 1981b), but the observed iron deposition may, on occasion, be an artifact of sampling and handling procedures. If anoxic water is membrane-filtered in the laboratory under aerobic conditions, then Fe(III) deposition on the membrane is rapid and often associated with the cell capsular material. On other occasions water has been allowed to stand under aerobic conditions before examining the bacteria (W.-D. Schmidt and Overbeck, 1984). Under such circumstances the deposition of iron may purely reflect the chemistry of the capsular material and may be unrelated to biological activity.

The water column of a eutrophic lake provides but one example of the development of "iron bacteria" along redox gradients and, as mentioned earlier, the scale of such events is controlled by the steepness of the gradient. When the water column is fully oxygenated, as, for example, during the winter months, then the redox gradient is confined to the sediment. Under such circumstances population distributions are compressed; *Leptothrix* is found only at the sediment–water interface, and *Ochrobium* between depths of 1.5 and 4.0 mm (Jones, 1981). Certain "iron bacteria" therefore exhibit benthic and planktonic stages in their population cycles and for many the transition may be mediated by the production of gas vacuoles (Walsby, 1981). The growth of "iron bacteria" may be restricted even further, namely to the surface film of the water. This usually occurs in swamps and ponds where the underlying water is anoxic, or at least experiences increased oxygen demand in the absence of light. Under such circumstances *Leptothrix* and *Pedomicrobium* are often the dominant organisms (Frølund, 1977; Ghiorse and Chapnick, 1983). In moving water the redox gradient is usually horizontal and may be represented in terms of distance, for the attached bacteria, or in units of time, for the planktonic forms. Common examples of such gradients include the discharges of land drains or the seepages of water-saturated soils into streams. Under such circumstances large populations of "iron bacteria" develop and zonation of populations may be observed. Unfortunately, although such sites are often accessible, little has been gained from them apart from the isolation of a few strains of *Leptothrix*. As mentioned earlier, the feature common to all communities of "iron bacteria" in the natural environment is the presence of a redox gradient dominated by the Fe^{2+}/Fe^{3+} couple. The inclusion of gradients in growth media often leads to the isolation of a wider variety of such bacteria (Hanert, 1981a; Jones, 1983).

5. Iron Acquisition

Apart from the previously mentioned strains of lactobacilli, all bacteria require iron as a component of electron transport systems and/or of cofactors. The concentration factor for iron in aquatic organisms is of the order of 10^6 and is matched only by that for gold and cobalt. In conventional growth media, 0.1 and 10 μM Fe might be regarded as low and high concentrations, respectively (Neilands, 1984). In natural waters the range is greater, 0.002–12.5 μM, but in the presence of oxygen the maximum concentration of free Fe^{3+} at neutral pH is 10^{-17} M. In the absence of a significant involvement of environmentally derived chelating agents (see discussion in Section 5.2), how then do bacteria acquire iron in freshwater? Much of our understanding of iron acquisition has come from studies of pathogenic bacteria and their need to overcome the iron-binding systems of their hosts, but clearly the results have applications and implications in the natural environment.

5.1. The Mechanism of Iron Acquisition—Siderophores

Bacteria possess low- and high-affinity systems for iron uptake; little is known about the former except that it appears to be widely distributed, since it permits growth after deletion of the high-affinity component. The high-affinity systems permit pathogenic bacteria to acquire sufficient iron in the presence of host iron-binding glycoproteins such as transferrin. The low solubility of Fe(III) implies that it is the high-affinity system that is also of importance in the natural environment. Under iron-limiting conditions bacteria produce siderophores, low-molecular-weight ligands specific for Fe(III), which permit its solubilization and transport. No such system has been found for other metals and this probably reflects the high concentration factor required for iron. Siderophores were originally divided into hydroxamate and catechol-based compounds, but such is the intensity of interest in this field that "new" siderophores are being reported with regularity. Although many of these function in conjunction with hydroxamate or catechol moieties, some, such as the α-hydroxyisovaleric acid siderophore in *Proteus mirabilis* (Evanylo *et al.*, 1984), are found in cells where such moieties are not detected. Clearly, our knowledge of siderophores is expanding rapidly and therefore this section will confine itself to those parts of current understanding that are relevant to the ecology of iron acquisition. A short and informative review of siderophores, ranging from their structure to the molecular genetics of control systems, is given by Neilands (1984).

In an iron-limited system, a lowered intracellular iron concentration triggers the two components involved in its further acquisition. These components, production of a siderophore and the necessary transport proteins in the cell outer membrane, appear to be under common control. Siderophores have been purified from several bacteria and are generally known by trivial names; e.g., enteric bacteria produce enterobactin and aerobactin, *Paracoccus denitrificans* produces parabactin; *Vibrio cholerae* produces vibriobactin; *Bacillus* sp., *Arthrobacter* sp., and *Anabaena* sp. produce schizokinen; *Pseudomonas* sp. produces pyochelin, pseudobactin, and ferribactin; *Actinomyces* sp. produces ferrioxamines; and *Mycobacterium* sp. produces mycobactin. Most are of low molecular weight (600–1000), possess a very high affinity for Fe(III) (stability constants of 10^{30} are the norm, although values of up to 10^{52} have been reported), and a low affinity for Fe(II). The majority of siderophores are produced extracellularly, although some, such as mycobactin, possess a lipid-soluble moiety and are located inside the cell envelope. The siderophore–Fe(III) complex is of such dimensions that it is too large to pass through the pores of the cell outer membrane by diffusion, hence the requirement for specific transport proteins. Some bacteria may produce more than one siderophore and several outer membrane transport proteins (Sigel and Payne, 1982; Williams *et al.*, 1984), whereas others, such as *E. coli*, may produce only the outer membrane proteins, and these are used to transport siderophores produced by other microorganisms, e.g., the ferrichrome produced by many fungi (Neilands, 1984). The requirement to carry the genetic information for only part of the iron acquisition system may be of advantage under certain growth-limiting conditions.

Siderophores may transport the iron to the bacterial cell surface or across the membrane into the cell. Once inside the cell, the iron is released, usually by reduction to Fe(II), and becomes available for metabolism. The siderophore may be decomposed or excreted for further use. The genes for siderophore production and uptake systems may be chromosomal or plasmid-borne. There is increased interest in the latter since the discovery that the plasmid coding for colicin production, pColV, in *E. coli* also codes for a siderophore system responsible for the enhanced virulence of the strains (Williams, 1979; Warner *et al.,* 1981). The antagonism of plant growth promoter bacteria to wilt fungi and take-all diseases has also been attributed to a plasmid-mediated pseudobactin, a high-affinity Fe(III) transport agent found in strains of *Pseudomonas.* It is argued that the bacteria deprive the plant pathogens of iron, thus allowing more successful germination and growth of the plant. The genes responsible have now been cloned (Moores *et al.,* 1984) and it is hoped that the better understanding of the molecular genetics of iron acquisition may be applied to improve agricultural and horticultural production.

Siderophores are produced by anaerobes as well as aerobes and by organisms other than bacteria. The siderophores of fungi and analogous systems of higher plants have not been considered here because this review is confined to freshwater systems. It is now known that algae may also possess iron acquisition systems and these are discussed in the final section of this chapter.

5.2. The Ecology of Iron Acquisition

In laboratory cultures it is possible to determine the system by which iron is acquired; in the field the situation is complicated by the variety of microorganisms and the presence of natural chelators derived from the catchment or produced *in situ.* There exist, however, certain situations in which siderophore activity may be demonstrated with some certainty. It is possible, for example, to detect the presence of iron chelators in nonwaterlogged soils, whereas such substances are absent or present in much lower quantities in subsurface water-saturated soils, where, presumably, iron is readily available in the Fe(II) form (Akers, 1981). Similarly, schizokinen, a siderophore produced by *Anabaena, Bacillus,* and *Arthrobacter* spp., is found in rice field soil immediately after annual flooding (Akers, 1983). In the aquatic systems the bacteria that might be expected to produce siderophores are those that must compete with higher animal transferrin and lactoferrin systems. Thus, the fish pathogens *Aeromonas salmonicida* and *Vibrio anguillarum* both possess high-affinity systems for Fe(III). The siderophores of *Aeromonas salmonicida* appear to be neither hydroxamate- nor catechol-based, in contrast to those found in *A. hydrophila.* As in other bacteria, iron limitation results in increased synthesis of membrane transport proteins, and in *A. salmonicida* two distinct mechanisms of iron acquisition have been found (Chart and Trust, 1983). The first is inducible and produces a low-molecular-weight, high-affinity system; the second is constitutive and requires contact with transferrin for transfer of iron to the bacterium. It is not difficult to see why a bacterium armed with two such systems is such a successful fish pathogen.

In freshwater systems it is the plankton that is most likely to be iron-limited, and it is in the water column that the process of iron acquisition is likely to be complicated by a variety of factors. Numerous substances are known to stabilize iron in a soluble or fine colloidal state (Cameron and Liss, 1984), and among these, humic acids are the most likely to be encountered in surface waters. Iron(III)-humic complexes are susceptible to photoreduction in acidic, humic-rich situations (Francko and Heath, 1982; Colliene, 1983) and it has now been shown that this mechanism also provides an adequate supply of iron for algae at neutral pH (Finden *et al.*, 1984). Among the photosynthetic organisms, cyanobacteria such as *Anabaena* produce schizokinen siderophores (Lammers and Sanders-Loehr, 1982) and the eucaryotic dinoflagellate *Procentrum minimum* produces a hydroxamate siderophore, procentrin (Trick *et al.*, 1983). Clearly these organisms are not dependent on indirect sources of iron, such as that obtained via photoreduction of Fe(III)–humic complexes. In the absence of direct evidence of siderophore production, studies on iron limitation have failed to demonstrate a common mechanism for acquisition. Anderson and Morel (1980) suggested that the coastal diatom *Thalassiosira weissflogii* was capable of extracellular reduction of Fe(III), although there was insufficient information to determine whether photoreduction might be involved in these experiments. In a later paper (Anderson and Morel, 1982), which took account of photoreduction, they concluded that the alga possessed a membrane-bound iron-complexing system which transferred iron to the cell. Biochemical analyses of the iron status of the cyanobacterium *Phormidium* sp. and the dinoflagellate *Gymnodinium microadriaticum* suggested that these components of coral reefs were iron-limited (Entsch *et al.*, 1983). Whether this is true of all marine phytoplankton is doubtful, since it appears that neritic species (i.e., those found in the off-shore zone) become iron-limited at far higher concentrations (10^{-7} M) than those that permitted maximal growth (10^{-9} M) of oceanic forms (Brand *et al.*, 1983). Perhaps this reflects adaptive mechanisms in those communities found in waters that contain insufficient organic material to provide adequate iron via photoreduction. The population of *Aphanizomenon flos-aquae* in Clear Lake, California, was shown to be iron-limited (Wurtsbaugh and Horne, 1983) and it is of interest to note that the lake is situated in a semiarid zone, which contributes little humic material to the inflow.

It will be clear from the foregoing that studies on iron acquisition in planktonic communities have largely been confined to algal and cyanobacterial populations. If the mechanism of photoreduction discussed by Finden *et al.* (1984) is common to most waters that receive adequate supplies of humic material, then it would follow that bacteria are unlikely to be iron-limited in the euphotic zone. Perhaps the hypolimnion of a deep oligotrophic lake is the most likely place to yield microorganisms that possess high-affinity iron acquisition systems. It is worth adding, as a final note, that bacteria that are found in the anoxic hypolimnia of eutrophic lakes may also require an iron acquisition system. Anaerobic bacteria also produces siderophores, but why are they required under conditions in which Fe(II) is stable? Although the hypolimnion of a eutrophic lake may contain Fe(II), this may not be available if sulfate reduction ensures that all the iron is precipitated as FeS. Sulfate-reducing bacteria have a high requirement for iron, which may be satisfied under the conditions described above by the possession of Fe(II)-selective sites in the lipopolysaccharide of their outer membrane (Bradley *et al.*, 1984).

ACKNOWLEDGMENTS. I would like to thank B. M. Simon and S. Gardener, who collaborated in the research on Fe(III) reduction, W. Davison and E. Tipping for valuable discussions and helpful comments, T. I. Furnass, who prepared the figures, and E. M. Evans, who typed the script. Research at the Freshwater Biological Association reported here was supported by the Natural Environment Research Council, U.K.

References

Akers, H. A., 1981, The effect of waterlogging on the quantity of microbial iron chelators (siderophores) in soil, *Soil Sci.* **132**:150–152.

Akers, H. A., 1983, Isolation of the siderophore schizokinen from soil of rice fields, *Appl. Environ. Microbiol.* **45**:1704–1706.

Anderson, M. A., and Morel, F. M., 1980, Uptake of Fe(II) by a diatom in oxic culture medium, *Mar. Biol. Lett.* **1**:263–268.

Anderson, M. A., and Morel, F. M., 1982, The influence of aqueous iron chemistry on the uptake of iron by the coastal diatom *Thalassiosira weissflogii, Limnol. Oceanogr.* **27**:789–813.

Archibald, F., 1983, *Lactobacillus plantarum,* an organism not requiring iron, *FEMS Microbiol. Lett.* **19**:29–32.

Aristovskaya, T. V., 1974, Genus *Seliberia,* in: *Bergey's Manual of Determinative Bacteriology,* 8th ed. (R. E. Buchanan and N. E. Gibbons, eds.), p. 160, Williams and Wilkins, Baltimore.

Aristovskaya, T. V., and Hirsch, P., 1974, Genus *Pedomicrobium,* in: *Bergey's Manual of Determinative Bacteriology,* 8th ed. (R. E. Buchanan and N. E. Gibbons, eds.), pp. 150–153, Williams and Wilkins, Baltimore.

Balashova, V. V., and Zavarzin, G. A., 1979, Anaerobic reduction of ferric iron by hydrogen bacteria, *Microbiology* **48**:733–778.

180   J. G. Jones

Blakemore, R. P., 1975, Magnetotactic bacteria, *Science* **190**:377–399.

Blakemore, R. P., Maratea, D., and Wolfe, R. S., 1979, Isolation and pure culture of a freshwater magnetotactic spirillum in chemically defined medium, *J. Bacteriol.* **140**:720–729.

Boone, D. R., and Bryant, M. P., 1980, Propionate degrading bacterium, *Syntrophobacter wolinii* sp. nov. gen. nov. from methanogenic ecosystems, *Appl. Environ. Microbiol.* **40**:626–632.

Bradley, G., Gaylarde, C. C., and Johnston, J. M., 1984, A selective interaction between ferrous ions and lipopolysaccharide in *Desulfovibrio vulgaris, J. Gen. Microbiol.* **130**:441–444.

Brand, L. E., Sunda, W. G., and Guillard, R. R. L., 1983, Limitations of marine phytoplankton reproductive rates by zinc, manganese and iron, *Limnol. Oceanogr.* **28**: 1182–1198.

Brock, T. D., 1978, The poisoned control in biogeochemical investigations, in: *Environmental Biogeochemistry and Geomicrobiology* (W. E. Krumbein, ed.), Vol. 3, pp. 717–725, Ann Arbor Science Publishers, Ann Arbor, Michigan.

Brock, T. D., and Gustafson, J., 1976, Ferric iron reduction by sulfur- and iron-oxidizing bacteria, *Appl. Environ. Microbiol.* **32**:567–571.

Caldwell, D. E., and Caldwell, S. J., 1980, Fine structure of *in situ* microbial iron deposits, *Geomicrobiol. J.* **2**:39–53.

Cameron, A. J., and Liss, P. S., 1984, The stabilization of "dissolved" iron in freshwaters, *Water Res.* **18**:179–185.

Chart, H., and Trust, T. J., 1983, Acquisition of iron by *Aeromonas salmonicida, J. Bacteriol.* **156**:758–764.

Collienne, R. H., 1983, Photoreduction of iron in the epilimnion of acidic lakes, *Limnol. Oceanogr.* **28**:83–100.

Cowen, J. P., and Silver, M. W., 1984, The association of iron and manganese with bacteria on marine macroparticulate material, *Science* **224**:1340–1342.

Cullimore, D. R., and McCann, A. E., 1977, The identification, cultivation and control of iron bacteria in ground water, in: *Aquatic Microbiology* (F. A. Skinner and J. M. Shewan, eds.), pp. 219–261, Academic Press, London.

Cunningham, C. R., and Davison, W., 1980, An opto-electronic sediment detector and its use in the chemical micro-profiling of lakes, *Freshwater Biol.* **10**:413–418.

Davison, W., and Seed, G., 1983, The kinetics of the oxidation of ferrous iron in synthetic and natural waters, *Geochim. Cosmochim. Acta* **47**:67–79.

Davison, W., and Tipping, E., 1984, Treading in Mortimer's footsteps: The geochemical cycling of iron and manganese in Esthwaite Water, *Freshwater Biol. Assoc. Annu. Rep.* **52**:91–101.

Davison, W., Heaney, S. I., Talling, J. F., and Rigg, E., 1981, Seasonal transformation and movements of iron in a productive English lake with deep-water anoxia, *Schweiz. Z. Hydrobiol.* **42**:196–224.

DeCastro, A. F., and Ehrlich, H. L., 1970, Reduction of iron oxide minerals by a marine *Bacillus. Antonie van Leeuwenhoek J. Microbiol. Serol.* **36**:317–327.

Dondero, N. C., 1975, The *Sphaerotilus–Leptothrix* group, *Annu. Rev. Microbiol.* **29**:407, 428.

Drake, C. H., 1965, Occurrence of *Siderocapsa treubii* in certain waters of Niederrhein, *Gewass. Abwass.* **39/40**:41–63.

Dubinina, G. A., 1976, Ecology of freshwater iron bacteria, *Biol. Bull.* **3**:473–489.

Dubinina, G. A., and Kuznetsov, S. I., 1976, The ecological and morphological characteristics of microorganisms in Lesnaya Lamba (Karelia), *Int. Rev. Ges. Hydrobiol.* **61**:1–19.

Dubinina, G. A., and Zhdanov, A. V., 1975, Recognition of the iron bacteria *"Siderocapsa"* as arthrobacters and description of *Arthrobacter siderocapsulatus* sp. nov., *Int. J. Syst. Bacteriol.* **25**:340–350.

Entsch, B., Sim, R. G., and Hatcher, B. G., 1983, Indications from photosynthetic components that iron is a limiting nutrient in primary producers on coral reefs, *Mar. Biol.* **73**:17–30.

Evanylo, L. P., Kadis, S., and Maudsley, J. R., 1984, Siderophore production by *Proteus mirabilis, Can. J. Microbiol.* **30**:1046–1051.

Finden, D. A. S., Tipping, E., Jaworski, G. H. M., and Reynolds, C. S., 1984, Light-induced reduction of natural iron (III) oxide and its relevance to phytoplankton, *Nature* **309**:783–784.

Francko, D. A., and Heath, R. T., 1982, UV-sensitive complex phosphorus: Associaton with dissolved humic material and iron in a bog lake, *Limnol. Oceanogr.* **27**:564–569.

Frankel, R. B., and Blakemore, R. P., 1984, Precipitation of Fe_3O_4 in magnetotactic bacteria, *Phil. Trans. R. Soc. Lond. B* **304**:567–574.

Frølund, A., 1977, The seasonal variation of the neuston of a small pond, *Bot. Tidsskrift.* **72**:45–56.

Gebers, R., 1981, Enrichment, isolation, and emended description of *Pedomicrobium ferrugineum* Aristovskaya and *Pedomicrobium manganicum* Aristovskaya, *Int. J. Syst. Bacteriol.* **31**:302–316.

Ghiorse, W. C., and Chapnick, S. D., 1983, Metal-depositing bacteria and the distribution of manganese and iron in swamp waters, *Ecol. Bull. (Stockholm)* **35**:367–376.

Ghiorse, W. C., and Hirsch, P., 1978, Iron and manganese deposition by budding bacteria, in: *Environmental Biogeochemistry and Geomicrobiology* (W. E. Krumbein, ed.), Vol. 3, pp. 897–909, Ann Arbor Science Publishers, Ann Arbor, Michigan.

Ghiorse, W. C., and Hirsch, P., 1979, An ultrastructural study of iron and manganese deposition associated with extracellular polymers of *Pedomicrobium*-like budding bacteria, *Arch. Microbiol.* **123**:213–226.

Gorlenko, V. M., Dubinina, G. A., and Kuznetsov, S. I., 1983, The ecology of aquatic microorganisms, in: *Die Binnengewässer. Einzeldarstellugen aus der Limnologie und ihren Nachbargebieten,* (H.-J. Elster and W. Ohle, eds.), pp. 1–252, E. Schweizerbart'sche, Stuttgart.

Gregory, E., Perry, R. S., and Staley, J. T., 1980, Characterization, distribution and significance of *Metallogenium* in Lake Washington, *Microb. Ecol.* **6**:125–140.

Hanert, H., 1968, Investigations on isolation, physiology, and morphology of *Gallionella ferruginea* Ehrenberg, *Arch. Mikrobiol.* **60**:348–376.

Hanert, H., 1970, Structure and growth of *Gallionella ferruginea* Ehrenberg in its natural habitat during the first 6 h of development, *Arch. Mikrobiol.* **75**:10–24.

Hanert, H., 1974, *In vivo* kinetics of individual development of *Gallionella ferruginea* in batch culture, *Arch. Microbiol.* **96**:58–74.

Hanert, H., 1981a, The genus *Gallionella,* in: *The Prokaryotes* (M. P. Starr, H. Stolp, H. G. Trüper, A. Balows, and H. G. Schlegel, eds.), pp. 509–515, Springer-Verlag, Berlin.

Hanert, H., 1981b, The genus *Siderocapsa* (and other iron- or manganese-oxidizing Eubacteria, in: *The Prokaryotes* (M. P. Starr, H. Stolp, H. G. Trüper, A. Balows, and H. G. Schlegel, eds.), pp. 1049–1059, Springer-Verlag, Berlin.

Heldal, M., and Tumyr, O., 1983, *Gallionella* from metalimnion in an eutrophic lake: Morphology and X-ray energy-dispersive microanalysis of apical cells and stalks, *Can. J. Microbiol.* **29**:303–308.

Hirsch, P., 1968, Biology of budding bacteria IV. Epicellular deposition of iron by aquatic budding bacteria, *Arch. Mikrobiol.* **60**:201–216.

Hirsch, P., 1974a, Genus *Clonothrix,* in: *Bergey's Manual of Determinative Bacteriology,* 8th ed. (R. E. Buchanan and N. E. Gibbons, eds.), p. 136, Williams and Wilkins, Baltimore.

Hirsch, P., 1974b, Genus *Crenothrix*, in: *Bergey's Manual of Determinative Bacteriology*, 8th ed. (R. E. Buchanan and N. E. Gibbons, eds.), pp. 135–136, Williams and Wilkins, Baltimore.

Hirsch, P., 1974c, Genus *Lieskeella*, in: *Bergey's Manual of Determinative Bacteriology*, 8th ed. (R. E. Buchanan and N. E. Gibbons, eds.), p. 134, Williams and Wilkins, Baltimore.

Hirsch, P., 1974d, Genus *Hyphomicrobium*, in: *Bergey's Manual of Determinative Bacteriology*, 8th ed. (R. E. Buchanan and N. E. Gibbons, eds.), pp. 148–150, Williams and Wilkins, Baltimore.

Hirsch, P., 1981, The genus *Toxothrix*, in: *The Prokaryotes* (M. P. Starr, H. Stolp, H. G. Trüper, A. Balows, and H. G. Schlegel, eds.), pp. 409–411, Springer-Verlag, Berlin.

Hirsch, P., and Skuja, H., 1974, Genus *Planctomyces*, in: *Bergey's Manual of Determinative Bacteriology*, 8th ed. (R. E. Buchanan and N. E. Gibbons, eds.), pp. 162–163, Williams and Wilkins, Baltimore.

Hirsch, P., and Zavarzin, G. A., 1974, Genus *Toxothrix*, in: *Bergey's Manual of Determinative Bacteriology*, 8th ed. (R. E. Buchanan and N. E. Gibbons, eds.), p. 120, Williams and Wilkins, Baltimore.

Jones, J. G., 1981, The population ecology of iron bacteria (Genus *Ochrobium*) in a stratified eutrophic lake, *J. Gen. Microbiol.* 125:85–93.

Jones, J. G., 1983, A note on the isolation and enumeration of bacteria which deposit and reduce ferric iron, *J. Appl. Bacteriol.* 54:305–310.

Jones, J. G., Gardener, S., and Simon, B. M., 1983, Bacterial reduction of ferric iron in a stratified eutrophic lake, *J. Gen. Microbiol.* 129:131–139

Jones, J. G., Gardener, S., and Simon, B. M., 1984a, Reduction of ferric iron by heterotrophic bacteria in lake sediments, *J. Gen. Microbiol.* 130:45–51.

Jones, J. G., Davison, W., and Gardener, S., 1984b, Iron reduction by bacteria: Range of organisms involved and metals reduced, *FEMS Microbiol. Lett.* 21:133–136.

Kelly, D. P., Norris, P. R., and Brierley, C. L., 1979, Microbiological methods for the extraction and recovery of metals, *Symp. Soc. Gen. Microbiol.* 29:263–308.

Kono, K., and Usami, S., 1982, Biological reduction of ferric iron by iron- and sulfur-oxidizing bacteria, *Agric. Biol. Chem.* 46:803–805.

Kucera, S., and Wolfe, R. S., 1957, A selective enrichment method for *Gallionella ferruginea*, *J. Bacteriol.* 74:344–349.

Kutuzova, R. S., 1974, Electron microscopic study of ooze overgrowths of an iron-oxidizing coccus related to *Siderococcus limoniticus* Dorff, *Microbiology* 43:237–241.

Kuznetsov, S. I., 1970, *The Microflora of Lakes and Its Geochemical Acivity*, University of Texas Press, Austin.

Lammers, P. J., and Sanders-Loehr, J., 1982, Active transport of ferric schizokinen in *Anabaena* sp., *J. Bacteriol.* 151:288–294.

Lascelles, J., and Burke, K. A., 1978, Reduction of ferric iron by L-lactate and D-L-glycerol-3-phosphate in membrane preparations from *Staphylococcus aureus* and interactions with the nitrate reductase system, *J. Bacteriol.* 134:585–589.

Lundgren, D. G., and Silver, M., 1980, Ore leaching by bacteria, *Annu. Rev. Microbiol.* 34:263–283.

Mah, R. A., 1982, Methanogenesis and methanogenic partnerships, *Phil. Trans. R. Soc. Lond. B.* 297:599–616.

McInerney, M. J., Bryant, M. P., and Pfennig, N., 1979, Anaerobic bacterium that degrades fatty acids in syntrophic association with methanogens, *Arch. Microbiol.* 122:129–135.

McInerney, M. J., Bryant, M. P., Hespell, R. B., and Costerton, J. W., 1981, *Syntrophomonas wolfei* gen. nov. sp. nov. an anaerobic, syntrophic, fatty acid-oxidizing bacterium, *Appl. Environ. Microbiol.* 41:1029–1039.

Moore, R. L., 1981, The genera *Hyphomicrobium, Pedomicrobium*, and *Hyphomonas*, in:

The Prokaryotes (M. P. Starr, H. Stolp, H. G. Trüper, A. Balows, and H. G. Schlegel), pp. 480–487, Springer-Verlag, Berlin.

Moores, J. C., Magazin, M., Ditta, G. S., and Leong, J., 1984, Cloning of genes involved in the biosynthesis of pseudobactin, a high-affinity iron transport agent of a plant growth-promoting *Pseudomonas* strain, *J. Bacteriol.* **157**:53–58.

Mortimer, C. H., 1941, The exchange of dissolved substances between mud and water in lakes: I and II, *J. Ecol.* **29**:280–329.

Mortimer, C. H., 1942, The exchange of dissolved substances between mud and water in lakes: III and IV, *J. Ecol.* **30**:147–201.

Mulder, E. G., 1964, Iron bacteria, particularly those of the *Sphaerotilus–Leptothrix* group, and industrial problems, *J. Appl. Bacteriol.* **27**:151–173.

Mulder, E. G., 1974, Genus *Leptothrix,* in: *Bergey's Manual of Determinative Bacteriology,* 8th ed. (R. E. Buchanan and N. E. Gibbons, eds.), pp. 129–133, Williams and Wilkins, Baltimore.

Mulder, E. G., and Deinema, M. H., 1981, The sheathed bacteria, in: *The Prokaryotes* (M. P. Starr, H. Stolp, H. G. Trüper, A. Balows, and H. G. Schlegel), pp. 425–440, Springer-Verlag, Berlin.

Mulder, E. G., and van Veen, W. L., 1974, Genus *Sphaerotilus,* in: *Bergey's Manual of Determinative Bacteriology,* 8th ed. (R. E. Buchanan and N. E. Gibbons, eds.), pp. 128–129, Williams and Wilkins, Baltimore.

Munch, J. C., and Ottow, J. C. G., 1977, Model experiments on the mechanism of bacterial iron-reduction in water logged soils, *Z. Pflanz. Dueng. Bodenkd.* **140**:549–562.

Munch, J. C., and Ottow, J. C. G., 1982, Effect of cell contact and iron(III) oxide form on bacterial iron reduction, *Z. Pflanz. Dueng. Bodenkd.* **145**:66–77.

Munch, J. C., and Ottow, J. C. G., 1983, Reductive transformation mechanism of ferric oxides in hydromorphic soils, *Ecol. Bull.* **35**:383–394.

Nealson, K., 1983, The microbial iron cycle, in: *Microbial Geochemistry* (W. E. Krumbein, ed.), pp. 159–190, Blackwell, Oxford.

Neilands, J. B., 1974, Siderophores of bacteria and fungi, *Microbiol. Sci.* **1**:9–14.

Nicholls, K. H., and Fung, D., 1982, Accumulation of iron in the cell walls of the two mono-specific freshwater genera *Catena* and *Dichotomosiphon* (Chlorophyceae), *Arch. Protistenkd.* **125**:209–214.

Novitsky, J. A., Scott, I. R., and Kepkay, P. E., 1981, Effects of iron, sulfur, and microbial activity on aerobic to anaerobic transitions in marine sediments, *Geomicrobiol. J.* **2**:211–223.

Nunley, J. W., and Krieg, N. R., 1968, Isolation of *Gallionella ferruginea* by the use of formalin, *Can. J. Microbiol.* **14**:385–389.

Obuekwe, C. O., Westlake, D. W. S., and Cook, R. D., 1981, Effect of nitrate on reduction of ferric iron by a bacterium isolated from crude oil, *Can. J. Microbiol.* **27**:692–697.

Ottow, J. C. G., 1970, Selection, characterization and iron-reducing capacity of nitrate reductaseless (nit⁻) mutants from iron reducing bacteria, *Z. Allg. Mikrobiol.* **10**:55–62.

Ottow, J. C. G., and Glathe, H., 1971, Isolation and identification of iron-reducing bacteria from gley soils, *Soil Biol. Biochem.* **3**:43–55.

Ottow, J. C. G., and Munch, J. C., 1978, Mechanisms of reductive transformations in the anaerobic microenvironment of hydromorphic soils, in: *Environmental Biogeochemistry and Geomicrobiology* (W. E. Krumbein, ed.), Vol. 2, pp. 483–491, Ann Arbor Science Publishers, Ann Arbor, Michigan.

Pfanneberg, T., and Fischer, W. R., 1984, An aerobic *Corynebacterium* from soil and its capability to reduce various iron oxides, *Zentralbl. Mikrobiol.* **139**:167–172.

Pringsheim, E. G., 1949, Iron bacteria, *Biol. Rev.* **24**:200–245.

Rogers, S. R., and Anderson, J. J., 1976, Measurement of growth and iron depositon in *Sphaerotilus discophorus, J. Bacteriol.* **126**:257–263.

Schmidt, J. M., and Starr, M. P., 1981, The *Blastocaulis-Planctomyces* group of budding and appendaged bacteria, in: *The Prokaryotes* (M. P. Starr, H. Stolp, H. G. Trüper, A. Balows, and H. G. Schlegel, eds.), pp. 496–504, Springer-Verlag, Berlin.

Schmidt, J. M., and Swafford, J. R., 1981, The genus *Seliberia*, in: *The Prokaryotes* (M. P. Starr, H. Stolp, H. G. Trüper, A. Balows, and H. G. Schlegel, eds.), pp. 516–519, Springer-Verlag, Berlin.

Schmidt, J. M., and Zavarzin, G. A., 1981, The genera *Caulococcus* and *Kusnezovia,* in: *The Prokaryotes* (M. P. Starr, H. Stolp,. H. G. Trüper, A. Balows, and H. G. Schlegel, eds.), pp. 529–530, Springer-Verlag, Berlin.

Schmidt, J. M., Sharp, W. P., and Starr, M. P., 1982, Metallic-oxide encrustations of the nonprosthecate stalks of naturally occurring populations of *Planctomyces bekefii, Curr. Microbiol.* 7:389–394.

Schmidt, W.-D., 1984, Die eisenbakterien des Plusssees. II. Morphologie und feinstruktur von *Siderocapsa geminata* (Skuja 1954/57), *Z. Allg. Mikrobiol.* 24:391–396.

Schmidt, W.-D., and Overbeck, J., 1974, Studies of "iron bacteria" from Lake Pluss. I. Morphology, fine structure and distribution of *Metallogenium* sp. and *Siderocapsa geminata, Z. Allg. Mikrobiol.* 24:329–339.

Shakhobova, N. N., 1981, Participation of *Arthrobacter* bacteria in the reduction of ferric compounds, *Isv. Akad. Nauk Tadzh, SSR, Ocd. Biol. Nauk* 1981(1):129–132.

Sigel, S. P., and Payne, S. M., 1982, Effect of iron limitation on growth, siderophore production, and expression of outer membrane proteins of *Vibrio cholerae, J. Bacteriol.* 150:148–155.

Sørensen, J., 1982, Reduction of ferric iron in anaerobic, marine sediment and interaction with reduction of nitrate and sulfate, *Appl. Environ. Microbiol.* 43:319–324.

Staley, J. T., and Bauld, J., 1981, The genus *Planctomyces,* in: *The Prokaryotes* (M. P. Starr, H. Stolp, H. G. Trüper, A. Balows, and H. G. Schlegel, eds.), pp. 505-508, Springer-Verlag, Berlin.

Svorcova, L., 1975, Iron bacteria of the genus *Siderocapsa* in mineral waters, *Z. Allg. Mikrobiol.* 15:553–557.

Svorcova, L., 1979, Diagnostik der Eisenbakterien der Familie Siderocapsaceae, *Arch. Hydrobiol.* 87:423–452.

Takai, Y., and Kamura, T., 1966, The mechanism of reduction in waterlogged paddy soil, *Folia Microbiol.* 11:304–313.

Thauer, R. K., Jungermann, K., and Decker, K., 1977, Energy conservation in chemotrophic anaerobic bacteria, *Bacteriol. Rev.* 41:100–180.

Tipping, E., and Cooke, D., 1982, The effects of adsorbed humic substances on the surface charge of geothite (α-FeOOH) in freshwaters, *Geochim, Cosmochim. Acta* 46:75–80.

Tipping, E., and Woof, C., 1983, Elevated concentrations of humic substances in a seasonally anoxic hypolimnion: Evidence for co-accumulation with iron, *Arch. Hydrobiol.* 98:137–145.

Tipping, E., Woof, C., and Cooke, D., 1981, Iron oxide from a seasonally anoxic lake, *Geochim. Cosmochim. Acta* 45:1411–1419.

Tipping, E., Woof, C., and Ohnstad, M., 1982, Forms of iron in the oxygenated waters of Esthwaite Water, U.K., *Hydrobiologia* 92:383–393.

Trick, C. G., Anderson, R. J., Gilliam, A., and Harrison, P. J., 1983, Procentrin, an extracellular siderophore produced by the marine dinoflagellate *Procentrum minimum, Science* 219:306–308.

Van Veen, W. L., Mulder, E. G., and Deinema, M. H., 1978, The *Sphaerotilus-Leptothrix* group of bacteria, *Microbiol. Rev.* 42:329–356.

Verdouw, H., and Dekkers, E. M. J., 1980, Iron and manganese in Lake Vechten (The Netherlands); Dynamics and role in the cycle of reducing power, *Arch. Hydrobiol.* 89:509–532.

Volker, H., Schweisfurth, R., and Hirsch, P., 1977, Morphology and ultrastructure of *Crenothrix polyspora* Cohn, *J. Bacteriol.* **131**:306–313.

Walker, J. C. G., 1984, Suboxic diagenesis in bonded iron formations, *Nature* **309**:340–342.

Walsby, A. E., 1981, Gas-vacuolate bacteria (apart form Cyanobacteria), in: *The Prokaryotes* (M. P. Starr, H. Stolp, H. G. Trüper, A. Balows, and H. G. Schlegel, eds.), pp. 441–447, Springer-Verlag, Berlin.

Walsh, F., and Mitchell, R., 1972a, A pH-dependent succession of iron bacteria, *Environ. Sci. Technol.* **6**:809–812.

Walsh, F., and Mitchell, R., 1972b, An acid tolerant iron-oxidizing *Metallogenium, J. Gen. Microbiol.* **72**:369–373.

Walsh, F., and Mitchell, R., 1973, Differentiation between *Gallionella* and *Metallogenium, Arch. Mikrobiol.* **90**:19–25.

Warner, P. J., Williams, P. H., Bindereif, A., and Neilands, J. B., 1981, ColV plasmid specified aerobactin synthesis by invasive strains of *Escherichia coli, Infect. Immunol.* **33**:540–545.

Williams, P. H., 1979, Novel iron uptake system specified by ColV plasmids: An important component in the virulence of invasive strains of *Escherichia coli, Infect. Immunol.* **26**:925–932.

Williams, P., Brown, M. R. W., and Lambert, P. A., 1984, Effect of iron deprivation on the production of siderophores and outer membrane proteins in *Klebsiella aerogenes, J. Gen. Microbiol.* **130**:2357–2365.

Wolfe, R. S., 1958, Cultivation, morphology and classification of the iron bacteria, *J. Am. Water Works Assoc.* **50**:1241–1249.

Woolfolk, C. A., and Whiteley, H. R., 1962, Reduction of inorganic compounds with molecular hydrogen by *Micrococcus lactilyticus, J. Bacteriol.* **84**:647–658.

Wurtsbaugh, W. A., and Horne, A. J., 1983, Iron in eutrophic Clear Lake, California: Its importance for algal nitrogen fixation and growth, *Can. J. Fish. Aquat. Sci.* **40**:1419–1429.

Zavarzin, G. A., 1974, Genus *Ochrobium*, in: *Bergey's Manual of Determinative Bacteriology*, 8th ed. (R. E. Buchanan and N. E. Gibbons, eds.), pp. 467–468, Williams and Wilkins, Baltimore.

Zavarzin, G. A., 1981, The genus *Metallogenium*, in: *The Prokaryotes* (M. P. Starr, H. Stolp, H. G. Trüper, A. Balows, and H. G. Schlegel, eds.), pp. 524–528, Springer-Verlag, Berlin.

Zavarzin, G. A., and Hirsch, P., 1974a, Genus *Gallionella*, in: *Bergey's Manual of Determinative Bacteriology*, 8th ed. (R. . Buchanan and N. E. Gibbons, eds.), pp. 160–161, Williams and Wilkins, Baltimore.

Zavarzin, G. A., and Hirsch, P., 1974b, Genus *Metallogenium*, in: *Bergey's Manual of Determinative Bacteriology*, 8th ed. (R. E. Buchanan and N. E. Gibbons, eds.), pp. 163–165, Williams and Wilkins, Baltimore.

5

The Influence of the Rhizosphere on Crop Productivity

J. M. WHIPPS and J. M. LYNCH

1. Introduction

The rhizosphere region is a variable zone containing a proliferation of microorganisms inside and outside the plant root. Many compounds are both taken up and passed out. Under normal growth conditions the rhizosphere exists because of the continuous loss of many forms of plant metabolites, which are rapidly utilized by microorganisms. Consequently, these rhizosphere microorganisms are in a position to affect both subsequent loss of material from the roots and nutrient uptake by the roots. In natural ecosystems an equilibrium develops between the plant and microorganisms that is affected only by the normal growth of plant and seasonal changes in the environment. However, in agriculture, man continually changes the normal equilibrium by manifold means (e.g., plant monoculture, herbicide, fungicide and pesticide treatments, fertilizer application, and cultivation), all of which modify subsequent plant growth and the associated rhizosphere biota. Because of the importance of agriculture, the majority of work on the rhizosphere and its effects on plant growth has involved research on crop plants and, although this has provided great insight into rhizosphere–plant interactions in these relatively few species, some care should be taken in extrapolating such results

J. M. WHIPPS • Agricultural Research Council, Letcome Laboratory, Wantage, Oxon OX12 9JT, England; *present address:* Plant Pathology and Microbiology Department, Glasshouse Crops Research Institute, Littlehampton, West Sussex BN17 6LP, England.
J. M. LYNCH • Plant Pathology and Microbiology Department, Glasshouse Crops Research Institute, Littlehampton, West Sussex BN17 6LP, England.

to all natural ecosystems. With this proviso, we attempt to show, first, the effect the plant has on development and maintenance of the rhizosphere and, second, the influence the rhizosphere has on plant physiology and consequently crop productivity, highlighting areas of research likely to be rewarding both scientifically and commercially in the future. We do not attempt a complete review of the literature, since there have been reviews on many aspects of rhizosphere biology in recent years (Barber, 1978; Hale, *et al.* 1978; Newman, 1978; Balandreau and Knowles, 1978; Hale and Moore, 1979; Bowen, 1979, 1980, 1982; Woldendorp, 1981; Foster and Bowen, 1982; Lynch, 1982, 1983; Subba Rao, 1982a; Suslow, 1982), but rather choose specific examples to illustrate our major points.

Many of the reviews on the rhizosphere have considered specifically the activities of saprophytes. However, symbionts, both biotrophs and necrotrophs, can be expected to make a major contribution to the rhizosphere biomass. Furthermore, many of the conceptual considerations, such as the analysis of carbon flow, are equally relevant to saprophytes and symbionts, both groups being chemoheterotrophs. We have therefore decided to include all these nutritional groups within our review of the rhizosphere.

2. Structure of the Rhizosphere

In the original concept the rhizosphere was a zone around the roots in which bacterial growth was stimulated by the release of nutrients, first described for legumes by Hiltner (1904). However, this description has been greatly refined and Balandreau and Knowles (1978) have described three distinct regions of the rhizosphere (Fig. 1): (1) the *endorhizosphere* is the epidermal/cortex region of the root, which commonly becomes colonized by microorganisms (Darbyshire and Greaves, 1973; von Bulow and Dobereiner, 1975; Dobereiner and Day, 1976; Old and Nicholson, 1975, 1978; Foster *et al.,* 1983); (2) the *rhizoplane* is the interface between the surface of the root and the soil; and (3) the *exorhizosphere* is the soil around the root where the microorganism proliferation occurs. We consider that the term *ectorhizosphere* should apply to this region, since it is already used for mycorrhiza and the source "ecto" is commonly used in describing other surface phenomena. The latter two regions are traditionally thought of as the rhizosphere, but it is better considered as a continuum of all these three regions, which changes in composition with root age. This is a particularly important concept because the rhizoplane *sensu stricto* may exist for only a very short while. Foster (1981, 1982), in studying this interface for various plant species by electron microscopy, noticed that the integrity of the plant cuticle was rapidly destroyed and

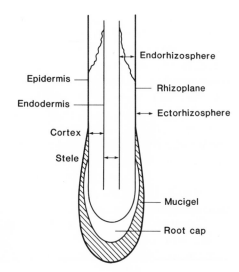

Epidermis

Endodermis

Cortex

Stele

Endorhizosphere

Rhizoplane

Ectorhizosphere

Mucigel

Root cap

Figure 1. Diagrammatic representation of the structure of the rhizosphere in and around roots.

the epidermal mucilages were mixed with soil colloids, soil organic matter, and bacterial and fungal polysaccharides to yield the mucigel. Mucigel is particularly important at the root tip, where a drop containing sloughed-off root cap cells is found (Samtsevitch, 1971, 1972). Once the cuticle has broken down, bacterial migration into the epidermal/cortex region can take place rapidly and the continuum is completed. With age, the structure of the cortex can break down to give roots with just a functional stele remaining bounded by a dead cortical zone (Holden, 1975; Henry and Deacon, 1981).

3. The Plant As the Driving Force for Rhizosphere Development and Maintenance

Plant roots behave as continuous sources of organic carbon compounds. These are probably lost as the result of the plant's need to grow roots to allow rapid uptake of water and inorganic nutrient rather than any direct carbon release mechanism. These processes are the controlling factors in microorganism development in and around the roots. In general, any increase in carbon loss from the plant would be expected to increase the biomass of the microbial population at least transiently. In some cases, however, higher plants release compounds that act as inhibitors of some microorganisms (Kapustka and Rice, 1976).

3.1. Loss of Organic Carbon Compounds

Almost every type of organic compound detected in plants has been found to be lost from roots (Rovira, 1969; Hale *et al.,* 1978). They are best considered as five different classes of compound: (1) exudates, water-soluble compounds which leak from roots without the expenditure of metabolic energy (e.g., sugars, amino acids, organic acids, vitamins, growth regulators); (2) secretions, low-molecular-weight compounds released by metabolic processes; (3) plant mucilages, polysaccharides of various qualities and origins; (4) mucigel, a complex product of the entire root–soil–microbial complex (see Section 2); and (5) lysates, compounds released from autolysis of old cells when the plasmalemma breaks down (Rovira *et al.* 1979).

Numerous studies have attempted to quantify this carbon loss from plant roots. Initially, these were carried out using large numbers of sterile and nonsterile plants in liquid culture, and although a large range of compounds was found, quantitatively the amounts found were small, between 0.1 and 4% of carbon equivalents in roots (Rovira and Davey, 1974). These experiments, however, took no account of CO_2 respired from roots and from the microorganisms associated wtih the roots, nor did they measure the insoluble carbon lost from the roots. They also neglected the effect of root restriction on carbon loss, for Barber and Gunn (1974) showed that, when barley roots were grown in 1-mm Ballotini beads compared with conventional solution culture, carbon loss was increased to 9% of the dry matter of the roots.

Growth of plants from seeds in soil or sand in a controlled environment containing $^{14}CO_2$ of constant specific activity, however, has provided a more sensitive and accurate system for analyzing carbon flow through plants and its subsequent fate in the growing medium (Whipps and Lynch, 1983). These systems enable all parts of the plant to be evenly labeled so that all carbon compounds lost from the roots, including insoluble material and respiratory CO_2, can be quantified. Such work has indicated that nonsterile roots lose more carbon than sterile root systems, and values of carbon lost from nonsterile, soil-grown roots are in the range of 12–40% of total net fixed carbon for cereals, depending on age and growing conditions (Barber and Martin, 1976; Sauerbeck *et al.,* 1976; Johnen and Sauerbeck, 1977; Martin, 1977a,b; Martin and Kemp, 1980; Whipps and Lynch, 1983; Whipps, 1984). Newman (1985) has suggested that values for respiration measured by these techniques are low and that quantities of carbon lost into the soil surrounding the roots are high and implied that these may have occurred because of problems with the procedure. However, unless dramatic improvements are made in this technically demanding experimental system, they still remain the best available estimates of carbon loss from roots in soil.

Many factors are known to affect carbon release both quantitatively and qualitatively, but, once again, the majority of work has been done in nutrient solution culture rather than soil, which must add an element of doubt to their relevance (Table I). Such effects may differ between plant species and are often dependent on plant age and developmental stage (Hale *et al.*, 1978; Rovira, 1979; Hale and Moore, 1979). It is evident from the information contained in Table I that the majority of effects are due to changes in the environment independent of man's activities, but the routine agricultural practice of applying foliar sprays of all kinds correspondingly produces changes in carbon release and, hence, changes in the rhizosphere microbiota (e.g., Balasubramanian and Rangaswami, 1969, 1973). In many cases the applied chemical itself is released and has a direct influence on associated microorganisms and other plants (Foy *et al.*, 1971; M. P. Greaves, personal communication). Coupland and Caseley (1979) showed that the herbicide glyphosate could be exuded from the roots of rhizomes of the weed *Agropyron repens,* but Penn and Lynch (1982) showed that the amounts released would be unlikely to affect a cereal crop. Unfortunately, little other recent research on this environmentally important subject has been published.

Extra research effort should also be directed to determining further the effects of pH, CO_2, and O_2 and their interrelationships, growth regulators (see Section 4.2.3), and environmental pollutants such as SO_2 and acid rain on rhizosphere biology. Interestingly, little or no research appears to have been carried out on the effects of foliar pathogens on exudation. Frič (1975) has shown that barley root cells become more permeable after powdery mildew infection of the leaves, but no direct estimate of carbon loss was obtained. This is even more surprising in view of the effect such pathogens have on translocation patterns within infected plants (Whipps and Lewis, 1981). It should of course be recognized that all the effects recorded in Table I are closely interrelated in the crop situation and an alteration in just one of these variables is likely to have consequences on some of the others. Hence, a resultant change in the rhizosphere population may not be a simple cause and effect relationship.

These quantitative and qualitative studies, however, give no impression of where on the roots carbon compounds are lost. Histological studies have shown that the root cap must provide much insoluble material (Samtsevich, 1965, 1971; Clowes, 1971), as do the older parts of the root system when they break down (Holden, 1975; Henry and Deacon, 1981). Exudation and secretion seem to take place in particular areas of the root surface. In a series of $^{14}CO_2$ pulse-chase experiments using wheat, the greatest quantity of carbon loss was found to occur at the root tips (corresponding to root cap sloughing), but the next greatest quantity was lost from the zone of elongation as diffusible exudates (McDougall, 1968,

Table I. Examples of Factors Affecting Release of Organic Compounds from Crop Plant Roots

Factor	Plant	Effect on release	Reference
Lowered temperature	Tomato	Decreased	Rovira (1969)
	Subterranean clover	Decreased	Rovira (1969)
	Maize	Increased	Vančura (1967)
	Cucumber	Increased	Vančura (1967)
	Wheat	Increased	Martin (1977a), Martin and Kemp (1980), Whipps (1984)
	Barley	Increased	Whipps (1984)
	Strawberry	Increased	Hussain and McKeen (1963)
Decreased day length	Wheat	Decreased	Barber and Martin (1976), Whipps (1984)
	Barley	Decreased	Barber and Martin (1976), Whipps (1984)
Decreased light intensity	Clover	Decreased	Rovira (1959)
	Tomato	Decreased	Rovira (1959)
Water stress	Wheat	Increased	Vančura (1967), Martin (1977b)
	Barley	Increased	Vančura (1967), Martin (1977b)
	Barley	Decreased	Shone et al. (1983)
	Lodgepole pine	Increased	C. P. P. Reid and Mexal (1977)
Increased pH	Wheat	Decreased[a]	McDougall (1970), Rovira and Ridge (1973)
	Pea clover	Increased[a]	Bonish (1973)
Anaerobiosis	Maize	Increased[b]	Grineva (1961)
	Peanut	Increased	Rittenhouse and Hale (1971)
	Wheat	Increased	Wiedenroth and Poskuta (1981)
Root injury/ impedance	Pea	Increased	Van Egeraat (1975)
	Wheat	Increased	Ayers and Thornton (1968)
	Barley	Increased	Barber and Gunn (1974)
	Soybean	Increased	D'Arcy (1982)
	Maize	Increased	Schonwitz and Ziegler (1982)
Changes in ionic concentration			
Decreased Ca^{2+}	Peanut	Increased	Shay and Hale (1973)
Decreased PO_4^{3-}	Pine	Increased	Bowen (1969)

Table I. (*continued*)

Factor	Plant	Effect on release	Reference
Changes in ionic concentration (*continued*)			
Decreased PO$_4^{3-}$ (*continued*)	Sorghum	Increased	Ratanayake, *et al.,* (1978), Graham *et al.* (1981)
	Citrus	Increased	Ratanayake *et al.* (1978), Graham *et al.* (1981)
Decreased K$^+$	Haricot beans	Increased	Trolldenier (1972)
Decreased N	Pine	Decreased	Bowen (1969)
Increased Na$^+$	Barley	Increased	Polonenko *et al.* (1983)
Ozone	Tomato	Decrease	McCool and Menge (1983)
Chemicals applied to foliage[c]	Sorghum	Variable	Balasubramanian and Rangaswami (1969, 1973)
	Sunhemp	Variable	Balasubramanian and Rangaswami (1969, 1973)
	Ragi	Variable	Balasubramanian and Rangaswami (1969, 1973)
	Tomato	Variable	Balasubramanian and Rangaswami (1969, 1973)
	Cajanus cajan	Variable	Sethunathan (1970a, b)
	Peanut	Variable	Hale, *et al.* (1977)
	Navy bean	Increased	Wyse, *et al.* (1976)
	Wheat	Variable	Jalili (1976), Jalili and Domsch (1975)
Growth regulators in root medium (e.g., cytokinin)	Peanuts	Variable	Thompson and Hale (1983)
Foliar saprophytes	Hyacynth bean	Increased	Bhat *et al.,* (1971)
Foliar infections	Squash	Increased	Magyarosy and Hancock (1974)
	Pea	Increased	Beute and Lockwood (1968)
	Barley	Increased	Paulech *et al.* (1981)
Root saprophytes	Maize	Increased	Vančura *et al.* (1977)
	Wheat	Increased	Barber and Martin (1976)
	Barley	Increased	
Root pathogens	Many	Variable	See review by Mitchell (1976)

[a]Indirect effect. pH may not be the actual controlling factor.
[b]Change in quality.
[c]A large range of chemicals was tried on each plant species; see individual references for details.

1970; McDougall and Rovira, 1970; Rovira, 1973). The preferential growth of microbial colonies on cell junctions has also been taken to imply microsites of exudation from which discrete colonies develop (Rovira and Campbell, 1974; Bowen and Foster, 1978; van Vuurde *et al.*, 1979). Only two experiments have attempted to relate quantitatively carbon supply from the roots and the associated bacterial growth (Barber and Lynch, 1977; Whipps and Lynch, 1983), but, unfortunately, neither of these was concerned with normal plant growth in soil, since it is difficult to quantify biomass in soil. However, Lynch and Panting (1981) demonstrated in samples taken from the field that the soil microbial biomass increased as roots grew in it and this biomass reached a maximum at the time of maximum root production. Quantification of carbon loss from a range of plant species in relation to microbial growth is an area in which research is greatly needed.

The carbon loss brings about the increase in numbers of microorganisms in the rhizosphere compared with the bulk soil, but such increases in biomass do not occur in all species. Specific classes of microorganisms are stimulated preferentially and these can change with time and vary with plant species and with plant mixtures. For instance, Vančura and Kunc (1977), using actidione to inhibit fungi and streptomycin to inhibit bacteria, found that bacterial respiration was greater than fungal respiration in the rhizosphere. Early work showed that some bacteria, such as *Pseudomonas* and other Gram-negative bacteria, dominated the rhizosphere, but this may have been partly due to the ability of these bacteria to grow on nonspecific media (Rovira and Davey, 1974). In gnotobiotic studies, however, *Pseudomonas* spp. were able to utilize at least 70% of the root carbon loss of wheat and maize, showing that they are suited to growth on the chemicals lost from the plant (Vančura, 1980). Fungi gradually become more prevelent with age of root and appear to be more specialized types, such as *Fusarium oxysporum, Cylindrocarpon radicola,* and *Trichoderma viride* (Parkinson *et al.*, 1963). Mycorrhizas are a special case and are discussed in Section 4.4.1. However, it is obvious that there are manifold interactions between bacteria and fungi on roots and these are dependent on many variables (Rambelli, 1973). In a survey of over 40 sites in England and Wales, Newman *et al.* (1981) found that the fungal-to-bacterial ratio (expressed as percentage cover of root surface of *Plantago lanceolata*) varied from 0.28 to 14.0. Subsequently, it was found found in a pot study, that the composition of bacteria and fungi (including mycorrhizas on the roots) of three grassland species was different, depending on species of plant and mixtures of plants with which it was growing (Lawley *et al.*, 1982; 1983). In a crop situation where monoculture occurs the situation is probably simpler and more predictable, and perhaps here is the place to start on the population dynamic studies called for by Bowen (1979, 1980).

In addition to providing substrates for general microbial growth, the carbon lost from roots can have much more selective effects. In soil many fungal propagules remain dormant until a root passes close by and are then stimulated to germinate due to a general release of nutrients, their subsequent metabolites, or more specific compounds. For instance, zoospores of *Pythium aphanidermatum* were attracted to a wide range of compounds released from pea roots (Royle and Hickman, 1964a,b), but zoospores of *Phytophthora cinnamoni* were attracted to ethanol produced by anaerobic respiration of *Eucalyptus* roots in wet soil (R. N. Allen and Newhook, 1973). There appears to be little evidence for specific chemotaxis being involved in the legume/*Rhizobium* association, because *Escherichia coli* was attracted by legume root exudates to the same extent as *R. leguminosarum* (Gaworzewska and Carlile, 1982). More specific attraction or interaction occurs with *Rhizobium* on legume root hairs and during the adhesion of *Agrobacterium* to roots, mediated by extracellular polysaccharides of the root (Dazzo, 1980).

The microbial metabolism of carbon compounds released by roots has implications for the analysis of plant physiological experiments that are conducted nonaxenically. A particular example is in phytoalexin studies, when it is usually necessary to extract large amounts of root material to obtain small amounts of the metabolite. In such a situation, how can it be certain that the metabolite comes from the plant and not the associated microbiota?

3.2. Uptake of Water and Nutrients

The role of microorganisms in ion uptake has been reviewed in detail by Barber (1978) and only examples of the types of influences that water and nutrient removal from the rhizosphere can have on the microbial population will be considered.

In broad terms, the microbial population around roots exists because of the carbon flow from the plant. As soon as the microorgansims begin to grow in the rhizosphere, however, they become competitors with the plant for the available water and inorganic nutrients and a balanced interdependent population becomes established. Axenic culture of plants is useful to study the underlying principles of this balance. For example, Turner and Newman (1984) found that a decreased nitrogen supply to grass plants reduced plant growth but had no effect on the numbers of *Flavobacterium* sp. and *Serratia marcescens* colonizing the roots, whereas phosphate deficiency increased the numbers of both bacterial species.

The microbial equilibrium of the rhizosphere is affected by the same environmental and other controlling factors mentioned above for carbon loss (Section 3.1). For instance, in time of drought both plant and rhizosphere microorganisms suffer water stress, with perhaps the rhizosphere organisms suffering more severely than those in the bulk soil because of the plant's demands. A similar situation may exist in soils low in certain essential elements, where competition between plant and associated microorganisms places a greater inhibitory effect on rhizosphere compared with bulk soil microorganisms. However, plant behavior can have more specific effects on this ecological niche. For instance, Marschner and Romheld (1983) showed that when maize plants were taking up nitrogen in the form of NH_4^+ the pH of the rhizosphere dropped, and while when they were taking up nitrogen as NO_3^- the pH rose. Such changes in pH alter the solubility and availability of other nutrients to plants. For example, *Brassica napus* grown in soil of low phosphate status supplied solely with nitrogen in the form of NO_3^- was found to lower the pH of the soil around the roots and this increased phosphate availability and uptake, whereas in soil amended with additional phosphate no pH drop was found (Grinstead *et al.,* 1982; Hedley *et al.,* 1982a,b, 1983). In low-phosphate soil, H^+ ions were released to maintain overall ionic balance and this was the mechanism for the pH drop, which did not occur in soils amended with phosphate. A similar effect was found for maize (Boero and Thien, 1979). Hence, all such pH-related changes in nutrient availability are still dependent on the type of plant, the type of soil, and, eventually, the types of microorganisms present (Section 4.2.2). Again, a change in any of the influencing factors on plant growth and metabolism will change the microbial population, which in turn may itself become another influencing factor.

4. Consequences of Rhizosphere Development on Root and Crop Growth—Saprophytic, Parasitic, and Mutualistic Microbial Effects

4.1. A Conceptual Note

Rhizosphere biologists have classically considered the nutrition of microorganisms that occur in the rhizosphere either as saprophytic or as parasitic and mutualistic symbioses (*sensu* Lewis, 1973; R. C. Cooke, 1977). However, in the last decade it has become apparent that many bacteria and fungi that inhabit the rhizosphere do not fall neatly into any of these categories and commonly change their nutrition with changes in the environment of the plant. For instance, the mutualistic bacteria of the *Rhizobium*–legume symbiosis may become necrotrophically parasitic on their host plant if the light intensity drops (Thornton, 1930; van Schevren, 1958). Similarly, some bacteria and fungi that naturally exist in the rhizosphere may become parasitic if root damage occurs or if the environment changes. For example, *Rhizoctonia sylvestris* and *Mycelium radicis atrovirens,* which form pseudomycorrhizas, become necrotrophically parasitic with low light intensity (Harley, 1969). Further, the idea that saprophytic microorganisms exist at all in the rhizosphere is a moot point, for in essence the continual supply of carbon compounds from the root generates the rhizosphere effect and so these microorganisms could well be regarded as neutral symbionts [organisms associated with a host upon which they are absolutely dependent, at least transiently, but on which they have no obvious deleterious or beneficial effect (R. C. Cooke, 1977)]. This may be particularly so when some bacteria subsequently enter the root cortex during the normal growth of plants (Darbyshire and Greaves, 1973; Old and Nicholson 1975, 1978). Some of the minor pathogens described by Salt (1979) may come into this category. Research on the physiology of these "neutral" parasitic symbionts in the root-inhabiting stage is much needed (Wilcox, 1983).

   The remainder of this review is separated for convenience into three sections describing the effects on plant growth of rhizosphere microorganisms and their directly generated regulatory and controlling mechanisms. These are arbitrarily categorized into saprophytic or symbiotic (parastic and mutualistic) microorganisms, but in reality such a classification may not be conceptually or nutritionally correct.

4.2. Saprophytic Effects

4.2.1. Soil Aggregation

   The structure and aggregate stability of the soil influences soil incrustation, water percolation, moisture content, tilth, aeration, temperature, microbial activity, and root penetration, and generally well-aggregated soils are easily worked and give better root growth conditions (Allison, 1973; Hattori, 1973; Lynch and Bragg, 1984). Aggregate stability is improved after growth of some crops, such as grasses and legumes (Clark, et al., 1967; G. W. Cooke and Williams, 1972) or decreased with others, such as wheat or maize (Page and Willard, 1946; Low, 1972), and may be related to changes in total soil organic matter in some cases or carbohydrates in others, depending on field conditions (Burns, 1981; Fletcher et al., 1980; J. B. Reid and Goss, 1981). However, there is evidence that *living* roots also exert an effect on soil aggregate stability. For instance, perennial ryegrass and lucerne were found to increase stability of sandy soils after only 28–42 days (Goss and Reid, 1979; J. B. Reid and Goss, 1980, 1981) and this effect could have been mediated by polysaccharide exudation from the plant directly and after alteration by microbial attack.

Maize, however, was found to decrease stability, but this effect was subsequently shown to be due to removal of organically bound Fe and Al by chelates exuded from maize roots that broke down organic matter–Fe or Al–mineral particle linkages (J. B. Reid et al., 1982). Using similar soil and conditions, however, E. Bragg and J. M. Lynch (unpublished results) were unable to observe the destabilizing effect of maize. The involvement of microorganisms in these stability effects is particularly important in the rhizosphere as the large numbers of bacteria present can, besides altering plant polysaccharides, also stabilize the soil directly by production of their own polysaccharides (Cheshire, 1979; Lynch, 1981, 1983). Saprophytic fungi could act as enmeshing agents and this could be particularly relevant in plants with high mycorrhizal infection (Tisdall and Oades, 1979). It could also be conceived that, for a short time, fungal infections such as take-all and other root and foot rots with a mycelium extending through the soil could act similarly. Microorganisms could also change the aggregation of soil particles around the roots by increasing or decreasing root hair frequency and/or number (Bowen and Rovira, 1961).

4.2.2. Nutrient Availability and Absorption

The effects of saprophytic rhizosphere microorganisms on nutrient uptake and availability have been reviewed recently (Barber, 1978; Woldendorp, 1981). The majority of work in this field has involved gnotobiotic studies in solution culture, and although these have yielded unequivocal evidence that bacteria and some fungi affect uptake of nutrients by roots, either positively or negatively, there are few equivalent studies in soil and these warrant further study. The microorganisms may influence uptake directly by competiton or indirectly by perhaps releasing H^+ ions or chelating agents, hormones, or toxins. These indirect processes could affect, on the small scale, permeability of the roots and, on the large scale, plant growth as a whole. Alteration in root morphology after inoculation in gnotobiotic studies has sometimes led to difficulties in interpretation of uptake data (e.g., Welte and Trolldenier, 1962; Williamson and Wyn Jones, 1973).

    4.2.2a. Phosphate. The uptake of phosphate by plant roots has been shown in solution culture to be dependent on many variables. Over short (30 min) periods, phosphate uptake and translocation in barley plants is promoted by bacteria, but over longer (24 hr) periods uptake is reduced (Barber et al., 1976). This effect is further complicated by the observation that phosphate uptake was stimulated by bacteria in young (6 day) and inhibited in older (12 day) seedlings. In soil, Benians and Barber (1974) showed that microorganisms decreased uptake and distribution of phosphate in barley seedlings and that this effect was probably due to competition for the available phosphate, much of which entered the nucleic acid fraction of the bacteria (Rovira and Bowen, 1966). Under other soil conditions bacteria may solubilize phosphate from insoluble sources by acid chelator production (Gerretson, 1948; Duff et al., 1963). However, it is in the rhizosphere rather than the bulk soil where such solubilization would benefit the plant. It is of significance that Odunfa and Oso (1978) have shown that a greater number of phosphate-solubilizing bacteria occurred in the rhizosphere of cowpea than in maize, and thus the effectiveness of such associated bacteria may differ between plant species. The work of Nye's group (see Section 3.2) indicates the importance of pH to the availability of phosphate to the plant. It is unknown if saprophytic fungi have any relevance to phosphate uptake by plants. This is in contrast to the overwhelming amount of data showing the importance of the mutualistic mycorrhizal fungi in enhancing phosphate supply to such infected plants (see Section 4.4.1). In most intensive cropping situations microbial action on phosphate uptake and availability would largely be negated by application of adequate fertilizer. However, in the Third World, where fertilization is not carried out to achieve maximum yield, such microbial effects would be much more important.

4.2.2b. Nitrogen. When 3-week-old barley plants growing in gnotobiotic solution culture were supplied with nitrogen as nitrate, bacteria were found to increase uptake (Barber, 1971). When supplied with ammonium chloride, however, uptake was decreased by the presence of bacteria. Similar physiological experiments have unfortunately not been carried out in sterile and nonsterile soil. Microorganisms in the rhizosphere must have quite significant effects on nitrogen availability and uptake for the plant, as they are intimately involved in the nitrogen cycle. As the rhizosphere is relatively rich in carbon compounds, the energy-consuming steps of the nitrogen cycle are likely to take place; i.e., immobilization within the organic nitrogen pool and associative nitrogen fixation as defined by Stewart et al. (1979). The rhizospheres of many plants have high numbers of associative nitrogen-fixers such as *Azospirillum* (Kosslak and Bohlool, 1983), but such nitrogen fixation has been calculated at best to supply only 15% of the nitrogen content of temperate cereals (Barber and Lynch, 1977). Similarly, Lethbridge and Davidson (1983a) have demonstrated that root-associated nitrogen fixation in wheat was negligible unless external sugar was supplied. From carbon release studies of wheat plants, Beck and Gilmour (1983) have found that associative nitrogen fixation is unlikely to be important in this species. In a soil study, J. M. Whipps (unpublished results) has found that 4-week-old maize seedlings, which have the C_4 photosynthetic pathway, lose between 17 and 21% of net fixed carbon from their roots, which is considerably lower than that for wheat and barley grown under similar conditions (Whipps, 1984) and implies that N_2 fixation in the rhizoplane-ectorhizosphere region of maize is relatively unimportant. This does not, however, preclude enhanced nitrogen fixation within the endorhizosphere. Under other environmental conditions with different crops, significant gains of nitrogen have been measured due to associative N_2 fixation. For instance, in temperate conditions lysimeter experiments have shown annual gains of 45 kg N ha^{-1} under mustard and 50–74 kg N ha^{-1} under perennial grasses and irises (Moore, 1966). In the tropics a range of values between 10 and 50 kg N ha^{-1} annually has been reported (Dobereiner et al., 1978; Burris et al., 1978a), although much controversy still remains over some values considerably in excess of these (von Bulow and Dobereiner, 1975; Dobereiner and Day, 1976), which may be due to problems of assay procedures (Stewart et al., 1979; Patriquin, 1982).

Nitrogen-15 incorporation studies are essential in such circumstances. In paddy fields, diazotrophic bacteria together with large numbers of cyanobacteria, both free-living and in association with the fern Azolla, are found, and here associative N_2-fixation may be much more significant to rice plants, with values of about 100 kg N ha^{-1} per crop (Peters and Calvert, 1982; Tirol et al., 1982). A more intimate relationship of associative N_2-fixing bacteria exists in some tropical grasses, such as Digitaria decumbens, maize, and sugar cane, many of which have a different (C_4) carbon metabolism in comparison with most temperate grasses, which have a C_3 type, and may represent primitive mutualistic symbioses (Dobereiner, 1974; Neyra and Dobereiner, 1977; Patriquin, 1982). Under these circumstances, where the bacteria are present within the living zone of the cortical cells, a more effective fixation and transfer of nitrogen may occur. However, the suggestion that C_4 plants such as maize could spare more photosynthate to support greater fixation has been widely questioned (Haystead and Sprent, 1981; Patriquin, 1982). For instance, work on Azotobacter by Jackson and Dawes (1976) has indicated that bacterial growth under N_2-fixing conditions is ATP- rather than carbon-limited because of the low partial pressure of oxygen required, and it could be that Azospirillum would be affected in the same way. Further, experiments of Neal et al. (1973) and Rennie and Larson (1979) with disomic substitution lines of wheat (a C_3 plant) indicated that root rot-susceptible lines (supposedly "leaky" in terms of nutrients), when inoculated with Azospirillum or Bacillus spp., did not fix nitrogen, but root rot-resistant lines, which were not "leaky," did fix N_2 showing again that a greater quantity of available carbon was not a prerequisite for improved N_2 fixation. Also, nitrogenase activity in associative N_2-fixing systems is very susceptible to changes in environmental variables as well as soil and nutritional factors (Balandreau, 1975; Patriquin, 1982) and may be constrained by these factors rather than photosynthate availability. A linked study quantifying carbon release, nitrogen fixation, and microbial numbers, as well as a histological examination locating the position in the rhizosphere where these activities are occurring, is badly needed for at least one C_4 plant, such as maize or a tropical grass, and one C_3 plant, such as wheat, to finally clarify the arguments related to the importance of associative nitrogen fixation to crop plants. However, care should be taken in such inoculation experiments because some bacteria may release growth hormones which directly stimulate plant growth in the absence of detectable N_2 fixation (Brown, 1976) and some inocula may decompose during the experimental period, releasing organic nitrogen, which is taken up without N_2 fixation occurring (Lethbridge and Davidson, 1983b).

Inoculation experiments, particularly in the field, have given inconsistent results with generally no or only slightly significant increases in yield and total nitrogen and often with no correlation to N_2 fixation (e.g., Albrecht et al., 1978; Burris, et al. 1978b; Barber et al., 1979; Rennie, 1980). These small increases in growth may be due to hormonal effects (Brown, 1974; Barea and Brown, 1974; R. L. Smith et al., 1976; Tien et al., 1979; Schank et al., 1981). Inoculation of Azospirillum together with vesicular-arbuscular mycorrhizas may enhance plant growth, particularly in phosphate–deficient soils, but this again may be a hormonal effect (Azcon et al., 1978; Raj et al., 1981; Barea et al., 1983). In view of the complexity of the environmental, nutritional, and soil effects on associative N_2-fixing systems (Patriquin, 1982), it is not surprising that consistent N gains have not been found.

Denitrification within the rhizosphere would be a way in which nitrogen is lost from the plant–microbial zone as N_2 and N_2O (M. S. Smith and Tiedje, 1979; Knowles, 1982), but the loss may be small (Burford et al., 1979). This process does not normally occur in the rhizosphere unless nitrate is added as fertilizer (Woldendorp, 1981), even though denitrifying bacteria may be enhanced in the rhizosphere (Woldendorp, 1963a,b). The significance of microorganisms to nitrogen uptake in crop plants may again be limited under cultural conditions where ample nitrogen is applied as fertilizer.

The passage of nitrogen derived from microbial metabolism into the plant root has seldom been considered. Clarholm (1983) has produced some evidence that protozoa can mediate the reaction (Fig. 2) and this hypothesis is worthy of further evaluation.

4.2.2c. Other Nutrients. Microorganisms may also influence the absorption and uptake of many other macro- and micronutrients. For instance, tomatoes grown aseptically for 3 weeks and then inoculated with *Trichoderma* and *Fusarium* showed a decrease in translocation of

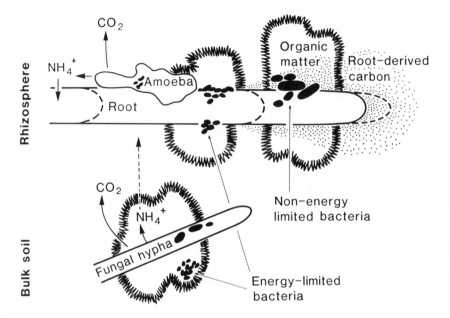

Figure 2. Model of proposed interactions in the rhizosphere and in the bulk soil. A root is growing in the soil from left to right. Under the influence of root-derived carbon (dots) bacteria utilizing organic matter are temporarily not energy-limited and start to mineralize nitrogen from the organic matter, which will be immediately immobilized in an increased bacterial biomass. The pulse of carbon is soon depleted and the bacterial production will be consumed by naked amebae that are attracted to the site. When digesting the bacteria, the protozoa release part of the bacterial nitrogen as ammonium on the root surface, where it can be taken up by the root. Below the root, in the bulk soil, a fungal hypha is decomposing organic material. Ammonium will be released as a waste product and it can diffuse toward the root as ammonium or, after nitrification, as NO_3^-. [From Clarholm (1983), with permission.]

^{35}S-labeled sulfate to the shoots (Subba Rao *et al.,* 1961). At anaerobic microsites on the roots or during anaerobiosis sulfate may be reduced to hydrogen sulfide by microorganisms, which has led to death of maize, citrus, and rice plants, particularly under conditions of high insolation, where exudation is increased (Vamos, 1959; Ford, 1965; Takai and Kamura, 1966; Dommergues *et al.,* 1969; Jacq and Dommergues, 1970). Uptake of potassium (or labeled ^{86}Rb) increased in aseptically grown maize, barley, and wheat seedlings following inoculation with microorganisms, but decreased in red clover (Trolldenier and Markwordt, 1962; Williamson and Wyn Jones, 1973). Interestingly, in a series of detailed experiments, uptake and translocation of manganese by barley roots were unequivocally shown to be increased by the presence of microorganisms irrespective of the fact that the microorganisms associated with the roots absorbed large quantities of manganese (Barber and Lee, 1974a,b). A soluble heat-labile factor produced by the microorganisms and destroyed by dialysis is thought to be responsible for this effect. Such effects may not be so clear in the soil, however. For instance, neutral to high pH and oxidative conditions favor the formation of tri- and tetravalent oxides, which are poorly available to plants, whereas under acid or anaerobic conditions divalent compounds are produced, sometimes to such a level that toxicity may be induced (Lohnis, 1951; R. Jones, 1972). Microorganisms may be able to alter solubility, for instance by pH change, and hence change availability of that nutrient to the plant.

4.2.3. Phytohormones

One of the most common and marked effects of inoculation of aseptically grown plants is an alteration in growth pattern of the roots or the plants as a whole (Swaby, 1942; Bowen and Rovira, 1961; Williamson and Wyn Jones, 1973; Lynch and Clark, 1984) and although many explanations have been put forward for such growth alterations, including changes in nutrient availability, it seems the most likely cause of such a widespread phenomenon is the production of plant growth regulators by rhizosphere organisms or alteration of the plant's own hormonal balance (Brown, 1974, 1975; Lynch, 1976; Hale, 1981). This view has come about because many investigators have shown that cultures of saprophytic rhizosphere microorganisms produce cytokinins, gibberellins, auxins, ethylene, and other substances. When the microorganisms are inoculated onto aseptically grown plants they can cause both increases and decreases of growth. These effects can be simulated by correct dosages of real samples of growth regulator (Brown et al., 1986; Brown, 1972; Barea and Brown, 1974). Hormonal production has also been shown to occur in mutualistic symbionts such as mycorrhiza and rhizobia (Kefford et al., 1960; Miller, 1967; Slankis, 1967; Crafts and Miller, 1974; M. F. Allen et al., 1980, 1982). In theory, such changes in growth or physiology associated with hormonal production should either increase or decrease yield, but no direct test of this hypothesis appears to have been carried out. Numerous attempts at increasing yield by inoculation of seeds or seedlings by appropriate bacterial and fungal cultures have been done in the hope that, by whatever mechanism (hormonal, nutrient availability, or biological control—[see Section 4.3.2a]), yields may be increased. Studies by Okon (1982) in Israel have shown that inoculation with Azospirillum benefited the growth and commercial yield of Zea mays, Sorghum bicolor, Setaria italica, and Panicum miliaceum grown under different environmental and soil conditions, at different levels of combined nitrogen, and in irrigated and nonirrigated plots of commercial size. Although inoculated plants contained more nitrogen than those that were not inoculated, the result does not verify that N_2 fixation was the action of the inoculant. A plausible explanation might be that the inoculants would promote nitrogen uptake by the plant and it is conceivable that a growth regulator effect could be involved. Clearly there is a need for much more fundamental study in the laboratory and in the glasshouse. Until this is accomplished it is probably premature to consider the technology of inoculum production, but Subba Rao (1982b) has given some guidelines.

4.2.4. Phytotoxins

In this section a distinction is made between toxins produced by saprophytic microorganisms in the rhizosphere and toxins produced by true pathogens as part of their disease syndrome (see Section 4.3). In the former category, toxins are produced by bacteria and fungi during their normal growth on substrates either derived from the living plant or from dead plant debris, such as straw. These compounds then deleteriously affect plant growth, but the microorganisms do not directly benefit from the damage to the roots by invading the roots. Such saprophytic microorganisms have been termed pseudopathogens (Drew and Lynch, 1980). Some toxins may have growth regulator effects (e.g., ethylene) or be derived from chemical reaction on minerals (e.g., H_2S), but these have been mentioned earlier and are not considered further.

In recent years much research has been carried out on toxin action; generally they are produced, and cause greatest harm to the plant, under anaerobic or waterlogged conditions, where metabolism of the toxins is restricted. Interpretation of these toxic affects in waterlogged soils is sometimes difficult because anaerobiosis *per se* can produce symptoms typical of waterlogging injury (Willey, 1970; Purvis and Williamson, 1972). Organic acids are commonly formed from decomposing crop residues and can quickly reach phytotoxic concentrations when they come into contact with growing roots (Cho and Ponnamperuma, 1971; Hollis and Rodriquez-Kabana, 1967; Lynch, 1978). In soils, although aromatic and longer chain fatty acids are more toxic than acetic acid in equivalent amounts, acetic acid is usually produced in greatest quantity and is the main cause for the increase in leakage of root cells induced by these toxins (Takijima, 1964; Lee, 1977; Lynch, 1980). Since the acids diffuse only slowly from their site of production, roots must come into contact with decomposing residues for a major effect to take place and in ploughed land this must cause minimal damage. However, in the direct-drilled situation where the drill slits may be lined with trash residues the problem is likely to be more significant.

In comparison to hormones and acids, relatively little is known of the production and action of other nonvolatile toxins, such as phenolics, but a large range of volatile compounds that may act as toxins, such as sulfur based chemicals, alcohols, and terpenes, have been found in soil and these have been reviewed (Lynch, 1976).

4.3. Parasitic Effects

4.3.1. Biology of Infection

Root-infecting pathogens cause damage to the plant and hence cause decreases in crop yield (Russell, 1981). Of a list of 50,000 parasitic and nonparasitic diseases in the U.S.A, only 6.7% are root diseases and of these 64% are root rots, 9% damping off diseases, 7% wilts, 6% seedling blights, 6% crown and foot rots, 3% root browning diseases, 3% root-infecting fungi, and 1% clubroots (Kommedahl and Windels, 1979). Despite such a low percentage of total infections, root infections are still of great economic importance in all parts of the world. The behavior of root pathogens has been well reviewed recently, indicating the upsurge in interest in this area (Garrett, 1970, 1979; Krupa and Dommergues, 1979; Schippers and Gams, 1979; Hornby and Fitt, 1981; Curl, 1982). Nutritionally they exist as necrotrophs, which kill living tissue in advance of themselves, such as root rots or vascular wilts, or as biotrophs, which derive their nutriment from living host cells, such as clubroot *(Plasmodiophora brassicae)* of crucifers, and may have varying capacities for saprophytic growth (Lewis, 1973; R. C. Cooke, 1977). Some may also be hemibiotrophs, since in some infections the point at which infected living cells die is not known (Luttrell, 1974; R. C. Cooke, 1977; R. C. Cooke and Whipps, 1980; Whipps and Lewis, 1981). The majority of root-infecting pathogens infect via wounds or breaks in the root surface (biotrophs are exceptions) and once in, depending on the organism, they may cause localized lesions with little or no root cell death or spread, causing massive tissue destruction or malformation, eventually leading to plant death. These correspond to the host-dominant and pathogen-dominant diseases of Kommedahl and Windels (1979). Virus diseases of plants can be carried by insects and microorganisms as vectors (Teakle and Hirucki, 1985). *Olpidium brassicae* acting as a vector of lettuce big vein virus is the only example of virus transmission by microorganisms that colonize the endorhizosphere and this could be a profitable area for further research. Viruses may form a special interaction with roots (Zeyen, 1979), but are not considered further here.

Aggressive necrotrophic pathogens such as some *Fusarium, Verticillium, Pythium, Phytophthora,* and *Erwinia* species enter the host roots and immediately cause an increase in permeability of host cells together with a general increase in respiration (Wheeler and Hanchey, 1968; Wheeler, 1978). These effects are brought about in a combination of ways. The organism may produce ˙cytolytic, pectolytic, or cellulolytic enzymes, or toxins that physically interfere with membrane integrity, all of which induce metabolite loss from cells (Lai *et al.,* 1968; Calonge *et al.,* 1969; Mount *et al.,* 1970; Wood *et al.,* 1972; Strobel, 1974). Once this has happened, cell death rapidly follows, often well in advance of the pathogen. In this way vascular wilts penetrate the stele and carry on the destruction throughout the plant. Less aggressive necrotrophs may become limited in their spread, either lacking the ability to produce large amounts of cytolytic metabolites or due to host resistance mechanisms, but little work has been done on this aspect of root infection in comparison with infections of the aerial parts.

Biotrophic root pathogens grow within living host tissues with little or no host cell death, at least initially. They may cause massive changes in root/hypocotyl structure, as with clubroot infection, or enter single cortical cells only, as with *Olpidium* spp. and *Ligniera* spp. infections (Salt, 1979). Little is known of the physiology of the latter group, but *Plasmodiophora* has been well studied and has all the attributes of biotrophy, i.e., it induces hyperplasia and hypertrophy of root cells and attracts nutrients mainly due to increased levels of phytohormones in the club, causes increased respiration within infected tissues, and changes nucleic acid metabolism in the parasitized region (Williams, 1966; Williams *et al.,* 1968, 1973; Keen and Williams, 1969; Bhattacharya and Williams, 1971; Dekhuijzen and Overeem, 1971; Butcher *et al.,* 1974).

4.3.2. Control of Disease

4.3.2a. Biocontrol. One of the most exciting areas for future research lies within the realm of biocontrol of plant diseases. At the moment only two biocontrol agents of soil-borne pathogens are commercially in use, *Peniophora gigantea* for control of *Heterobasidium annosum* on conifers (Rishbeth, 1975) and strain 84 of *Agrobacterium radiobacter* for controlling crown gall caused by *A. tumefaciens* (Moore and Warren, 1979), but there are many other possible organisms under investigation for control of pests as well as fungal and bacterial pathogens. Biological control had largely been neglected before the 1970s because of technical problems associated with the work, but interest increased as problems with chemical control and cultural practices became apparent. This has been reflected by the upsurge in the number of reviews on the subject (Baker and Cook, 1974; Bruehl, 1975; Charudattan, 1978; Gindrat, 1979a,b; Mangenot and Diem, 1979; Schroth and Hancock, 1981, 1982; Kerr, 1982; Cook and Baker, 1983; Deacon, 1983; Linderman *et al.,* 1983).

Essentially, biological control strategies can be divided into three basic types. The first is *competition,* where the biological agent competes for nutrients or space, often involving antibiosis or mycostasis. The second is *induced resistance* (hypovirulence), where either a nonpathogenic strain or a closely related species infects the host, sensitizing it to resist subsequent attack by the pathogen. The third is *predation* or *parasitism,* where direct infection and growth by the control agent occurs at the expense of the parasitized pathogen. None of these are mutually exclusive and a successful antagonist may have attributes of each strategy. These processes are carried out by bacteria, actinomycetes, and fungi (Baker and Cook, 1974). Traditionally, because of competition aspects, bacteria have always been assumed to play the major biocontrolling role in the rhizosphere, but the flood of recent research into biological controlling agents in recent years has shown this to be too narrow a view, as indicated in Table II. To obtain a more comprehensive list of biological control organisms, reference should be made to the reviews mentioned above.

1. Competition. The two biological control organisms in commercial use today are based upon the competition strategy. The fungus *Heterobasidium annosum* causes root decay and heart rot of conifers. It infects stumps of felled conifers by both roots and the cut surface of the trunk and it is able to use this as a food base from which infection can spread to healthy roots of living trees. The fungus *Peniophora gigantea* is routinely inoculated onto stumps of felled conifers and this fungus rapidly spreads through the stumps and prevents subsequent infection of the stump by *H. annosum* (Rishbeth, 1975). The second case involves the rhizosphere organism *Agrobacterium tumefaciens,* which causes crown gall of many economically important crops, such as stone fruits, pome

Table II. Recent Examples of Biological Control by Antagonistic Organisms in Natural or Artificial Conditions

Pathogen	Antagonist	Reference
Bacterial diseases		
Agrobacterium tumefaciens	*Agrobacterium radiobacter*	Moore and Warren (1979)
Erwinia carotovora	Mixed bacteria	Kloepper *et al.* (1980c)
Fungal diseases		
Pythium ultimum	*Pythium oligandrum*	Vesely (1977, 1979), Al-Hamdani *et al.* (1983)
Pythium ultimum	*Pseudomonas fluorescens*	Howell and Stipanovic (1980)
Pythium spp.	*Laetisaria* sp.	Martin *et al.* (1983)

(continued)

Table II. (*continued*)

Pathogen	Antagonist	Reference
Pythium sp., *Phytophthora* sp.	Actinomycetes	Keast and Tonkin (1983)
Phytophthora cinnamomi	*Pseudomonas* sp., *Streptomyces* sp.	Malajczuk (1979)
Fusarium oxysporum f. sp. *dianthi*	*Arthrobacter* sp., *Serratia liquifaciens*	Sneh (1981)
F. oxysporum f. sp. *lini*	*Pseudomonas fluorescens, P. putida*	Kloepper *et al.* (1980a), Scher and Baker (1982)
F. oxysporum f. sp. *lini*	*Pseudomonas* sp.	Vandenburgh *et al.* (1983)
F. oxysporum f. sp. *pisi*	*Corynebacterium*	Opgenorth and Endo (1983)
Verticillium sp.	*Penicillium chrysogenum, Trichoderma viride*	Dutta (1981)
Ceratocystis ulmi	*Pseudomonas* sp.	Scheffer (1983)
Gaeumannomyces graminis	*Gaeumannomyces graminis, Phialophora radicicola*	Deacon (1976, 1981), Wong and Southwell (1980), Wong (1981)
G. graminis	*Bacillus mycoides*	Campbell and Faull (1979)
G. graminis	*Pseudomonas fluorescens*	Cook and Rovira (1976), Kloepper *et al.* (1980a)
	Pseudomonas sp.	Weller (1983), Weller and Cook (1983)
G. graminis	*Streptomyces lavendula*	Tschudi and Kern (1979)
G. graminis	Amebae	Chakraborty (1983), Chakraborty *et al.* (1983)
Sclerotinia sclerotiorum	*Sporodesmium sclerotivorum*	Uecker *et al.*, (1978), Ayers and Adams (1979)
Sclerotium cepivorum	*Coniothyrium minitans*	Ahmed and Tribe (1977)
	Bacillus subtilus	Utkhede and Rahe (1983)
Sclerotium rolfsii	*Trichoderma harzianum*	Wells *et al.* (1972), Backman and Rodriguez-Kabana (1975), Chet *et al.* (1979), Abd-El-Moity and Shatla (1981)
Sclerotium rolfsii	*Pseudomonas aeroginosa*	Brathwaite and Cunningham (1982)
Rhizoctonia solani	*Rhizoctonia solani*	Casthano and Butler (1978)
	Laetisaria sp.	Odvody *et al.* (1980)
	Trichoderma harzianum	Chet *et al.* (1979)
Rhizoctonia sp.	*Bacillus* sp., *Streptomyces* sp.	Merriman *et al.* (1974)
	Trichoderma hamatum, T. harzianum	Nelson *et al.* (1983)
Nematode infections		
Aphelenchus avenae	*Nematotoctonus* sp.	Giuma *et al.* (1973)
Dictylenchus myceliophagus	*Arthrobotrys* sp.	Mankau (1980)
Meloidogyne sp.	*Bacillus penetrans*	Stirling and Wachtel (1980)
Meloidogyne sp.	*Dactyella oviparasitica*	Stirling *et al.* (1979)
Heterodera avenae	*Nematophthora gynophila*	Kerry and Crump (1980)

fruits, grapevines, chrysanthemums, and dahlias (Moore and Warren, 1979). It infects through wounds in the roots, and the T-DNA from the tumor-inducing (Ti) plasmid of the bacterium enters the host cell and causes it to divide out of control, which results in galling. The disease is controlled by inoculating roots with strain 84 of *A. radiobacter.* This bacterium produces an antibiotic, Agrocin 84, which kill strains of *A. tumefaciens* that produce unique chemicals called nopalines. Other strains, which produce related chemicals called octapines, are not controlled by Agrocin 84 because the genes coding for sensitivity to Agrocin 84 are localized on the nopaline Ti plasmid. This control system has worked well for over 10 years, but there are now signs of breakdown of control (Moore and Warren, 1979; Panagopoulas *et al.,* 1979). This could be disastrous if the pathogenic strains produced by transconjugation have gained the ability to produce Agrocin 84 and thus become immune to it.

Recently there has been a great deal of interest in certain strains of *Pseudomonas fluorescens* controlling root diseases in suppressive soil in that, if inoculated onto seeds or seedlings, they could control diseases in conducive soil. This came about from the work of Schroth's group who found that strain B10 of *P. fluorescens* produced a siderophore, an iron-chelating agent which binds Fe^{3+} so that it makes iron unavailable to other microorganisms. Pathogens such as *Fusarium oxysporum* and *Gaeumannomyces graminis* thus deprived of iron, were unable to grow and infect their hosts. Addition of 50 μM Fe^{3+} EDTA allowed such suppressive soil to become conducive again (Kloepper *et al.,* 1980b,c; Schroth and Hancock, 1982). *Pseudomonas putida* also controls *Fusarium* wilt by a similar mechanism (Scher and Baker, 1982). A major question that arises is whether the siderophore-producing rhizosphere organisms limit the availability of iron to the plant. There is also evidence that the availability of iron, siderophore production, and phytotoxin production are interrelated in some cases and that this influences the uptake of iron by plants (Marschner, 1978; Hemming *et al.,* 1982). This warrants further study.

Antibiosis and fungistasis are well documented and easily demonstrated phenomena in the soil and rhizosphere (Lockwood, 1977). Generally, fungi, actinomycetes, or bacteria can be shown to produce antibiotics in cultures and from this it has been implied that such factors are important in the soil, but good *direct* evidence for this is sparse. It is likely, however, that any factor that weakens a potential parasite will add to overall biological control because in the rhizosphere the intense competition would then predispose the susceptible pathogen to autolysis. For instance, *Bacillus subtilis,* one of the most widely cited biological control agents has been shown to be effective in the field and the laboratory by antibiotic production (Aldrich and Baker, 1970; Broadbent *et al.,* 1971;

Utkhede and Rahe, 1983), and a strain of *Pseudomonas fluorescens* controls *Pythium ultimum* by production of a specific antibiotic, pyoluteorin, to the extent that *P. fluorescens* has been used as a seed inoculum (Howell and Stipanovic, 1980). A factor intimately linked here is that in the bulk soil many pathogenic spores or resting bodies (e.g., sclerotia) remain dormant until stimulated to germinate by the presence of a root either due to specific compounds or to a general increase in the availability of nutrients, thus overriding any antibiosis present (Griffin and Roth, 1979). Much work remains to be done to clarify the controlling factors of pathogen germination and infection in the rhizosphere.

Ectomycorrhiza and endomycorrhiza are well-documented mutualistic symbioses generally known to improve phosphate and perhaps nitrogen uptake by plants (see Section 4.4.1), but they may also have a competitive biocontrol role. Ectomycorrhiza could act against bacteria, fungi, and nematodes by antibiotic production, by acting as a mechanical barrier to infection, by stimulating synthesis of host antimicrobial compounds, and by changing the rhizosphere microbiota to resist pathogens (Marx, 1975; Schenk, 1981). Vesicular-arbuscular mycorrhizas could also act by some of these mechanisms, but the evidence is sparse and subject to much controversy (Ross, 1972; Krupa and Nylund, 1971; Schonbeck and Dehne, 1977; Schonbeck, 1979; Bartschi *et al.,* 1981; Dehne, 1982; J. H. Graham and Menge, 1982; Zambolin and Schenck, 1983). Recently ectomycorrhizal fungi have been shown to produce siderophores (Szanizslo *et al.,* 1981) and, besides acting as biological control agents in a way similar to *Pseudomonas,* could again aid the plant with its iron nutrition (Cline *et al.,* 1982; Orlando and Neilands, 1982; Powell *et al.,* 1982).

2. Hypovirulence and induced resistance. Hypovirulence and induced resistance are well-documented phenomena for aerial pathogens. For instance, *Endothia parasitica,* which causes chestunut blight or canker (Grente and Sauret, 1969), can be controlled by inoculation with a nonvirulent but highly competitive (hypovirulent) race of the same pathogen, which outgrows the virulent race. Similarly, plants may be given induced resistance by inoculation with an organism that stimulates resistance to a subsequent challenge with a pathogen. Such mechanisms may be important in the phenomenon of take-all decline (TAD), where, after 5 years of continual cropping, the incidence of take-all, caused by *Gaeumannomyces graminis,* decreases to a low, constant amount. Numerous speculations have been made about this (Hornby, 1979, 1983; Deacon, 1981; Rovira and Wildermuth, 1981; Schneider, 1982), but there appear to be two main schools of thought. The first holds that a gradual buildup of antagonistic organisms (fungal, bacterial, and animal), eventually restricts the pathogen. *Pseudomonas* spp. have been particularly implicated here (Smiley, 1979). The second suggests that hypovirulent forms of *G. graminis* develop, or the closely related *Phialophora radicicola* increases in importance, and these infect the roots and cause little or no damage, but induce a defense reaction to pathogenic races of *G. graminis* (Balis, 1970; Wong, 1975; Deacon, 1973, 1976; Sivasithamparam and Parker, 1980). Such control effects are subject to mediation by interaction among the fungus, clay type, and antagonsitic flora (Stotzky and Burns, 1982; Campbell and Ephgrave, 1983).

3. Hyperparasitism or predation. There have been many examinations of possible biocontrol agents that directly parasitize pathogens, and examples are included in Table II [see Gindrat (1979a,b) and Papavizas and Lumsden (1980) for recent reviews]. These range from necrotrophic mycoparasites, such as *Trichoderma harzianum*, which attacks *Rhizoctonia solani* and *Sclerotium rolfsii* (Chet *et al.*, 1979), to pathogenic bacteria, such as *Bacillus mycoides,* which infects *Gaeumannomyces graminis* (Campbell and Faull, 1979), to parasitic amebae on *G. graminis* (Chakraborty, 1983), and bacteria, fungi, and viruses parasitic on nematodes (Mankau, 1980). However, for the majority of biocontrol organisms the detailed interactions have only rarely been elucidated, since obtaining control of disease is generally considered more important than understanding the mechanism. In the examples where the mechanism of biocontrol have been examined in detail, the work has been carried out in laboratory culture and the mode of action found on Petri dishes may not actually be the mechanism occurring in the soil. Here, antibiotic production or competition may be more important than direct parasitism. In many cases, putative biocontrol organisms detected in pot or Petri dish experiments fail when applied in agricultural situations. For instance, inoculation of peas with *Trichoderma hamatum* gave protection from *Pythium* infection in some soils. In soils of low iron content, however, biocontrol was lost due to competition for the available iron by resident fluorescent pseudomonads (Hubbard *et al.*, 1983). Subsequent provision of iron allowed the biocontrol to be effective. This is an interesting comparison to the biocontrol mode of action of the pseudomonads of Schroth's group (Section 4.3.2a1). Potatoes inoculated with "growth-promoting rhizobacteria" did give a large increase in yield in pot trials, but were gradually replaced on the roots through the growing season by the resident microbiota, casting doubt on its application in the field, where competition and stress are likely to be more important (Kloepper *et al.*, 1980a; Kloepper *et al.*, 1980c). Similarly, Howie and Echandi (1983), using potato, obtained smaller increases in yield by inoculation with rhizobacteria in the field compared with pot trials and again the inoculated bacteria decreased in frequency with time. However, there is evidence that such problems of maintaining biocontrol around roots are being overcome (Weller, 1983; Weller and Cook, 1983) and may depend upon a thorough understanding of the initial interaction between the biocontrol agent and the host plant and how this is subsequently related to the manifold other variables, such as soil type and cultivation system.

4.3.2b. Plant Breeding. Plant breeding is a classic strategy to obtain an increase in yield by improving the desired characteristics of the plant. This procedure would, of course, include breeding resistance to plant diseases and has been developed for all the major groups of pathogens that attack crop plants, including fungi, bacteria, mycoplasmas, viruses, and invertebrate, and vertebrate pests (Russell, 1978). Unfortunately, attempts at breeding resistance to many root pathogens have not been as successful or widespread as that for aerial diseases and, although resistant varieties to some soil-borne diseases have been developed (e.g., *Fusarium oxy-sporum,* and *Cochliobolus sativus*), no good resistance to the important cereal pathogen *Gaeumannomyces graminis* has been found (D. G. Jones and Clifford, 1978). It is of interest in this context that chromosome substitution in wheat can give rise to changes in the normal rhizosphere microbiota. In a series of experiments, wheat plants susceptible to *Cochliobolus sativus* had resistance conferred on them by substitution of chromosome 5B from a resistant variety. This changed the quantity and quality of the rhizosphere population in the substituted variety, as well as forming a population of which 20% were antibiotic to *C. sativus,* whereas none had been before (T. G. Atkinson *et al.,* 1974; Larson and Atkinson, 1970; Neal, 1971; Neal *et al.,* 1970, 1973). It was suggested that these changes may be related to changes in exudation pattern. The possibility of manipulation of the rhizosphere microbiota by plant breeding to control disease has been overlooked and it is likely that it is an important component and occurs unobserved in breeding programs, where yield is all that is measured in the end.

Gilmour *et al.* (1978) showed that rice genotype was a factor in the amount of nitrogenase activity that could be associated with roots. The older cultivars generally had the most activity and modern cultivars the least. Nitrogenase activity was probably significant before the extensive use of nitrogen fertilizers and where the plant had to fight for survival ecologically.

Elliott and Lynch (1984) demonstrated that the more recent wheat cultivars bred in the Pacific Northwest of the U.S. appeared to be more prone to invasion by growth-inhibitory pseudomonads, which were recovered from the endorhizosphere of plants growing in the presence of straw residues. Plant breeding is usually done in the absence of straw and this should be considered if the effects of these bacteria are proven to be of general significance. One problem of interpretation is that the effects appear to be highly variable (Elliott and Lynch, 1985). These pseudo-monads contain unstable plasmids (R. Wheatcroft, L. F. Elliott, and J. M. Lynch, unpublished results) and this may contribute to the variabliity of the bacterial effects. Many of the nonpathogenic and pathogenic bacteria of the rhizosphere are expected to contain plasmids (Coplin, 1982) and this is likely to be a most profitable area for study. Can bacteria be introduced into the rhizosphere with stable plasmid characters that will benefit plants?

    4.3.2c. Chemical or Heat Sterilization. Recently a review (Powlson, 1975) and a whole book (Mulder, 1979) have been devoted to the subject of soil disinfestation, and so only a few salient points will be made here concerning the importance of these procedures in controlling root disease. Three methods of soil sterilization are in use: (1) Chemical fumigation, using compounds such as methyl bromide or chloropicrin, which are sterilants in their own right, or compounds such as Dazomet, which decompose in the soil to release the active ingredient. While these are used on fields, their main use is in the glasshouse industry because of cost. They also have the further problem of toxicity to man. (2) Heat treatment, which may take the form of air–steam mixtures at various temperatures or, in special cases, oven drying. Again this is costly and can only be used on high-return crops. A modified form of heat treatment has been developed in Israel, where under high solar radiation the soil is covered by plastic sheeting, which raises the temperature of the soil sufficiently to kill many pathogens (Katan *et al.*, 1976; Grinstein *et al.*, 1979; Pullman *et al.*, 1979). Although this is an excellent technique environmentally, it is restricted to areas of high insolation. (3) γ-Irradiation, has been used under special experimental circumstances, although it is impractical on a commercial scale. All these procedures, besides killing or greatly reducing populations of pathogenic organisms, also lower populations of other microorganisms initially, as well as altering the physical and chemical properties of the soil, particularly by increasing the availability of nutrients. Subsequently, large increases in numbers of microorganisms occur utilizing the available carbon, but these gradually form an equilibrium again. In the majority of treatments, plants grow much better in sterilized soil than in untreated soil, but in some circumstances, where a pathogen returns in the absence of the normal resident microbiota, disease can be much worse (Baker and Cook, 1974). This has led to the procedure of inoculating sterilized soil with a small amount of non-sterilized soil that does not contain pathogens. Similarly, mycorrhizal establishment after inoculation into sterilized soil can be slower due to the greater availability of nutrients (Jakobsen and Anderson, 1982), which would have serious effects for routine production of mycorrhizal stock. These procedures are only satisfactory in the short term, because during exposure to the air and with cultivation the normal microbial equilibrium returns. Thus, sterilization represents a continual financial drain. The glasshouse industry has obviated the problem of soil pathogens to some extent by opting for soil-less growing systems, such as nutrient film, glass fiber block, or "soil-less compost," which are all pathogen-free initially. However, even these systems can have epiphytotics of *Phytophthora* and *Pythium,* for instance, with no protection from a resident antagonistic rhizosphere microbiota. In these situations the plants must be destroyed and the whole growing system sterilized if no chemical or environmental answer can be found. More research is needed into the rhizosphere microbiota of these artificial systems, with particular emphasis on the possibility of inoculation with "beneficial" bacteria and fungi.

    4.3.2d. Cultivation. Cultivation is an all-encompassing term and numerous books have been published on this subject alone (e.g., Davies *et al.,* 1972; Phillips and Young, 1973; Anonymous, 1978; H. P. Allen, 1981). In this section, however, discussion is restricted to those points salient to the control of disease. In agriculture worldwide the main form of control practiced on soil-borne diseases is crop rotation, although in Western countries monoculture is becoming the norm. Crop rotation can have the benefit of decreasing the use of fertilizers and pesticides, and the extra cost of carrying out this type of cultivation is of little consequence in countries where labor is plentiful or the value of the land is very high because of its scarcity; e.g., Switzerland (Gindrat, 1979a). There are examples of disease decline in monoculture, which makes the low continual loss of yield acceptable, for instance, with take-all of cereals (Shipton, 1977) and common scab of potatoes (Weinhold *et al.,* 1964), but for the majority of crops in monoculture a steady increase in disease and a decrease in yield is found. Rotations of 2 or 3 years away from susceptible grasses and cereals are long enough to control foot rot caused by *Cercosporella herpotrichoides* or take-all, as the crop residues containing the pathogen are decomposed by soil microorganisms. These rotations may include other crops or a fallow period. Deep ploughing is useful for the control of some pathogens, such as *Sclerotium rolfsii* and *Sclerotinia sclerotiorum,* where the persistent sclerotia are removed from potential host plants near the soil surface, depriving the fungus of a food base. Other cultivation techniques may cause desiccation of pathogenic propagules (Sequeira, 1958) and direct heat or chemical sterilization can eradicate pathogens, as discussed above (Section 4.3.2c).

    The mechanical procedures, such as ploughing, discing, and harrowing, involved in preparing the planting crops in the soil during rotation can help control disease directly, but the inclusion of organic (and inorganic) amendments during these stages can further increase plant yield by providing nutrients, improving soil structure, and controlling disease.

These additions may consist of animal manure of all kinds or green cover crops, commonly legumes, which provide plenty of available carbon, nitrogen, and micronutrients when incorporated. Such amendments stimulate large populations of microorganisms and, in general, pathogenic propagules are either stimulated to germinate so that they cannot then compete and soon die (Owens *et al.,* 1969), or dormancy is increased so that they are then decomposed by the active microbiota (Adams *et al.,* 1968). Care must be taken in choosing the correct amendment to control a specific disease, however, as leguminous residues, for instance, are unsatisfactory for controlling *Rhizoctonia solani* on rape (Seidel, 1970).

One aspect that should be mentioned here, although not directly involved in soil-borne disease control, is the application of chemicals to soil or foliage, such as fertilizers, herbicides, pesticides, fungicides, or growth regulators. In the majority of cases it is thought likely that they exert little influence on pathogenic organisms in the soil. However, changes in pH in the soil caused by nitrogen in the form of NH_4^+-fertilizer application can decrease the severity of take-all, providing nitrogen in the form of NO_3^- is also available (Smiley, 1975), and urea and other foliar sprays can change the rhizosphere microbiota markedly (Balasubramanian and Rangaswami, 1973; Vaidehi, 1973; Rovira and Davey, 1974; Vrany, 1974) and offer an opportunity for controlling root disease which should be followed up.

4.4. Mutualistic Effects

4.4.1. Mycorrhizas

Mycorrhizal interaction is an area where great research effort has been made in the last decade with a view to increasing plant yield. Fundamental knowledge has been obtained, but only now are some of the benefits of the research being seen in agriculture (Hayman, 1982; Menge, 1983). Mycorrhizal fungi can be simply classified as ecto- or endomycorrhizal, depending on the growth form of the fungus on the root. In the former a distinct fungal sheath or mantle is found, but in the latter, which can be subdivided into ericaceous, orchidaceous, and vesicular-arbuscular mycorrhizas (VAM), growth characteristically occurs within the plant root and a thin fungal sheath is rarely found. Although this is a gross simplification and more detailed ideas on mycorrhizal classification have recently been discussed by Read (1983), only the morphologically distinct ectomycorrhizas and VAM groups are important in most agricultural situations, and so further discussion is limited to these two groups. Ectomycorrhizas are typically formed with trees or other woody species, but VAMs are found in most Phanaerogams and are consequently potentially important both economically and ecologically (Gerdemann, 1968). All aspects of the structure and function of mycorrhizas have been reviewed in great detail (e.g., Harley, 1969; Sanders *et al.,* 1975; Hayman, 1978, 1982; Marx and Krupa, 1978; D. Atkinson *et al.,* 1983; Harley and Smith, 1983; Scannerini and Bonfante-Fasolo, 1983) and so only the properties of these fungi that can influence plant growth are mentioned. It is of some note that ectomycorrhizas and VAM infections, which are so different in morphology, have extremely similar physiological roles.

The most striking feature of mycorrhizal infection in normal ecosystems is that the infected plants tend to grow better than nonmycorrhizal plants, particularly in phosphate-deficient soils (Hayman, 1978; Safir, 1980). In some cases, for instance, many tree species, the ectomycorrhizal association is obligatory for normal plant development to occur. A range of mechanisms are involved with this improved growth, but undoubtedly the most common is that of enhanced phosphate uptake. This is a conglomerate effect of the fungal mycelium in the soil exploiting a greater volume of soil for nutrient uptake than the roots alone could do and accumulating the phosphate within the fungal mycelium, which then acts as a reservoir of phosphate as well as providing a rapid transfer system to the plant (Ali, 1969; Gray and Gerdemann, 1969; Harley, 1969; Sanders and Tinker, 1973; Sanders *et al.,* 1977; Cox *et al.,* 1980; Strullu *et al.,* 1982). It should be remembered, however, that although much of the work of mycorrhizas throughout the world has been done on phosphate-deficient soils and is therefore directly applicable to the agricultural system in that soil, in Western agriculture, where any phosphate deficiency is obviated by fertilizer application, mycorrhizas may not have such an important effect. Mycorrhizas can also increase uptake of other nutrients, such as potassium, sulfate, copper, and zinc, although data on increases in nitrogen uptake are more equivocable (Gerdemann, 1964; Bowen, 1973; Cooper and Tinker, 1978; Lambert *et al.,* 1979; Chambers *et al.,* 1980; Bowen and Smith, 1981; Ames *et al.,* 1983; Buwalda *et al.,* 1983; France and Reid, 1983; Gildon and Tinker, 1983a,b; Harley and Smith, 1983; C. P. P. Reid *et al.,* 1983). Interestingly, mycorrhizas also enable plants to establish and grow in harsh conditions, such as heavily polluted or eroding sites, or areas with high variability of temperature or adverse pH (Schramm, 1966; Marx, 1980; Gildon and Tinker, 1983a,b; Menge, 1983). There are also several reports that mycorrhizal plants are less sensitive to periods of drought than nonmycorrhizal plants and that mycorrhizal plants play a role in the water relations of plants (Safir *et al.,* 1972; Duddridge *et al.,* 1980; M. F. Allen *et al.,* 1981; Hardie and Leyton, 1981; M. F. Allen and Boosalis, 1983; Parke *et al.,* 1983). As mentioned before (Section 4.3.2a), mycorrhizas may also increase disease resistance in plants by several possible mechanisms, one of which (hormone production) may also stimulate plant growth directly (Slankis, 1973; M. F. Allen *et al.*, 1980, 1982). In legumes infected with VAM, increased nodulation and nitrogen fixation results from improved phosphorus nutrition (Schenk and Hinson, 1973; S. E. Smith and Daft, 1977; S. E. Smith *et al.*, 1979; Munns and Mosse, 1980) and this synergistic effect on yield and, thus, protein nitrogen could have real significance for agriculture, emphasizing the need for further research on mycorrhizal–legume relationships.

In return for all these potentially beneficial effects, the plant supplies most if not all the carbon and vitamins needed by the mycorrhizal fungus. Ectomycorrhizas may well obtain some of their carbon requirement from external hyphae decomposing soil organic matter, but generally, by nature of their obligate relationship, ectomycorrhizas and VAMs must obtain the majority of the carbon from the host (Lewis and Harley, 1965a–c; C. P. P. Reid and Woods, 1969; Ho and Trappe, 1973; Bevege *et al.*, 1975; France and Reid, 1983). The importance of the carbon supply is further indicated by the fact that when the available carbohydrate within the plant is decreased, by shading, for instance, mycorrhizal associations tend to become necrotic (Harley, 1969) and, under conditions where nutrients are nonlimiting, VAM infection can decrease plant growth (Mosse, 1973; Sparling and Tinker, 1978; Stribley *et al.*, 1980; Buwalda and Goh, 1982). J. H. Graham *et al.* (1981) have found that VAM infection of sorghum decreases exudation under low-phosphate conditions and this may represent an economy measure of the mycorrhizal plant and could be one of the reasons for the changes in rhizosphere microbiota found around mycorrhizal roots (Rambelli, 1973). However, whether the altered translocation pattern and carbon utilization by the mycorrhizal fungus represents a significant drain on the plant's yield capabilities is the subject of some controversy. Ectomycorrhizas have been estimated to utilize between 25 and 40% of the total net photosynthate of their host trees (Tranquillini, in Harley, 1969; Newman, 1978) and VAMs less than 10% of the total fixed carbon in *Vicia faba* and leeks, although enhanced photosynthesis or increased "leafiness" may offset the effect of the carbon drain in VAM plants (Paul and Kucey, 1981; Kucey and Paul, 1982; Snellgrove *et al.*, 1982). These would appear to be high prices to pay for the benefits discussed above, but in some cases may be acceptable to the plant, as the carbon loss may occur anyway with yearly root turnover (Fogel and Hunt, 1979; Fogel, 1983; St. John and Coleman, 1983) or represent a loss of excess photosynthate because plant growth but not photosynthesis is constrained in some way (Whipps, 1984). It is likely that improvements in crop yield would sooner be obtained if the source–sink relationship together with the values of carbon lost from mycorrhizal roots could be better equated with the beneficial effects of mycorrhizal infection. This would enable subsequent selection of the beneficial trait required on a carbon cost basis. This is particularly important in view of the ease with which many cultural treatments can carry out the role of mycorrhizas in Western agriculture (see Section 4.3.2d). In many agricultural systems mycorrhizal infections may be self-limiting anyway. For instance, Jensen and Jacobsen (1980), who investigated wheat and barley grown in fields with soil containing varying soluble phosphate content, found that shoot phosphate content differed little between sites. On low-phosphate sites mycorrhizal infection was much greater, compensating for low available soluble phosphate in the soil. Although there has been much investigation of the potential of mycorrhizal inoculants (Hayman, 1982; Menge, 1983), the only uses in commercial practice are following fumigation of soils used to grow citrus and in some conifer nurseries. We consider that there are good opportunities for their use with other hardy nursery stock and possibly some protected crops, but more development is required before widespread use of mycorrhizal inoculations can be made sensibly in nutrient-poor sites and sites of reclamation.

4.4.2. Symbiotic Nitrogen Fixation

There are three possible groups of associations between nitrogen-fixing organisms and roots of plants, but this section will be limited to discussion of the *Rhizobium*–legume symbiosis. This is because the nonlegume, N_2-fixing, nodule-forming symbioses, which are restricted to woody trees and shrubs, are not important in most agricultural systems even though they may be members of a dominant flora in some geographic localities (Akkermans, 1978; Becking, 1982). The associative N_2-fixing grass–bacteria relationships have been mentioned earlier under saprophytic effects (Section 4.2.2b), although some authors consider these as primitive mutualistic symbioses (Dobereiner, 1974). Numerous reviews and monographs have been published on all aspects of nitrogen fixation (Bergersen, 1978; Hewitt and Cutting, 1979; Sprent, 1970; Phillips, 1980; Bauer, 1981; Haystead and Sprent, 1981; Atkins and Rainbird, 1982; Beringer, 1982; Gibson and Jordan, 1983; Vance, 1983).

Symbiotic N_2 fixation can amount to several hundred kg N ha^{-1} annually (Sprent, 1979; Phillips, 1980). Most legumes are grown principally in the semiarid tropics either for grain or as forage/pasture species and are commonly used as first line protein in the developing countries, unlike in Western agriculture, where they are often used as animal feedstuffs. The ability to fix N_2 also makes it more advantageous to grow legumes in the poorer countries, where application of large amounts of nitrogenous fertilizers to all crops is impossible economically. In more advanced countries nitrogen fertilizer until recently was inexpensive and gave the farmer flexibility of cropping, which legumes alone do not. Because of this, legumes may not be an economic crop in the West, particularly when the nitrogen that is fixed does not become immediately available, as with pasture species such as clover. However, as energy costs increase, the use of legumes as a crop will presumably increase.

It is unlikely that N_2 fixation alone would provide all the nitrogen requirement of legumes in the field and evidence from experiments with soybean, *Phaseolus vulgaris,* and *Pisum arvense* suggest that the proportion of plant nitrogen derived from nitrate or N_2 fixation differs with growth stage and environmental conditions (Oghoghorie and Pate, 1971, 1972; Harper, 1974; Garcia and Hanway, 1976; P. H. Graham and Halliday, 1977). The process of N_2 fixation has a cost in terms of carbohydrate subsequently not available to the plant for growth, but it is unlikely in most field situations that carbon availability would be the limiting factor of legume growth and productivity. During periods of most vigorous N_2 fixation, values between 4 and 9 g C respired g N^{-1} fixed have been obtained for the nodulated root systems of several legume species, although values outside this range have been recorded, depending on methodology and environmental conditions (Phillips, 1980; Haystead and Sprent, 1981; Minchin *et al.,* 1981; Atkins and Rainbird, 1982). The theoretical minimum requirement for the nitrogenase reaction is about 2 g C g N^{-1} (Bulen and Le Compte, 1966; Hardy and Havelka, 1975) and measurement of nodule respiration alone gives about 3–4 g C g N^{-1} (Ryle *et al.,* 1979), and so, on these grounds, it appears that half the nodule respiration is associated with the nitrogenase reaction and half with NH_4 assimilation and nodule growth and maintenance. Measurements of total carbon distribution in non-nodulated or nodulated plants growing with or without nitrate have indicated that approximately 10% less of net fixed carbon is available for growth in N_2-fixing plants than in non-nodulated plants and this appears to be due solely to supplying the microsymbiont with energy (Haystead and Sprent, 1981; Minchin *et al.,* 1981; Warembourg *et al.,* 1982). These analyses may be further complicated by the presence of mycorrhizal infections on the roots, which can modify the carbon distribution pattern (Pang and Paul, 1980; Paul and Kucey, 1981; Kucey and Paul, 1982).

In order to optimize or improve the growth and efficiency of legumes, numerous suggestions, mostly related to removing known limitations to symbiotic nitrogen fixation, have been made: (1) integration of genetically altered *Rhizobium* strains (Johnston and Beringer, 1979; Beringer, 1982); (2) screening for more efficient rhizobia and those adapted to adverse conditions (e.g., high concentrations of toxic chemicals, low pH, water stress, extremes of temperature); (3) screening for host genotypes with better characteristics for prolonged nitrogen fixation (e.g., delayed leaf senescence, increased photosynthetic efficiency, decreased photorespiration, earlier establishment of root nodules, and extension of the period of N_2 fixation into the pod filling stage), and (4) selection of mycorrhizas and bacteria to enhance plant and nodule growth (Stewart *et al.,* 1979; Phillips, 1980; Azcon-Aguilar and Barea, 1981; Grimes and Mount, 1984). All these ideally should be tailored to fit the crop and prevailing environmental conditions. In reality, however, the best hope for improving legumes in general could be to concentrate on a single or few specific crops, work out the easiest and simplest manipulations that can be done to give *repeatable* increases in yield, and extrapolate from these, rather than working on a broad front on many species. Work at the fundamental level should still carry on, because it is here that any major, novel breakthrough would come.

The preparation and use of artificially produced *Rhizobium* inoculum is well underway throughout the world (Subba Rao, 1982b) and it remains to be seen whether the successes already made in some environments where competition from naturally occurring rhizobia was lacking or where conditions favored the process of inoculation (i.e., new plant species in a new area with alien *Rhizobium* strains) (Nutman, 1975) can be built on in the future. Much fundamental knowledge is already present and it may well be that all that is needed is its application by agronomists to exploit this approach to its full potential.

5. Future Research Directions

The work reviewed in this chapter indicates the important role that rhizosphere microorganisms have in the growth of plants. They are intimately involved in all stages of uptake and release of nutrients by plants, and they modify the immediate soil environment and directly influence plant growth by both chemical effects (e.g., toxin or hormone production), and by forming parasitic and mutualistic symbioses. For each plant and soil, all rhizosphere microorganisms interact with each other and with the plant generally, establishing a slowly changing population during the life of the plant that is affected by environmental variables and cultural practices. It is evident that these interactions are the key to understanding and utilizing rhizosphere biology to improve crops, and from this standpoint only by working across a whole range of traditional research disciplines as a team can advances be made. There will always be a place for the pure plant physiologist, microbiologist, plant pathologist, soil scientist, plant breeder, etc.; but by integrating these spheres of interest in a field situation with the interests of agronomists who understand the everyday problems of growing crops, a complete attack on a specific crop problem can be made. Thus, to obtain the best yields, the correct cropping sequence combined with the correct soil amendment and the correct chemical and biological treatment for any particular environmental situation is required. This is the concept of integrated agriculture and some of the variables are listed in Table III. The chemical and cultural groups of methods are well established and although new chemicals and possible new cropping systems may be developed, it is unlikely that these alone would give any major breakthrough in agriculture. However, in the biological group, research is intense and already rewards for this work are being seen. It is in this area that most scientifically stimulating and financially rewarding results appear to lie.

In the future, the single most important problem to be solved for the biological group is the production of viable inocula capable, when present in the soil, of maintaining the required role in the rhizosphere. This is difficult because the nature of rhizospheres is so varied. If this problem can be solved, the wealth of laboratory and pot experiments with a host of biological agents (mycorrhizas, rhizobia, associative nitrogen-fixers, growth-promoting rhizobacteria, etc.) could be usefully employed. Without these attributes, a microorganism, no matter how successful it might be at its putative role in the laboratory, is of no use agronomically. Perhaps the glasshouse industry is the obvious place to start these studies, because the soil (if used) commonly has a less variable microbial population, due to sterilization techniques, and the environment is controlled to an extent.

Great interest has been generated by the possibility of genetic manipulation of both plants and microorganisms to improve plant growth (Kado and Lurquin, 1982), but to date there has been no clearly proven commercial success of these techniques.

Another tool in the research on these microbial aids to plant growth will be an understanding of the development and maintenance of micro-

Table III. Methods to Increase Plant Growth and Productivity[a]

Chemical	Cultural	Biological
Fungicides	Cultivation (ploughing, etc.)	Disease control
Herbicides	Water	Hormone production
Pesticides	Crop rotation	Increase nutrient uptake
Growth regulators	Mixing crop varieties (species)	Decrease nutrient loss
Disinfectants (+ other sterilization techniques)	Manuring	Breeding
Fertilizers		Bioherbicides

[a]For any plant–soil–environment interrelationship all these controlling factors are interdependent and no one set of methods should be considered on its own.

bial populations around roots. Several mathematical models of the growth of microorganisms in the rhizosphere have been developed with respect to saprophytic and symbiotic organisms (Newman and Watson, 1979; Gilligan, 1979, 1983; Ferriss, 1981, 1983; Buwalda *et al.,* 1982; Drury *et al.,* 1983; Sanders *et al.,* 1983) and a synthesis of these ideas might help predict what will happen to inoculants on roots under a range of conditions. However, implicit in this concept is a greater *quantitative* knowledge of microbial populations in the rhizosphere and their relationship to carbon loss from roots and how the various controlling factors influence these populations. This sort of information (Newman and Watson, 1977; Bowen, 1979, 1980; Whipps and Lynch, 1983) is much needed and is likely to have immediate relevance in crop productivity.

References

Abd-El-Moity, T. H., and Shatla, M. N., 1981, Biological control of white rot disease of onion *(Sclerotium cepivorum)* by *Trichoderma harzianum, Phytopathol. Z.* **100**:29–35.

Adams, P. B., Papavizas, G. C., and Lewis, J. A., 1968, Survival of root-infecting fungi in soil. III. The effect of cellulose amendment on chlamydospore germination of *Fusarium solani* f. sp. *phaseoli* in soil, *Phytopathology* **58**:373–377.

Ahmed, A. H. M., and Tribe, H. T., 1977, Biological control of white rot of onion *(Sclerotium cepivorum)* by *Coniothyrium minitans, Plant Pathol.* **26**:75–78.

Akkermans, A. D. L., 1978, Root nodule symbioses in non-leguminous N$_2$-fixing plants, in: *Interactions between Non-Pathogenic Soil Microorganisms and Plants* (Y. R. Dommergues and S. V. Krupa, eds.), pp. 335–372, Elsevier, Amsterdam.

Albrecht, S. L., Okon, Y., and Burris, R. M., 1978, Effect of light and temperature on the association between *Zea mays* and *Spirillum lipoferum, Plant Physiol.* **60**:528–531.

Aldrich, J., and Baker, R., 1970, Biological control of *Fusarium roseum* f. sp. *dianthi* by *Bacillus subtilis, Plant Disease Reporter* **54**:446–448.

Al-Hamdani, A. M., Lutchmeah, R. S., and Cooke, R. C., 1983, Biological control of *Pythium ultimum*-induced damping-off by treating cress seed with the mycoparasite *Pythium oligandrum, Plant Pathol.* **32**:449–454.

Ali, B., 1969, Cytochemical and autoradiographic studies of mycorrhizal roots of *Nardus, Arch. Mikrobiol.* **68**:236–245.

Allen, H. P., 1981, *Direct Drilling and Reduced Cultivations,* Farming Press, Ipswich.

Allen, M. F., and Boosalis, M. G., 1983, Effects of two species of VA mycorrhizal fungi on drought tolerance of winter wheat, *New Phytol.* **93**:67–76.

Allen, M. F., Moore, T. S.,and Christensen, M., 1980, Phytohormone changes in *Bouteloua gracilis* infected by vesicular-arbuscular mycorrhizae. I. Cytokinin increases in the host plant, *Can. J. Bot.* **58**:371–374.

Allen, M. F., Smith, W. K., Moore, T. S., and Christensen, M., 1981, Comparative water relations and photosynthesis of mycorrhizal and non-mycorrhizal *Bouteloua gracilis* HBK Lag ex Steud, *New Phytol.* **87**:677–685.

Allen, M. F., Moore, T. S., and Christensen, M., 1982, Phytohormone changes in *Bouteloua gracilis* infected by vesicular-arbuscular mycorrhizae. II. Altered levels of gibberellin-like substances and abscisic acid in the host plant, *Can. J. Bot.* **60**:468–471.

Allen, R. N., and Newhook, F. J., 1973, Chemotaxis of zoospores of *Phytopthora cinnamoni* to ethanol in capillaries of soil pore dimensions, *Trans. Br. Mycol. Soc.* **61**:287–302.

Allison, F. E., 1973, *Soil Organic Matter and its Role in Crop Production*, Elsevier, Amsterdam.

Ames, R. N., Reid, C. P. P., Porter, L. K., and Cambardella, C., 1983, Hyphal uptake and transport of nitrogen from two ¹⁵N-labelled sources by *Glomus mosseae*, a vesicular-arbuscular mycorrhizal fungus, *New Phytol.* **95**:381–396.

Anonymous, 1978, *Maximizing Yields of Crops*, Proceedings of a symposium organized by Agricultural Development and Advisory Service and the Agricultural Research Council, HMSO, London.

Atkins, C. A., and Rainbird, R. M., 1982, Physiology and biochemistry of biological nitrogen fixation in legumes, in: *Advances in Agricultural Microbiology* (N. S. Subba Rao, ed.), pp. 25–51, Butterworth Scientific, London.

Atkinson, D., Bhat, K. K. S., Coutts, M. P., Mason, P. A., and Read, D. J. (eds.), 1983, Tree root systems and their mycorrhizas, *Plant Soil* **71**:1–525.

Atkinson, T. G., Neal, J. L., and Larson, R. I., 1974, Root rot reaction in wheat: Resistance not mediated by rhizosphere or laimosphere antagonists, *Phytopathology* **64**:97–101.

Ayers, W. A., and Adams, P. B., 1979, Mycoparasitism of sclerotia of *Sclerotinia* and *Sclerotium* species by *Sporodesmium sclerotivorum*, *Can. J. Microbiol.* **25**:17–23.

Ayers, W. A., and Thornton, R. M., 1968, Exudation of amino acids by intact and damaged roots of wheat and peas, *Plant Soil* **28**:193–207.

Azcon, R., Azcon-G de Aguilar, C., and Barea, J. M., 1978, Effects of plant hormones present in bacterial cultures on the formation and responses to VA endomycorrhizas, *New Phytol.* **80**:359–364.

Azcon-Aguilar, C., and Barea, J. M., 1981, Field inoculation of *Medicago* with VA mycorrhizae and *Rhizobium* in phosphate-fixing agricultural soil, *Soil Biol. Biochem.* **13**:19–22.

Backman, P. A., and Rodriguez-Kabana, R. A., 1975, A system for the growth and delivery of biological control agents to the soil, *Phytopathology* **65**:819–821.

Baker, K. F., and Cook, R. J., 1974, *Biological Control of Plant Pathogens*, Freeman, San Fransisco.

Balandreau, J., 1975, Mesure de l'activité nitrogénasique des microorganismes fixateurs libres d'azote de la rhizosphere de quelques graminées, *Rev. Ecol. Biol. Sol* **12**:273–290.

Balandreau, J., and Knowles, R., 1978, The rhizosphere, in: *Interactions between Non-Pathogenic Soil Microorganisms and Plants* (Y. R. Dommergues and S. V. Krupa, eds.), pp. 243–268, Elsevier, Amsterdam.

Balasubramanian, A., and Rangaswami, G., 1969, Studies on the influence of foliar nutrient sprays on the root exudation pattern in four crop plants, *Plant Soil* **30**:210–220.

Balasubramanian, A., and Rangaswami, G., 1973, Influence of foliar application of chemicals on the root exudation and rhizosphere microflora of *Sorghum vulgare* and *Crotalaria juncea*, *Folia Microbiol.* **18**:492–498.

Balis, C., 1970, A comparative study of *Phialophara radicicola*, an avirulent fungal root parasite of grasses and cereals, *Ann. Appl. Biol.* **66**:59–73.

Barber, D. A., 1971, The influence of microorgansims on the assimilation of nitrogen by plants from soil and fertilizer sources, in: *Nitrogen-15 in Soil–Plant Studies*, pp. 91–101, International Atomic Energy Authority, Vienna.

Barber, D. A., 1978, Nutrient uptake, in: *Interactions between Non-Pathogenic Soil Microorganisms and Plants* (Y. R. Dommergues and S. V. Krupa, eds.), pp. 131–162, Elsevier, Amsterdam.

Barber, D. A., and Gunn, K. B., 1974, The effect of mechanical forces on the exudation of organic substances by roots of cereal plants grown under sterile conditions, *New Phytol.* **73**:39–45.

Barber, D. A., and Lee, R. B., 1974a, The effect of microorganisms on the absorption of manganese by plants, *New Phytol.* **73**:97–106.

Barber, D. A., and Lee, R. B., 1974b, Effects of microbial products on the absorption of manganese by barley, *Agric. Res. Council Letcombe Lab. Annu. Rep. 1973*, **1974**:31–33.

Barber, D. A., and Lynch, J. M., 1977, Microbial growth in the rhizosphere, *Soil Biol. Biochem.* **9**:305–308.

Barber, D. A., and Martin, J. K., 1976, The release of organic substances by cereal roots into soil, *New Phytol.* **76**:69–80.

Barber, D. A., Bowen, G. D., and Rovira, A. D., 1976, Effects of microorganisms on absorption and distribution of phosphate in barley, *Aust. J. Plant Physiol.* **3**:801–808.

Barber, L. E., Russell, S. A., and Evans, H. J., 1979, Inoculation of millet with *Azospirillum*, *Plant Soil* **52**:49–57.

Barea, J. M., and Brown, M. E., 1974, Effects on plant growth produced by *Azotobacter paspali* related to synthesis of plant growth regulating substances, *J Appl. Bacteriol.* **40**:583–593.

Barea, J. M., Bonis, A. F., and Olivares, J., 1983, Interactions between *Azospirillum* and VA mycorrhiza and their effects on growth and nutrition of maize and ryegrass, *Soil Biol. Biochem.* **15**:705–709.

Bartschi, H., Gianinazzi-Pearson, V., and Vegh, I., 1981, Vesicular-arbuscular mycorrhiza formation and root rot disease *(Phytophthora cinnamomi)* development in *Chamaecyparis lawsoniana, Phytopathol. Z.* **102**:213–218.

Bauer, W. D., 1981, Infection of legumes by rhizobia, *Annu. Rev. Plant Physiol.* **32**:407–449.

Beck, S. M., and Gilmour, C. M., 1983, Role of wheat root exudates in associative nitrogen fixation, *Soil Biol. Biochem.* **15**:33–38.

Becking, J. H., 1982, Nitrogen fixation in nodulated plants other than legumes, in: *Advances in Agricultural Microbiology* (N. S. Subba Rao, ed.), pp. 89–110, Butterworth Scientific, London.

Benians, G. J., and Barber, D. A., 1974, The uptake of phosphate by barley plants from soil under aseptic and non-sterile conditions, *Soil Biol. Biochem.* **6**:195–200.

Bergersen, F. J., 1978, Physiology of legume symbiosis, in: *Interactions between Non-Pathogenic Soil Microorganisms and Plants* (Y. R. Dommergues and S. V. Krupa, eds.), pp. 305–333, Elsevier, Oxford.

Beringer, J. E., 1982, Microbial genetics and biological nitrogen fixation, in: *Advances in Agricultural Microbiology* (N. S. Subba Rao, ed.), pp. 3–23, Butterworth Scientific, London.

Beute, M. K., and Lockwood, J. L., 1968, Mechanism of increased root rot in virus-infected peas, *Phytopathology* **58**:1643–1651.

Bevege, D. I., Bowen, G. D., and Skinner, M. F., 1975, Comparative carbohydrate physiology of ecto- and endomycorrhizas, in: *Endomycorrhizas* (F. E. Sanders, B. Mosse, and P. B. Tinker, eds.), pp. 149–174, Academic Press, London.

Bhat, J. V., Limaye, K. S., and Vasantharajan, V. N., 1971, The effect of the leaf surface microflora on the growth and root exudation of plants, in: *Ecology of Leaf Surface Microorganisms* (T. F. Preece and C. H. Dickinson, eds.), pp. 581–595, Academic Press, London.

Bhattacharya, P. K., and Williams, P. H., 1971, Microfluorometric quantitation of nuclear proteins and nucleic acids in cabbage root hair cells infected by *Plasmodiophora brassicae, Physiol. Plant Pathol.* **1**:167–175.

Boero, G., and Thien, S., 1979, Phosphatase activity and phosphorus availability in the rhizosphere of corn roots, in: *The Soil–Root Interface* (J. L. Harley and R. S. Russell, eds.), pp. 231–242, Academic Press, London.

Bonish, P. M., 1973, Cellulase and red clover exudates, *Plant Soil* **38**:307–314.

Bowen, G. D. 1969, Nutrient status effects on loss of amides and amino acids from pine roots, *Nature (London)* **211**:665–666.

Bowen, G. D., 1973, Mineral nutrition of ectomycorrhizae, in: *Ectomycorrhizae* (G. C. Marks and T. T. Kozlowski, eds.), pp. 151–203, Academic Press, London.

Bowen, G. D., 1979, Integrated and experimental approaches to the study of growth of organisms around roots, in: *Soil-Borne Plant Pathogens* (B. Schippers and W. Gams, eds.), pp. 209–227, Academic Press, London.

Bowen, G. D., 1980, Misconceptions, concepts and approaches in rhizosphere biology, in: *Contemporary Microbial Ecology* (D. C. Ellwood, J. N. Hedger, M. J. Latham, J. M. Lynch, and J. M. Slater, eds.), pp. 283–304, Academic Press, London.

Bowen, G. D., 1982, The root–micoorganism ecosystem, in: *Biological and Chemical Interactions in the Rhizosphere, Proceedings of a Symposium of Swedish Natural Science Research Council 1981*, pp. 3–42, Sudt Offset, Stockholm.

Bowen, G. D., and Foster, R. C., 1978, Dynamics of microbial colonization of plant roots, in *Proceedings Symposium on Soil Microbiology and Plant Nutrition* (W. J. Broughton and C. K. John, eds.), pp. 231–256, University Press, Malaysia.

Bowen, G. D., and Rovira, A. D., 1961, The effects of micro-organisms on plant growth. 1. Development of roots and root hairs in sand and agar, *Plant Soil* **15**:166–188.

Bowen, G. D., and Smith, S. E., 1981, The effects of mycorrhizas on nitrogen uptake by plants, in: *Terrestrial Nitrogen Cycles: Processes, Ecosystem Strategies and Management Impacts* (F. E. Clark and T. Rosswall, eds.), Bulletin No. 33, pp. 237–247, Swedish Natural Science Research Council, Stockholm.

Brathwaite, C. W. D., and Cunningham, H. G. A., 1982, Inhibition of *Sclerotium rolfsii* by *Pseudomonas aeruginosa, Can. J. Bot.* **60**:237–239.

Broadbent, P., Baker, K. F., and Waterworth, Y., 1971, Bacteria and actinomycetes antagonistic to fungal root pathogens in Australian soils, *Aust. J. Biol. Sci.* **24**:925–944.

Brown, M. E., 1972, Plant growth substances produced by micro-organisms of soil rhizosphere, *J. Appl. Bacteriol.* **35**:443–451.

Brown, M. E., 1974, Seed and root bacterization, *Annu. Rev. Phytopathol.* **12**:181–197.

Brown, M. E., 1975, Rhizosphere microorganisms—Opportunists, bandits or benefactors, in: *Soil Microbiology* (N. Walker, ed.), pp. 21–38, Butterworth Scientific, London.

Brown, M. E., 1976, Role of *Azotobacter paspali* in association with *Paspalum notatum, J. Appl. Bacteriol.* **40**:341–348.

Brown, M. E., Jackson, R. M., and Burlingham, S. K., 1968, Effects produced on tomato plants, *Lycopersicum esculentum*, by seed or root treatment with gibberellic acid and indol-3-yl-acetic acid, *J. Exp. Bot.* **19**:544–552.

Bruehl, G. W. (ed.), 1975, *Biology and Control of Soil-Borne Plant Pathogens*, American Phytopathological Society, St. Paul, Minnesota.

Bulen, W. A., and Le Compte, J. R., 1966, The nitrogenase system from *Azotobacter:* Two-enzyme requirement for N_2 reduction, ATP-dependent H_2 evolution and ATP hydrolysis. *Proc. Natl. Acad. Sci. USA* **56**:979–986.

Burford, J. R., Dowdell, R. J., Crees, R., and Hall, K. C., 1979, Soil aeration and denitrification, *Agric. Res. Council Letcombe Lab. Annu. Rep. 1978,* **1979**:26.

Burns, R. G., 1981, Microbial adhesion to soil surfaces; Consequences for growth and enzyme activities, in: *Microbial Adhesion to Surfaces* (R. C. W. Berkeley, J. M. Lynch, J. Melling, P. R. Rutter, and B. Vincent, eds.), pp. 249–262, Ellis Horwood, Chichester.

Burris, R. H., Albrecht, S. L., and Okon, Y., 1978a, Physiology and biochemistry of *Spirillum lipoferum*, in: *Proceedings of the International Symposium on the Limitations and Potentials of Biological Nitrogen Fixation in the Tropics, Brazil* (J. Dobereiner, R. H. Burris, and A. Hollaender, eds.), pp. 303–315, Plenum Press, New York.

Burris, R. H., Okon, Y., and Albrecht, S. L., 1978b, Properties and reactions of *Spirillum lipoferum, Ecol. Bull.* **26**:353–363.

Butcher, D. N., El-Tigani, S., and Ingram, D. S., 1974, The role of indole glucosinolates in the clubroot disease of the cruciferae, *Physiol. Plant Pathol.* **4**:127–140.

Buwalda, J. G., and Goh, K. M., 1982, Host–fungus competition for carbon as a cause of growth depression in vesicular-arbuscular mycorrhizal ryegrass, *Soil Biol. Biochem.* **14**:103–106.

Buwalda, J. G., Ross, G. J. S., Stribley, D. P., and Tinker, P. B., 1982, The development of endomycorrhizal root systems III. The mathematical representation of the spread of vesicular-arbuscular mycorrhizal infection in root systems, *New Phytol.* **91**:669–682.

Buwalda, J. G., Stribley, D. P., and Tinker, P. B., 1983, Increased uptake of anions by plants with vesicular-arbuscular mycorrhizas, *Plant Soil* **71**:463–467.

Calonge, F. O., Fielding, S. M., Byrde, R. J. W., and Akinrefon, O. A., 1969, Changes in ultrastructure following fungal invasion and the possible relevance of extracellular enzymes, *J. Exp. Bot.* **20**:350–357.

Campbell, R., and Ephgrave, J. M., 1983, Effect of bentonite clay on the growth of *Gaeumannomyces graminis* var. *tritici* and on its interactions with antagonistic bacteria, *J. Gen. Microbiol.* **129**:771–778.

Campbell, R., and Faull, J. L., 1979, Biological control of *Gaeumannomyces graminis:* Field trials and the ultrastructure of the interaction between the fungus and a successful antagonistic bacterium, in: *Soil-Borne Plant Pathogens* (B. Schippers and W. Gams, eds.), pp. 603–609, Academic Press, London.

Castanho, B., and Butler, E. E., 1978, *Rhizoctonia* decline: Studies on hypovirulence and potential use in biological control, *Phytopathology* **68**:1511–1514.

Chakraborty, S., 1983, Population dynamics of amoebae in soils suppressive and non-suppressive to wheat take-all, *Soil Biol. Biochem.* **15**:661–664.

Chakroborty, S., Old, K. M., and Warcup, J. H., 1983, Amoebae from take-all suppressive soil which feeds on *Gaeumannomyces graminis tritici* and other soil fungi, *Soil Biol. Biochem.* **15**:17–24.

Chambers, C. A., Smith, S. E., and Smith, F. A., 1980, Effects of ammonium and nitrate ions on mycorrhizal infection, nodulation and growth of *Trifolium subterraneum, New Phytol.* **85**:47–62.

Charudattan, R., 1978, *Biological Control Projects in Plant Pathology—A Directory,* Institute of Food and Agricultural Sciences, University of Florida, Tampa.

Cheshire, M. V., 1979, *Nature and Origin of Carbohydrates in Soils,* Academic Press, London.

Chet, I., Hadar, Y., Elad, Y., Katan J., and Henis, Y., 1979, Biological control of soil-borne plant pathogens by *Trichoderma harzianum,* in: *Soil-Borne Plant Pathogens* (B. Schippers and W. Gams, eds.), pp. 585–591, Academic Press, London.

Cho, D. Y., and Ponnamperuma, F. N., 1971, Influence of soil temperature on the chemical kinetics of flooded soils and the growth of rice, *Soil Sci.* **112**:184–194.

Clarholm, M., 1983, Dynamics of Soil Bacteria in Relation to Plants, Protozoa and Inorganic Nitrogen, Report no. 17, Department of Microbiology, Swedish University of Agricultural Sciences, Uppsala.

Clark, A. L., Greenland, D. J., and Quirk, J. P., 1967, Changes in some physical properties of the surface of an impoverished red-brown earth under pasture, *Aust. J. Soil Res.* **5**:59–68.

Cline, G. R., Powell, P. E., Szaniszlo, P. J., and Reid, C. P. P., 1982, Comparison of the abilities of hydroxamic, synthetic, and other natural organic acids to chelate iron and other ions in nutrient solution, *Soil Sci. Soc. Am. J.* **46**:1158–1164.

Clowes, F. A. L., 1971, The proportion of cells that divide in root meristems of *Zea mays* L., *Ann. Bot.* **35**:249–261.

Cook, R. J., and Baker, K. F., 1983, *The Nature and Practice of Biological Control of Plant Pathogens,* American Phytopathological Society, St. Paul, Minnesota.

Cook. R. J., and Rovira, A. D., 1976, The role of bacteria in the biological control of *Gaeumannomyces graminis* by suppressive soils, *Soil Biol. Biochem.* **8**:269–273.

Cooke, G. W., and Williams, R. J. B., 1972, Problems with cultivation and soil structure at Saxmundham, in: *Rothamsted Report 1971, Part 2,* pp. 122–142, Lawes Agricultural Trust, Harpenden, U.K.

Cooke, R. C., 1977, *The Biology of Symbiotic Fungi,* Wiley, London.

Cooke, R. C., and Whipps, J. M., 1980, The evolution of modes of nutrition in fungi parasitic on terrestrial plants, *Biol. Rev.* **55**:341–362.

Cooper, K. M., and Tinker, P. B., 1978, Translocation and transfer of nutrients in vesicular-arbuscular mycorrhizas II. Uptake and translocation of phosphorus, zinc and sulphur, *New Phytol.* **81**:43–52.

Coplin, D. L., 1982, Plasmids in plant pathogenic bacteria, in: *Phytopathogenic Prokaryotes,* Vol. 2 (M. S. Mount and G. H. Lacy, eds.), pp. 255–280, Academic Press, New York.

Coupland, D., and Caseley, J. C., 1979, Presence of ^{14}C activity in root exudates and guttation fluid from *Agropyron repens* treated with ^{14}C-labelled glyphosate, *New Phytol.* **83**:17–22.

Cox, G., Moran, K. J., Sanders, F., Nockolds, C., and Tinker, P. B., 1980, Translocation and transfer of nutrient in vesicular-arbuscular mycorrhizas III, Polyphosphate granules and phosphorus translocation, *New Phytol.* **84**:645–659.

Crafts, C. B., and Miller, C. D., 1974, Detection and identification of cytokinins produced by mycorrhizal fungi, *Plant Physiol.* **54**:586–588.

Curl, E. A., 1982, The rhizosphere: Relation to pathogen behavior and root disease, *Plant Dis.* **66**:624–630.

Darbyshire, J. F., and Greaves, M. P., 1973, Bacteria and protozoa in the rhizosphere, *Pestic. Sci.* **4**:349–360.

D'Arcy, A. L., 1982, Etude des exsudats racinaires de Soja et de Lentille 1. Cinetique d'exsudation des composés phénoliques, des amino acides et des sucres, au cours des premiers jours de la vie des plantules, *Plant Soil* **68**:399–403.

Davies, D. B., Eagle, D. J., and Finney, J. B., 1972, *Soil Management,* Farming Press, Ipswich.

Dazzo, F. B., 1980, Microbial adhesion to plant surfaces, in: *Microbial Adhesion to Surfaces* (J. M. Lynch, J. Melling, P. R. Rutter, and B. Vincent, eds.), pp. 311–328, Ellis Horwood, Chichester.

Deacon, J. W., 1973, *Phialophora radicicola* and *Gaeumannomyces graminis* on roots of grasses and cereals, *Trans. Br. Mycol. Soc.* **61**:471–485.

Deacon, J. W., 1976, Biological control of the take-all fungus *Gaeumannomyces graminis* by *Phialophora radicicola* and similar fungi, *Soil Biol. Biochem.* **8**:275–283.

Deacon, J. W., 1981, Ecological relationships with other fungi—Competitors and hyperparasites, in: *Biology and Control of Take-all* (M. J. C. Asher and P. J. Shipton, eds.), pp. 75–101, Academic Press, London.

Deacon, J. W., 1983, *Microbial and Plant Pests and Diseases,* Van Nostrand Reinhold, Workingham, U.K.

Dehne, M. W., 1982, Interaction betwen vesicular-arbuscular mycorrhizal fungi and plant pathogens, *Phytopathology* **72**:1115–1119.

Dekhuijzen, H. M., and Overeem, J. C., 1971, The role of cytokinins in clubroot formation, *Physiol. Plant Pathol.* **1**:151–161.

Dobereiner, J., 1974, Nitrogen-fixing bacteria in the rhizosphere, in: *The Biology of Nitrogen Fixation* (A. Quispel, ed.), pp. 26–120, North-Holland, Amsterdam.

Dobereiner, J., and Day, J. M., 1976, Associative symbioses in tropical grasses: Characterization of micro-organisms and dinitrogen-fixing sites, in *Proceedings 1st International Symposium on Nitrogen Fixation* (W. E. Newton and C. J. Nyman, eds.), pp. 518–538, Washington State University Press, Pullman, Washington.

Dobereiner, J., Burris, R. H., Hollaender, A., Franco, A. A., Neyra, C. A., and Scott, D. B. (eds.), 1978, *Proceedings of the International Symposium on the Limitations and Potentials of Biological Nitrogen Fixation in the Tropics, Brazil,* Plenum Press, New York.

Dommergues, Y., Combremont, R., Beck, G., and Ollat, C., 1969, Note préliminaire concernant la sulfato-réduction rhizospherique dans un sol salin tunisien, *Rev. Ecol. Biol. Sol* **6:**115–129.

Drew, M. C., and Lynch, J. M., 1980, Soil anaerobiosis, microorganisms, and root function, *Annu. Rev. Phytopathol.* **18:**37–66.

Drury, R. E., Baker, R., and Griffin, G. J., 1983, Calculating the dimensions of the rhizosphere, *Phytopathology* **73:**1351–1354.

Duddridge, J., Malibari, A., and Read, D. J., 1980, Structure and function of mycelial rhizomorphs with special reference to their role in water transport, *Nature* **287:**834–836.

Duff, R. B., Webley, D. M., and Scott, R. O., 1963, Solubilization of minerals and related materials by 2-ketogluconic acid-producing bacteria, *Soil Sci.* **95:**105–114.

Dutta, B. K., 1981, Studies on some fungi isolated from the rhizosphere of tomato plants and the consequent prospect for the control of *Verticillium* wilt, *Plant Soil* **63:**209–216.

Elliott, L. F., and Lynch, J. M., 1984, Pseudomonads as a factor in the growth of winter wheat (*Triticum aestivum* L.), *Soil Biol. Biochem.* **16:**69–71.

Elliott, L. F., and Lynch, J. M., 1985, Plant growth-inhibitory pseudomonads colonizing winter wheat (*Triticum aestivum* L.) roots, *Plant Soil* **84:**57–65.

Ferriss, R. S., 1981, Calculating rhizosphere volume, *Phytopathology* **71:**1229–1231.

Ferriss, R. S., 1983, Calculating the dimensions of the rhizosphere—A response, *Phytopathology* **73:**1355–1357.

Fletcher, M. F., Latham, M. J., Lynch, J. M., and Rutter, P. R., 1980, Characteristics of interfaces and their role in microbial attachment, in: *Microbial Adhesion to Surfaces* (R. C. W. Berkeley, J. M. Lynch, J. Melling, P. R. Rutter, and B. Vincent, eds.), pp. 67–78, Ellis Harwood, Chichester.

Fogel, R., 1983, Root turnover and productivity of coniferous forests, *Plant Soil* **71:**75–85.

Fogel, R., and Hunt, G., 1979, Fungal and arboreal biomass in a western Oregon Douglas fir ecosystem: Distribution patterns and turnover, *Can. J. For. Res.* **9:**245–256.

Ford, H. W., 1965, By-products from bacteria are toxic to citrus roots under flooded conditions, *Florida Field Rep.* **4:**8–12.

Foster, R. C., 1981, The ultrastructure and histochemistry of the rhizosphere, *New Phytol.* **89:**263–273.

Foster, R. C., 1982, The fine structure of epidermal cell mucilages of roots, *New Phytol.* **91:**727–740.

Foster, R. C., and Bowen, G. D., 1982, Plant surfaces and bacterial growth: The rhizosphere and rhizoplane, in: *Phytopathogenic Prokaryotes,* Vol. 1 (M. S. Mount and G. H. Lacy, eds.), pp. 159–185, Academic Press, New York.

Foster, R. C., Rovira, A. D., and Cock, T. W., 1983, *Ultrastructure of the Root–Soil Interface,* American Phytopathological Society, St. Paul, Minnesota.

Foy, C. L., Hurt, W., and Hale, M. G., 1971, Root exudation of plant growth regulators, in: *Biochemical Interactions among Plants,* pp. 75–85, National Academy of Science, Washington D.C.

France, R. C., and Reid, C. P. P., 1983, Interactions of nitrogen and carbon in the physiology of ectomycorrhizae, *Can. J. Bot.* **61**:964–984.

Frič, F., 1975, Translocation of ^{14}C-labelled assimilates in barley plants infected with powdery mildew (*Erysiphe graminis* f. sp. *hordei* Marchal), *Phytopathol. Z.* **84**:88–95.

Garcia, L. R., and Hanway, J. J., 1976, Foliar fertilization of sybeans during the seed-filling period, *Agron. J.* **68**:653–657.

Garrett, S. D., 1970, *Pathogenic Root-Infecting Fungi*, Cambridge University Press, Cambridge.

Garrett, S. D., 1979, The soil–root interface in relation to disease, in: *The Soil–Root Interface* (J. L. Harley and R. S. Russell, eds.), pp. 301–313, Academic Press, London.

Gaworzewska, E. T., and Carlile, M. J., 1982, Positive chemotaxis of *Rhizobium leguminosarum* and other bacteria towards root exudates of legumes and other plants, *J. Gen. Microbiol.* **128**:1179–1188.

Gerdemann, J. W., 1964, The effect of mycorrhizas on the growth of maize *Mycologia* **56**:342–349.

Gerdemann, J. W., 1968, Vesicular-arbuscular mycorrhiza and plant growth, *Annu. Rev. Phytopathol.* **6**:397–418.

Gerretsen, F. C., 1948, The influence of microorganisms on the phosphate intake by the plant, *Plant Soil* **1**:51–85.

Gibson, A. H., and Jordon, D. C., 1983, Ecophysiology of nitrogen-fixing systems, in: *Encyclopedia of Plant Physiology*, Vol. 12C. *Physiological Plant Ecology* III. *Responses to the Chemical and Biological Environment* (O. L. Lange, P. S. Nobel, C. B. Osmund, and M. Ziegler, eds.), pp. 301–390, Springer-Verlag, Berlin.

Gildon, A., and Tinker, P. B., 1983a, Interactions of vesicular-arbuscular mycorrhizal infection and heavy metals in plants. I. The effects of heavy metals on the development of vesicular-arbuscular mycorrhizas, *New Phytol.* **95**:247–261.

Gildon, A., and Tinker, P. B., 1983b, Interactions of vesicular-arbuscular mycorrhizal infections and heavy metals in plants. II. The effects of infection on uptake of copper, *New Phytol.* **95**:262–268.

Gilligan, C. A., 1979, Modelling rhizosphere infection, *Phytopathology* **69**:782–784.

Gilligan, C. A., 1983, Modelling of soil-borne pathogens, *Annu. Rev. Phytopathol.* **21**:45–64.

Gilmour, J. T., Gilmour, C. M., and Johnston, T. H., 1978, Nitrogenase activity in rice plant root systems, *Soil Biol. Biochem.* **10**:261–264.

Gindrat, D., 1979a, Biocontrol of plant disease by inoculation of fresh wounds, seeds and soil with antagonists, in: *Soil-Borne Plant Pathogens* (B. Schippers and W. Gams, eds.), pp. 537–551, Academic Press, London.

Gindrat, D., 1979b, Biological soil disinfestation, in: *Soil Disinfestation* (D. Mulder, ed.), pp. 253–287, Elsevier, Amsterdam.

Giuma, A. Y., Hackett, A. M., and Cooke, R. C., 1973, Thermostable nematotoxins produced by germinating conidia of some endozoic fungi, *Trans. Br. Mycol. Soc.* **60**:49–56.

Goss, M. J., and Reid, J. B., 1979, Influence of perennial ryegrass roots on aggregate stability, *Agric. Res. Council Letcombe Lab. Annu. Rep. 1978*, **1979**:24–25.

Graham, J. H., and Menge, J. A., 1982, Influence of vesicular-arbuscular mycorrhizae and soil phosphate on take-all disease of wheat, *Phytopathology* **72**:95–98.

Graham, J. H., Leonard, R. T., and Menge, J. A., 1981, Membrane-mediated decrease in root exudation responsible for phosphorus inhibition of vesicular-arbuscular mycorrhiza formation, *Plant Physiol.* **68**:548–552.

Graham, P. H., and Halliday, J., 1977, Inoculation and nitrogen fixation in the genus *Phaseolus*, in: Exploiting the Legume—*Rhizobium* Symbiosis in Tropical Agriculture (J. M.

Vincent, A. S. Whitney and J. Bose, eds.), *Misc. Publ. College Tropical Agric. Univ. Hawaii* **145**:313–334.

Gray, L. E., and Gerdemann, J. W., 1969, Uptake of phosphorus-32 by vesicular-arbuscular mycorrhizae, *Plant Soil* **30**:415–422.

Grente, J., and Sauret S., 1969, L'hypovirulence exclusive phénomène original en pathologie végétale, *C. R. Acad. Sci. Paris* **268**:2347–2350.

Griffin, G. J., and Roth, D. A., 1979, Nutritional aspects of soil mycostasis, in: *Soil-Borne Plant Pathogens* (B. Schippers and W. Gams, eds.), pp. 79–96, Academic Press, London.

Grimes, H. D.,and Mount, M. S., 1984, Infuence of *Pseudomonas putida* on nodulation of *Phaseolus vulgaris, Soil Biol. Biochem.* **16**:27–30.

Grineva, G. M., 1961, Excretion by plant roots during brief periods of anaerobiosis, *Sov. Plant Physiol.* **8**:549–552.

Grinsted, M. J., Hedley, M. J., White, R. E., and Nye, R. W., 1982, Plant induced changes in the rhizosphere of rape (*Brassica napus* var. Emerald) seedlings. I. pH change and the increase in P concentration in the soil solution, *New Phytol.* **91**:19–29.

Grinstein, A., Orion, D., Greenberger, A., and Katan, J., 1979, Solar heating of the soil for the control of *Verticillium dahliae* and *Pratylenchus thornei* in potatoes, in: *Soil-Borne Plant Pathogens* (B. Schippers and W. Gams, eds.), pp. 431–438, Academic Press, London.

Hale, M. G., 1981, Plant growth regulators and the rhizosphere ecosystem in: *Proceedings Plant Growth Regulators Society America (8th)*, pp. 256–261.

Hale, M. G., and Moore, L. D., 1979, Factors affecting root exudation II: 1970–1978, *Adv. Agron.* **31**:93–124.

Hale, M. G., Orcutt, D. M., and Moore, L. D., 1977, GA₃ and 2,4,-D effects on free sterol and fatty acid content of peanut, *Plant Physiol. Suppl.* **59**:30.

Hale, M. G., Moore, L. D., and Griffin, G. J., 1978, Root exudates and exudation, in: *Interations between Non-pathogenic Soil Microorganisms and Plants* (Y. R. Dommergues and S. V. Krupa, eds.), pp. 163–204, Elsevier, Amsterdam.

Hardie, K., and Leyton, L., 1981, The influence of vesicular-arbuscular mycorrhiza on growth and water relations of red clover. I. In phosphate deficient soil, *New Phytol.* **89**:599–608.

Hardy, R. W. F., and Havelka, U. D., 1975, Nitrogen fixation research: A key to world food, *Science* **188**:633–643.

Harley, J. L., 1969, *The Biology of Mycorrhiza,* Leonard Hill, London.

Harley, J. L., and Smith, S. E., 1983, *Mycorrhizal Symbiosis,* Academic Press, London.

Harper, J. E., 1974, Soil and symbiotic nitrogen requirements for optimum soybean production, *Crop Sci.* **14**:255–260.

Hattori, T., 1973, *Microbial Life in the Soil,* Marcel Dekker, New York.

Hayman, D. S., 1978, Endomycorrhizae, in: *Interations between Non-pathogenic Soil Microorganisms and Plants* (Y. R. Dommergues and S. V. Krupa, eds.), pp. 401–442, Elsevier, Amsterdam.

Hayman, D. S., 1982, Practical aspects of vesicular-arbuscular mycorrhiza, in: *Advances in Agricultural Microbiology* (N. S. Subba Rao, ed.), pp. 325–373, Butterworth Scientific, London.

Haystead, A., and Sprent, J. I., 1981, Symbiotic nitrogen fixation, in: *Physiological Processes Limiting Plant Productivity* (C. B. Johnson, ed.), pp. 345–364, Butterworth Scientific, London.

Hedley, M. J., Nye, P. H., and White, R. E., 1982a, Plant-induced changes in the rhizosphere of rape (*Brassica napus* var. Emerald) seedlings II. Origin of the pH change, *New Phytol.* **91**:31–44.

Hedley, M. J., White, R. E., and Nye, P. H., 1982b, Plant-induced changes in the rhizosphere of rape (*Brassica napus* var. Emerald) seedlings III. Changes in the *L* value, soil phosphate fraction and phosphatase activity, *New Phytol.* **91**:45–56.

Hedley, M. J., Nye, P. H., and White, R. E., 1983, Plant-induced changes in the rhizosphere of rape (*Brassica napus* var. Emerald) seedlings IV. The effect of rhizosphere phosphorus status on the pH, phosphatase activity and depletion of soil phosphorus fractions in the rhizosphere and on the cation–anion balance in the plants, *New Phytol.* **95**:69–82.

Hemming, B. C., Orser, C., Jacobs, D. L., Sands, D. C., and Strobel, G. A., 1982, The effects of iron on microbial antagonism by fluorescent pseudomonads, *J. Plant Nutr.* **5**:683–702.

Henry, C. M., and Deacon, J. W., 1981, Natural (non-pathogenic) death of the cortex of wheat and barley seminal roots, as evidenced by nuclear staining with acridine orange, *Plant Soil* **60**:255–274.

Hewitt, E. J., and Cutting, C. V. 1979, *Nitrogen Assimilation of Plants*, Academic Press, London.

Hiltner, L., 1904, Über neuere Erfahrungen und Probleme auf dem Gebiet der bodenbakteriologie und unter besonderer Berücksichtigung der Gründüngung und Brache, *Arb. Dtsch. Landw. Ges. Berl.* **98**:59–78.

Ho, I., and Trappe, J. M., 1973, Translocation of ^{14}C from *Festuca* plants to their endomycorrhizal fungi, *Nature New Biol.* **244**:30–31.

Holden, J., 1975, Use of nuclear staining to assess rates of cell death in cortices of cereal roots, *Soil Biol. Biochem.* **7**:333–334.

Hollis, J. P., and Rodriquez-Kabana, R., 1967, Fatty acids in Louisiana rice fields, *Phytopathology* **57**:841–847.

Hornby, D., 1979, Take-all decline: A theorist's paradise, in: *Soil-Borne Plant Pathogens* (B. Schippers and W. Gams, eds.), pp. 133–156, Academic Press, London.

Hornby, D., 1983, Suppressive Soils, *Annu. Rev. Phytopathol.* **21**:65–85.

Hornby, D., and Fitt, B. D. L., 1981, Effects of root-infecting fungi on structure and function of cereal roots, in: *Effects of Disease on the Physiology of the Growing Plant* (P. G. Ayres, ed.), pp. 101–130, Cambridge University Press, Cambridge.

Howell, C. R., and Stipanovic, R. D., 1980, Suppression of *Pythium ultimum*-induced damping-off of cotton seedlings by *Pseudomonas fluorescens* and its antibiotic, pyroluteorin, *Phytopathology* **70**:712–715.

Howie, W. J., and Echandi ,E., 1983, Rhizobacteria: Influence of cultivar and soil type on plant growth and yield of potato, *Soil Biol. Biochem.* **15**:127–132.

Hubbard, J. P., Harmon, G. E., and Hadar, Y., 1983, Effect of soil-borne *Pseudomonas* spp. on the biological control agent, *Trichoderma hamatum*, on pea seeds, *Phytopathology* **73**:655–659.

Hussain, S. S., and McKeen, W. E., 1963, Interactions between strawberry roots and *Rhizoctonia fragariae*, *Phytopathology* **53**:541–545.

Jackson, F. A., and Dawes, E. A., 1976, Regualtion of the tricarboxylic acid cycle and poly-β-hydroxybutyric metabolism in *Azotobacter beijerinckii* grown under nitrogen or oxygen limitation, *J. Gen. Microbiol.* **97**:303–312.

Jacq, V., and Dommergues, Y., 1970, Influence de l'intensité d'eclairement et de l'âge de la plante sur la sulfato-réduction rhizospherique, *Zentralbl. Bakteriol. Parasitenkd. Infektionskr.* **125**:661–669.

Jakobsen, I., and Anderson, A. J., 1982, Vesicular-arbuscular mycorrhiza and growth in barley: Effects of irradiation and heating of soil, *Soil Biol. Biochem.* **14**:171–178.

Jalili, B. L., 1976, Biochemical nature of root exudates in relation to root rot of wheat III. Carbohydrate shifts in response to foliar treatments, *Soil Biol. Biochem.* **8**:127–129.

Jalili, B. L., and Domsch, K. H., 1975, Effect of systemic fungitoxicants on the development of endotrophic mycorrhiza, in: *Endomycorrhizas* (F. E. Sanders, B. Mosse, and P. G. Tinker, eds.), pp. 619–626, Academic Press, London.

Jensen, A., and Jacobsen, I., 1980, The occurrence of vesciular-arbuscular mycorrhiza in barley and wheat grown in some Danish soils with different fertilizer treatments, *Plant Soil* **55**:403–414.

Johnen, B. G., and Sauerbeck, D. R., 1977, A tracer technique for measuring growth, mass and microbial breakdown of plant roots during vegetation, in: *Soil Organisms As Components of Ecosystems* (V. Lohm and T. Persson, eds.), *Ecol. Bull. (Stockholm)* **25**:366–373.

Johnston, A. W. B., and Beringer, J. E., 1979, Genetics of the *Rhizobium*–legume symbiosis, in: *Nitrogen Assimilation of Plants* (E. J. Hewitt and C. V. Cutting, eds.), pp. 67–72, Academic Press, London.

Jones, D. G., and Clifford, B. C., 1978, *Cereal Diseases—Their Pathology and Control*, BASF, Ipswich.

Jones, R., 1972, Comparative studies of plant growth and distribution in relation to water-logging VI. The effect of manganese in the growth of dune and duneslack plants, *J. Ecol.* **60**:141–145.

Kado, C. I., and Lurquin, P. F., 1982, Prospects of genetic engineering in agriculture, in: *Phytopathogenic Prokaryotes*, Vol. 2 (M. S. Mount and G. H. Lacy, eds.), pp. 303–325, Academic Press, New York.

Kapustka, L. A., and Rice, E. L., 1976, Acetylene reduction (N_2-fixation) in soil and old field succession in central Oklahoma, *Soil Biol. Biochem.* **8**:497–503.

Katan, J., Greenberger, A., Alon, H., and Grinstein, A., 1976, Solar heating by polyethylene mulching for the control of diseases caused by soil-borne pathogens, *Phytopathology* **66**:683–688.

Keast, D., and Tonkin, C., 1983, Antifungal activity of Western Australian soil actinomycetes against *Phytophthora* and *Pythium* species and a mycorrhizal fungus, *Laccaria laccata, Aust. J. Biol. Sci.* **36**:191–203.

Keen, N. T., and Williams, P. H., 1969, Translocation of sugars into infected cabbage tissues during clubroot development, *Plant Physiol.* **44**:748–754.

Kefford, N. P., Brockwell, J., and Zwar, J. A., 1960, The symbiotic synthesis of auxin by legumes and nodule bacteria and its role in nodule development, *Aust. J. Biol. Sci.* **13**:456–467.

Kerr, A., 1982, Biological control of soil-borne microbial pathogens and nematodes, in: *Advances in Agricultural Microbiology* (N. S. Subba Rao, ed.), pp. 429–463, Butterworth Scientific, London.

Kerry, B. R., and Crump, D. M., 1980, Two fungi parasites on females of cystnematodes (*Heterodera* spp.), *Trans. Br. Mycol. Soc.* **74**:119–125.

Kloepper, J. W., Schroth, M. N., and Miller, T. D., 1980a, Effects of rhizosphere colonization by plant growth-promoting rhizobacteria on potato plant development and yield, *Phytopathology* **70**:1078–1082.

Kloepper, J. W., Leong, J., Teintze, M., and Schroth, M. N., 1980b, *Pseudomonas* siderophores: A mechanism explaining disease-suppressive soils, *Curr. Microbiol.* **4**:317–320.

Kloepper, J. W., Leong, J., Teinze, M., and Schroth, M., 1980c, Enhanced plant growth by siderophores produced by plant growth-promoting bacteria, *Nature* **286**:885–886.

Knowles, R., 1982, Denitrification in soils, in: *Advances in Agricultural Microbiology* (N. S. Subba Rao, ed.), pp. 243–266, Butterworth Scientific, London.

Kommedhal, T., and Windels, C. E., 1979, Fungi: Pathogen or host dominance in disease, in: *Ecology of Root Pathogens* (S. V. Krupa and Y. R. Dommergues, eds.), pp. 1–103, Elsevier, Amsterdam.

Kosslak, R. M., and Bohlool, B. B., 1983, Prevalence of *Azospirillum* spp. in the rhizosphere of tropical plants, *Can. J. Microbiol.* **29:**649–652.

Krupa, S. V., and Dommergues, Y. R. (eds.), 1979, *Ecology of Root Pathogens,* Elsevier, Amsterdam.

Krupa, S., and Nylund, J.-E., 1971, Studies on ectomycorrhizae of pine. III. Growth inhibition of two root pathogenic fungi by volatile organic constituents of ectomycorrhizal root systems of *Pinus sylvestris* L., *Eur. J. For. Pathol.* **2:**88–94.

Kucey, R. M. N., and Paul, E. A., 1982, Carbon flow, photosynthesis, and N_2 fixation in mycorrhizal and nodulated faba beans (*Vicia faba* L.), *Soil Biol. Biochem.* **14:**407–412.

Lai, M., Weinhold, A. R., and Hancock, J. G., 1968, Permeability changes in *Phaseolus aureus* associated with infection by *Rhizoctonia solani, Phytopathology* **58:**240–245.

Lambert, D. H., Baker, D. E., and Cole, H., 1979, The role of mycorrhizae in the interactions of phosphorus with zinc, copper and other elements, *Soil Sci. Soc. Am. J.* **43:**976–980.

Larson, R. I., and Atkinson, T. G., 1970, A cytogenetic analysis of reaction to common root in some hard red spring wheats, *Can. J. Bot.* **48:**2067.

Lawley, R. A., Newman, E. I., and Campbell, R., 1982, Abundance of endomycorrhizas and root-surface microorganisms on three grasses grown separately and in mixtures, *Soil Biol. Biochem.* **14:**237–240.

Lawley, R. A., Campbell, R., and Newman, E. I., 1983, Composition of the bacterial flora of the rhizosphere of three grassland plants grown separately and in mixtures, *Soil Biol. Biochem.* **15:**605–607.

Lee, R. B., 1977, Effects of organic acids on the loss of ions from barley roots, *J. Exp. Bot.* **28:**578–587.

Lethbridge, G., and Davidson, M. S., 1983a, Root-associated nitrogen-fixing bacteria and their role in the nitrogen nutrition of wheat estimated by ^{15}N isotope dilution, *Soil Biol. Biochem.* **15:**365–374.

Lethbridge, G., and Davidson, M. S., 1983b, Microbial biomass as a source of nitrogen for cereals, *Soil Biol. Biochem.* **15:**375–376.

Lewis, D. H., 1973, Concepts in fungal nutrition and the origin of biotrophy, *Biol. Rev.* **61:**218–220.

Lewis, D. H., and Harley, J. L., 1965a, Carbohydrate physiology of mycorrhizal roots of beech. I. Identity of endogenous sugars and utliization of exogenous sugars, *New Phytol.* **64:**224–237.

Lewis, D. H., and Harley, J. L., 1965b, Carbohydrate physiology of mycorrhizal roots of beech. II. Utilization of exogenous sugars by uninfected and mycorrhizal roots, *New Phytol.* **64:**238–255.

Lewis, D. H., and Harley, J. L., 1965c, Carbohydrate physiology of mycorrhizal roots of beech. III. Movement of sugars between host and fungus, *New Phytol.* **62:**256–269.

Linderman, R. G., Moore, L. W., Baker, K. F., and Cooksey, D. A., 1983, Strategies for detecting and characterizing systems for biological control of soil-borne plant pathogens, *Plant Dis.* **67:**1058–1064.

Lockwood, J. L., 1977, Fungistasis in soils, *Biol. Rev.* **52:**1–43.

Lohnis, M. P., 1951, Manganese toxicity in field and market garden crops, *Plant Soil* **3:**193–222.

Low, A. J., 1972, The effect of cultivation on the structure and other characteristics of grassland and arable soils (1945–1970), *J. Soil Sci.* **23:**363–380.

Luttrell, E. S., 1974, Parasitism of fungi on vascular plants, *Mycologia* **66:**1–15.

Lynch, J. M., 1976, Products of soil micro-organisms in relation to plant growth, *CRC Crit. Rev. Microbiol.* **5:**67–107.

Lynch, J. M., 1978, Production and phytotoxicity of acetic acid in anaerobic soils containing plant residues, *Soil Biol. Biochem.* **10:**131–135.

Lynch, J. M., 1980, Effects of organic acids on the germination of seeds and growth of seed-
lings, *Plant Cell Environ.* **3**:255–259.
Lynch, J. M., 1981, Promotion and inhibition of soil aggregate stabilization by micro-organ-
isms, *J. Gen. Microbiol.* **126**:371–375.
Lynch, J. M., 1982, The rhizosphere, in: *Experimental Microbial Ecology* (R. G. Burns and
J. M. Slater, eds.), pp. 395–411, Blackwells, London.
Lynch, J. M., 1983, Interactions between bacteria and plants in the root environment, in:
Bacteria and Plants (M. E. Rhodes-Robert and F. A. Skinner, eds.), pp. 1–23, Academic
Press, London.
Lynch, J. M., and Bragg, E., 1984, Microorganisms and soil aggregate stability, in: *Advances
in Soil Sciences,* Vol. 2. (B. A. Stewart, ed.), pp. 133–172, Springer-Verlag, New York.
Lynch, H. M., and Clark, S. J., 1984, Effects of microbial colonization of barley (*Hordeum
vulgare* L.) roots on seedling growth, *J. Appl Bacteriol.* **56**:47–52.
Lynch, J. M., and Panting, L. M., 1981, Measurement of the microbial biomass in intact
cores of soil, *Microb. Ecol.* **7**:229–234.
Magyarosy, A. C., and Hancock, J. G., 1974, Association of virus-induced changes in lai-
mosphere microflora and hypocotyl exudation with protection to *Fusarium* stem rot,
Phytopathology **64**:994–1000.
Malajczuk, N., 1979, Biocontrol of *Phytophthora cinnamomi* in eucalyptus and avocados
in Australia, in: *Soil-Borne Plant Pathogens* (B. Schippers and W. Gams, eds.), pp. 635–
652, Academic Press, London.
Mangenot, F., and Diem, H. G., 1979, Fundamentals of biological control, in: *Ecology of
Root Pathogens* (S. V. Krupa and Y. R. Dommergues, eds.), pp. 207–625, Elsevier,
Amsterdam.
Mankau, R., 1980, Biological control of nematode pests by natural enemies, *Annu. Rev.
Phytopathol.* **18**:415–440.
Marschner, H., 1978, Role of the rhizosphere in iron nutrition of plants, *Iran. J. Agric. Res.*
6:69–80.
Marschner, H., and Romheld, V. 1983, *In vivo* measurement of root-induced pH changes
at the soil–root interface: Effect of plant species and nitrogen source, *Z. Pflanzenphy-
siol.* **111**:241–251.
Martin, J. K., 1977a, Factors influencing the loss of organic carbon from wheat roots, *Soil
Biol. Biochem.* **9**:1–7.
Martin, J. K., 1977b, Effect of soil moisture on the release of organic carbon from wheat
roots, *Soil Biol. Biochem.* **9**:303–304.
Martin, J. K., and Kemp, J. R., 1980, Carbon loss from roots of wheat cultivars, *Soil Biol.
Biochem.* **12**:551–554.
Martin, S. B., Hoch, H. O., and Abawi, G. S., 1983, Population dynamics of *Laetisaria
arvalis* and low-temperature *Pythium* spp. in untreated and pasturised beet field soils,
Phytopathology **73**:1445–1449.
Marx, D. H., 1975, The role of ectomycorrhizae in the protection of pine from root infection
by *Phytophthora cinnamomi*, in: *Biology and Control of Soil-Borne Plant Pathogens*
(G. W. Bruehl, ed.), pp. 112–115, American Phytopathological Society, St. Paul,
Minnesota.
Marx, D. H., 1980, Ectomycorrhizal fungus inoculations: A tool for improving forestation
practices, in: *Tropical Mycorrhiza Research* (P. Mikola, ed.), pp. 13–71, Clarendon
Press, Oxford.
Marx, D. H., and Krupa, S. V., 1978, Ectomycorrhiza, in: *Interactions between Non-patho-
genic Soil Micro-organisms and Plants* (Y. R. Dommergues and S. V. Krupa, eds.), pp.
373–400, Elsevier, Amsterdam.
McCool, P. M., and Menge, J. A., 1983, Influence of ozone on carbon partitioning in

tomato: Potential role of carbon flow in regulation of the mycorrhizal symbiosis under conditions of stress, *New Phytol.* **94**:241–247.

McDougall, B. M., 1968, The exudation of ^{14}C-labeled substances from roots of wheat seedlings, in: *Transactions of the Ninth International Congress of Soil Science Adelaide,* pp. 647–655.

McDougall, B. M., 1970, Movement of ^{14}C-photosynthate into the roots of wheat seedlings and exudation of ^{14}C from intact roots, *New Phytol.* **69**:37–46.

McDougall, B. M., and Rovira, A. D., 1970, Sites of exudation of ^{14}C-labelled compounds from wheat roots, *New Phytol.* **69**:999–1003.

Menge, J. A., 1983, Utilization of vesicular-arbuscular mycorrhizal fungi in agriculture, *Can. J. Bot.* **61**:1015–1024.

Merriman, P. R., Price, R. D., Kollmorgen, F., Piggott, T., and Ridge, E. H., 1974, Effect of seed inoculation with *Bacillus subtilis* and *Streptomyces griseus* on the growth of cereals and carrots, *Aust. J. Agric. Res.* **25**:219–276.

Miller, C. O., 1967, Zeatin and zeatin riboside from a mycorrhizal fungus *(Rhizopogon roseolus), Science* **157**:1055–1057.

Minchin, F. R., Summerfield, R. J., Hadley, P., Roberts, E. H., and Rawsthorne, S., 1981, Carbon and nitrogen nutrition of nodulated roots of grain legumes, *Plant Cell Environ.* **4**:5–26.

Mitchell, J. E., 1976, The effect of roots on the activity of soil-borne plant pathogens, in: *Physiological Plant Pathology. Encyclopedia of Plant Physiology New Series,* Vol. 4 (R. Heitefuss and P. H. Williams, eds.), pp. 104–128, Springer-Verlag, Berlin.

Moore, A. W., 1966, Non-symbiotic nitrogen fixation in soil and soil–plant systems, *Soils Fertil.* **29**:113–128.

Moore, L. W., and Warren, G., 1979, *Agrobacterium radiobacter* strain 84 and biological control of crown gall, *Annu. Rev. Phytopathol.* **17**:163–179.

Mosse, B., 1973, Plant growth responses to vesicular-arbuscular mycorrhiza.IV. In soil given additional phosphate, *New Phytol.* **72**:127–136.

Mount, M. S., Bateman, D. F., and Basham, M. G., 1970, Induction of electrolyte loss, tissue maceration, and cellular death of potato tissue by an endopolygalacturonate trans-eliminase, *Phytopathology* **60**:924–931.

Mulder, D., 1979 (ed.), *Soil Disinfestation,* Elsevier, Amsterdam.

Munns, D. N., and Mosse, B., 1980, Mineral nutrition of legume crops, in: *Advances in Legume Science* (R. J. Summerfield and A. H. Bunting, eds.), pp. 115–125, HMSO, London.

Neal, J. L., 1971, A simple method for enumeration of antibiotic producing microorganisms in the rhizosphere, *Can. J. Microbiol.* **17**:1143–1145.

Neal, J. L., Atkinson, T. G., and Larson, R. I., 1970, Changes in the rhizosphere microflora of spring wheat induced by disomic substitution of a chromosome, *Can. J. Microbiol.* **16**:153–158.

Neal, J. L., Larson, R. I., and Atkinson, T. G., 1973, Changes in rhizosphere populations of selected physiological groups of bacteria related to substitution of specific pairs of chromosomes in spring wheat, *Plant Soil* **39**:209–212.

Nelson, E. B., Kuter, G. A., and Hoitink, H. A. J., 1983, Effects of fungal antagonists and compost age on suppression of rhizoctonia damping-off in container media amended with composted hardwood bark, *Phytopathology* **73**:1457–1462.

Newman, E. I., 1978, Root microorganisms: Their significance in the ecosystem, *Biol. Rev.* **53**:511–554.

Newman, E. I., 1985, The rhizosphere: Carbon sources and microbial populations, in: *Ecological Interactions in Soil: Plants, Microbes and Animals* (A. H. Fitter, D. Atkinson, D. J. Read, and M. B. Usher, eds.), pp. 107–121, Blackwell Scientific, Oxford.

Newman, E. I., and Watson, A., 1977, Microbial abundance in the rhizosphere: A computer model, *Plant Soil* **48**:17–56.

Newman, E. I., Heap, A. J., and Lawley, R. A., 1981, Abundance of mycorrhizas and root-surface microorganisms of *Plantago lanceolata* in relation to soil and vegetation: A multi-variate approach, *New Phytol.* **89**:95–108.

Neyra, C. A., and Dobereiner, J., 1977, Nitrogen fixation in grasses, *Adv. Agron.* **29**:1–38.

Nutman, P. S., 1975, *Rhizobium* in the soil, in: *Soil Microbiology* (N. Walker, ed.), pp. 111–131, Butterworth Scientific, London.

Odunfa, V. S. A., and Oso, B. A., 1978, Bacterial populations in the rhizosphere soils of cowpea and sorghum, *Rev. Ecol. Biol. Sol* **15**:413–420.

Odvody, G. N., Boosalis, M. G., and Kerr, C. D., 1980, Biological control of *Rhizoctonia solani* with a soil-inhabiting basidiomycete, *Phytopathology* **70**:655–658.

Oghoghorie, C. G. O., and Pate, J. S., 1971, The nitrate stress syndrome of the nodulated field pea (*Pisum arvense* L.), *Plant Soil* **1971**(special volume):185–202.

Oghoghorie, C. G. O., and Pate, J. S., 1972, Exploration of the nitrogen transport system of a nodulated legume using ^{15}N, *Planta* **104**:35–49.

Okon, Y., 1982, Field inoculation of grasses with *Azospirillum*, in: *Biological Nitrogen Fixation Technology for Tropical Agriculture* (P. H. Graham and S. C. Harris, eds.), pp. 459–483, Centro Internacional de Agricultura Tropical, Cali, Colombia.

Old, K. M., and Nicholson, T. H., 1975, Electron microscopical studies of the microflora of roots of sand dune grass, *New Phytol.* **74**:51–58.

Old, K. M., and Nicholson, T. H., 1978, The root cortex as part of a microbial continuum, in: *Microbial Ecology* (M. W. Loutit and J. A. R. Miles, eds.), pp. 291–294, Springer-Verlag, Berlin.

Opgenorth, D. C., and Endo, R. M., 1983, Evidence that antagonistic bacteria suppress fusarium wilt of celery in neutral and alkaline soils, *Phytopathology* **73**:703–708.

Orlando, J. A., and Neilands, J. B., 1982, Ferrichrome compounds as a source of iron for higher plants, in: *Chemistry and Biology of Hydroxamic Acids* (K. Horst, ed.), pp. 123–129, S. Karger, Basel.

Owens, L. D., Gilbert, R. G., Griebel, G. E., and Menzies, J. D., 1969, Identification of plant volatiles that stimulate microbial respiration and growth in soil, *Phytopathology* **59**:1468–1472.

Page, J. B., and Willard, C. J., 1946, Cropping systems and soil properties, *Soil Sci. Soc. Am. Proc.* **11**:81–88.

Panagopoulos, C. G., Psallidas, P. G., and Alivizatos, A. S., 1979, Evidence of a breakdown in the effectiveness of biological control of crown gall, in: *Soil-Borne Plant Pathogens* (B. Schippers and W. Gams, eds.), pp. 569–578, Academic Press, London.

Pang, P. C., and Paul, E. A., 1980, Effects of vesicular-arbuscular mycorrhizae on ^{14}C and ^{15}N distribution in nodulated faba beans, *Can. J. Soil Sci.* **60**:241–250.

Papavizas, G. C., and Lumsden, R. D., 1980, Biological control of soil-borne fungal propagules, *Annu. Rev. Phytopathol.* **18**:389–413.

Parke, J. L., Linderman, R. G., and Black, C. M., 1983, The role of ectomycorrhizas in drought tolerance of Douglas-fir seedlings, *New Phytol.* **95**:83–95.

Parkinson, D., Taylor, G. S., and Pearson, R., 1963, Studies on the fungi in the root region. I. The development of fungi on young roots, *Plant Soil* **19**:332–349.

Patriquin, D. G., 1982, New developments in grass–bacteria associations, in: *Advances in Agricultural Microbiology* (N. S. Subba Rao, ed.), pp. 139–190, Butterworth Scientific, London.

Paul, E. A., and Kucey, R. M. N., 1981, Carbon flow in plant microbial associations, *Science* **213**:473–474.

Paulech, C., Fric, F., Minarcic, P., Priehradny, S., and Vizarova, G., 1981, Response of barley roots to infection by the parasitic fungus *Erysiphe graminis* DC, *Plant Soil* 63:119–121.

Penn, D. J., and Lynch, J. M., 1982, Toxicity of glyphosate applied to roots of barley seedlings, *New Phytol.* 90:51–55.

Peters, G. A., and Calvert, M. E., 1982, The *Azolla–Anabaena* symbioses, in: *Advances in Agricultural Microbiology* (N. S. Subba Rao, ed.), pp. 191–218, Butterworth Scientific, London.

Phillips, D. A., 1980, Efficiency of symbiotic nitrogen fixation in legumes, *Annu. Rev. Plant Physiol.* 31:29–49.

Phillips, S. M., and Young, H. M., 1973, *No-Tillage Farming*, Reiman, Milwaukee.

Polonenko, D. R., Dumbroff, E. B., and Mayfield, C. I., 1983, Microbial responses to salt-induced osmotic stress. III. Effects of stress on metabolites in the roots, shoots and rhizosphere of barley, *Plant Soil* 73:211–225.

Powell, P. E., Szaniszlo, P. J., Clive, G. R., and Reid, C. P. P., 1982, Hydroxamate siderophores in the iron nutrition of plants, *J. Plant Nutr.* 5:653–673.

Powlson, D. S., 1975, Effects of biocidal treatments on soil organisms, in: *Soil Microbiology. A Critical Review* (N. Walker, ed.), pp. 193–224, Butterworth Scientific, London.

Pullman, G. S., DeVay, J. E., Garber, R. H., and Weinhold, A. R., 1979, Control of soil-borne fungal pathogens by plastic tarping of soil, in: *Soil-Borne Plant Pathogens* (B. Schippers and W. Gams, eds.), pp. 439–446, Academic Press, London.

Purvis, A. C., and Williamson, R. E., 1972, Effects of flooding and gaseous composition of the root environment on growth of corn, *Agron. J.* 64:674–678.

Raj, J., Bagyaraj, D. J., and Manjunath, A., 1981, Influence of soil inoculation with vesicular-arbuscular mycorrhizae and a phosphate dissolving bacterium on plant growth and ^{32}P-uptake, *Soil Biol. Biochem.* 13:105–108.

Rambelli, A., 1973, The rhizosphere of mycorrhizae, in: *Ectomycorrhizae, Their Ecology and Physiology* (G. C. Marks and T. T. Kozlowski, eds.), pp. 299–349, Academic Press, London.

Ratnayake, M., Leonard, R. T., and Menge, J. A., 1978, Root exudation in relation to supply of phosphorus and its possible relevance to mycorrhizal formation, *New Phytol.* 81:543–552.

Read, D. J., 1983, The biology of mycorrhiza in the Ericales, *Can. J. Bot.* 61:985–1004.

Reid, C. P. P., and Mexal, J. G., 1977, Water stress effects on root exudation by lodgepole pine, *Soil Biol. Biochem.* 9:417–422.

Reid, C. P. P., and Woods, F. W., 1969, Translocation of C^{14}-labelled compounds in mycorrhizae and its implications in interplant nutrient cycling, *Ecology* 50:179–187.

Reid, C. P. P., Kidd, F. A., and Ekwebelam, S. A., 1983, Nitrogen nutrition, photosynthesis and carbon allocation in ectomycorrhizal pine, *Plant Soil* 71:415–432.

Reid, J. B., and Goss, M. J., 1980, Changes in aggregate stability of a sandy loam effected by growing roots of perennial ryegrass *(Lolium perenne), J. Sci. Food Agric.* 31:325–328.

Reid, J. B., and Goss, M. J., 1981, Effect of living roots of different plant species on the aggregate stability of two arable soils, *J. Soil Sci.* 32:521–541.

Reid, J. B., Goss, M. J., and Robertson, P. D., 1982, Relationship between the decreases in soil stability effected by the growth of maize roots and changes in organically bound iron and aluminum, *J. Soil Sci.* 33:397–410.

Rennie, R. J., 1980, ^{15}N-Isotope dilution as a measure of dinitrogen fixation by *Azospirillum* associated with maize, *Can. J. Bot.* 58:21–24.

Rennie, R. J., and Larson, R. I., 1979, Dinitrogen fixation associated with disomic chromosome substitution lines of spring wheat, *Can. J. Bot.* 57:2771–2775.

Rishbeth, J., 1975, Stump inoculation: A biological control of *Fomes annosus*, in: *Biology and Control of Soil-Borne Plant Pathogens* (G. W. Bruehl, ed.), pp. 158–162, American Phytopathological Society, St. Paul, Minnesota.

Rittenhouse, R. L., and Hale, M. G., 1971, Loss of organic compounds from roots II. Effect of O_2 and CO_2 tension on release of sugars from peanut roots under axenic conditions, *Plant Soil* **35**:311–321.

Ross, J. P., 1972, Influence of *Endogone* mycorrhizae on *Phytophthora* rot of soybean, *Phytopathology* **62**:896–897.

Rovira, A. D., 1959, Root excretions in relation to the rhizosphere effect IV. Influence of plant species, age of plant, light, temperature and calcium nutrition on exudation, *Plant Soil* **11**:53–64.

Rovira, A. D., 1969, Plant root exudates, *Bot. Rev.* **35**:35–57.

Rovira, A. D., 1973, Zones of exudation along plant roots and spatial distribution of microorganisms in the rhizosphere, *Pestic. Sci.* **4**:361–366.

Rovira, A. D., 1979, Biology of the soil–root interface, in: *The Soil–Root Interface* (J. L. Harley and R. S. Russell, eds.), pp. 145–60, Academic Press, London.

Rovira, A. D., and Bowen, G. D., 1966, Phosphate incorporation by sterile and non-sterile plant roots, *Aust. J. Biol. Sci.* **19**:1167–1169.

Rovira, A. D., and Campbell, R., 1974, Scanning electron microscopy of microorganisms on the roots of wheat, *Microb. Ecol.* **1**:15–23.

Rovira, A. D., and Davey, C. B., 1974, Biology of the rhizosphere, in: *The Plant Root and Its Environment* (E. W. Carson, ed.), pp. 153–204, University Press of Virginia, Charlottesville, Virginia.

Rovira, A. D., and Ridge, E. M., 1973, Exudation of ^{14}C-labelled components from wheat roots: Influence of nutrients, microorganisms and added organic compounds, *New Phytol.* **72**:1081–1087.

Rovira, A. D., and Wildermuth, G. B., 1981, The nature and mechanisms of suppression, in: *Biology and Control of Take-All* (M. J. C. Asher and P. J. Shipton, eds.), pp. 385–415, Academic Press, London.

Rovira, A. D., Foster, R. C., and Martin, J. K., 1979, Note on terminology: Origin, nature and nomenclature of the organic materials in the rhizosphere, in: *The Soil–Root Interface* (J. L. Harley and R. Scott Russell, eds.), pp. 1–4, Academic Press, London.

Royle, D. J., and Hickman, C. J., 1964a, Analysis of factors governing *in vitro* accumulation of zoospores of *Pythium aphanidermatum* on roots I. Behaviour of zoospores, *Can. J. Microbiol.* **10**:151–162.

Royle, D. J., and Hickman, C. J., 1964b, Analysis of factors governing *in vitro* accumulation of zoospores of *Pythium aphanidermatum* on roots II. Substances causing response, *Can. J. Microbiol.* **10**:201–219.

Russell, G. E., 1978, *Plant Breeding for Pest and Disease Resistance,* Butterworth Scientific, London.

Russell, G. E., 1981, Disease and crop yield: The problems and prospects for agriculture, in: *Effects of Disease on the Physiology of the Growing Plant* (G. P. Ayres, ed.), pp. 1–11, Cambridge University Press, Cambridge.

Ryle, G. J. A., Powell, C. E., and Gordon, A. J., 1979, The respiratory costs of nitrogen fixation in soyabeans, cowpea, and white clover. 1. Nitrogen fixation and the respiration of the nodulated root, *J. Exp. Bot.* **30**:135–144.

Safir, G. R., 1980, Vesicular-arbuscular mycorrhizae and crop productivity, in: *The Biology of Crop Productivity* (P. S. Carlson, ed.), pp. 231–252, Academic Press, London.

Safir, G. R., Boyer, J. S., and Gerdemann, J. W., 1972, Nutrient status and mycorrhizal enhancement of water transport in soybean, *Plant Physiol.* **49**:700–703.

St. John, T. V., and Coleman, D. C., 1983, The role of mycorrhizae in plant ecology, *Can. J. Bot.* **61**:1005–1014.

Salt, G. A., 1979, The increasing interest in "minor pathogens," in: *Soil-borne Plant Pathogens* (B. Schippers and W. Gams, eds.), pp. 289–312, Academic Press, London.

Samtsevich, S. A., 1965, Active secretions of plant roots and their significance, *Sov. Plant Physiol.* **12**:731–740.

Samtsevich, J. A., 1971, Root excretions of plants, An important source of humus formation in the soil, *Trans. Int. Symp. Humus et Planta V* (B. Novak, J. Macura, M. Kutilek, J. Pokorna-Kozova, and V. Tichy, eds.), pp. 147–153, Prague.

Samtsevich, S. A., 1972, Effect of plant cover and soil cultivations on the number of microorganisms and content of organic substances in the soil, *Symp. Biol. Hung.* **11**:41–48.

Sanders, F. E., and Tinker, P. B., 1973, Phosphate flow into mycorrhizal roots, *Pestic. Sci.* **4**:385–395.

Sanders, F. E., Mosse, B., and Tinker, P. B., (eds.), 1975, *Endomycorrhizas,* Academic Press, London.

Sanders, F. E., Tinker, P. B., Black, R. B. L., and Palmerley, S. M., 1977, The development of endomycorrhizal root systems I. Spread of infection and growth promoting effects with four species of vesicular-arbuscular endophyte, *New Phytol.* **78**:257–268.

Sanders, F. E., Buwalda, J. G., and Tinker, P. B., 1983. A note on modelling methods for studies of ectomycorrhizal systems, *Plant Soil* **71**:507–512.

Sauerbeck, D. R., Johnen, B. G., and Six, R., 1976, Atmung, Abbau and Ausscheidungen von Weizenwurzeln im Laufe Ihrer Entwicklung, *Landwirtsch. Forsch. Sonderh.* **32**:49–58.

Scannerini, S., and Bonfante-Fasolo, P., 1983, Comparative ultrastructural analysis of mycorrhizal associations, *Can. J. Bot.* **61**:917–943.

Schank, S. C., Weter, K. L., and Macrae, I. C., 1981, Plant yield and nitrogen content of a digitgrass in response to *Azospirillum* inoculation, *Appl. Environ. Microbiol.* **41**:343–345.

Scheffer, R. J., 1983, Biological control of Dutch Elm disease by *Pseudomonas* species, *Ann. Appl. Biol.* **103**:21–30.

Schenk, N. C., 1981, Can mycorrhizae control root disease?, *Plant Dis.* **65**:230–234.

Schenk, N. C., and Hinson, K., 1973, Response of nodulating and non-nodulating soybeans to a species of *Endogone* mycorrhiza, *Agron. J.* **65**:849–850.

Scher, F. M., and Baker, R. R., 1982, Effect of *Pseudomonas putida* and a synthetic iron chelator on induction of soil suppressiveness to *Fusarium* wilt pathogens, *Phytopathology* **72**:1567–1573.

Schippers, B., and Gams, W. (eds.), 1979, *Soil-Borne Plant Pathogens,* Academic Press, London.

Schneider, R. W., 1982 (ed.), *Suppressive Soils and Plant Disease,* American Phytopathological Society, St. Paul, Minnesota.

Schonbeck, F., 1979, Endomycorrhiza in relation to plant diseases, in: *Soil-Borne Plant Pathogens* (B. Schippers and W. Gams, eds.), pp. 271–280, Academic Press, London.

Schonbeck, F., and Dehne, H. W., 1977, Damage to mycorrhizal and non-mycorrhizal cotton seedlings by *Thielaviopsis basicola, Plant Dis. Rep.* **61**:266–267.

Schonwitz, R., and Ziegler, H., 1982, Exudation of water-soluble vitamins and of some carbohydrates by intact roots of maize seedlings (*Zea mays* L.) into a mineral nutrient solution, *Z. Pflanzenphysiol.* **707**:7–14.

Schramm, J. R., 1966, Plant colonization studies on black wastes from anthracite mining in Pennsylvania, *Trans. Am. Phil. Soc.* **56**:1–194.

Schroth, M. N., and Hancock, J. G., 1981, Selected topics in biological control, *Annu. Rev. Microbiol.* **35**:453–476.

Schroth, M. N., and Hancock, J. G., 1982, Disease-suppressive soil and root-colonizing bacteria, *Science* **216**:1376–1381.

Seidel, D., 1970, Pflanzen in ihren Auswirkungen auf phytopathogene Bodenpilze VI. *Rhizotonia solani* Kuhn, *Zentralbl. Bakteriol. Parasitenkd. Infektionskr. Abt. II* **120**:49–59.

Sequeira, L., 1958, Bacterial wilt of bananas: Dissemination of the pathogen and control of the disease, *Phytopathology* **48**:64–69.

Sethunathan, N., 1970a, Foliar sprays of growth regulators and rhizosphere effect in *Cajanus cajan* Millsp. I: Quantitative changes, *Plant Soil* **33**:62–70.

Sethunathan, N., 1970b, Foliar sprays of growth regulators and the rhizosphere effect in *Cajanus cajan* Millsp. II. Qualitative changes in the rhizosphere and certain metabolic changes in the plant, *Plant Soil* **33**:71–80.

Shay, F. J., and Hale, M. G., 1973, Effect of low levels of calcium on exudation of sugars and sugar derivatives from intact peanut roots under axenic conditions, *Plant Physiol.* **51**:1061–1063.

Shipton, P. J., 1977, Monoculture and soil-borne plant pathogens, *Annu. Rev. Phytopathol.* **15**:387–407.

Shone, M. G. T., Whipps, J. M., and Flood, A. V., 1983, Effects of localized and overall water stress on assimilate partitioning in barley between shoots, roots and root exudates, *New Phytol.* **95**:625–634.

Sivasithamparam, K., and Parker, C. A., 1980, Effect of certain isolates of soil fungi on take-all of wheat, *Aust. J. Bot.* **28**:421–427.

Slankis, V., 1967, Renewed growth of ectotrophic mycorrhizae as an indication of an unstable symbiotic relationship, in: *Proceedings of the 14th Congress of the International Forest Research Organisation, Munich,* Vol. 5, pp. 84–99.

Slankis, V., 1973, Hormonal relationships in mycorrhizal developments, in: *Ectomycorrhizae* (G. C. Marks and T. T. Kozlowski, eds.), pp. 231–298, Academic Press, London.

Smiley, R. W., 1975, Forms of nitrogen and the pH in the root zone and their importance to root infections, in: *Biology and Control of Soil-Borne Plant Pathogens* (G. W. Bruehl, ed.), pp. 55–62, American Phytopathological Society, St. Paul, Minnesota.

Smiley, R. W., 1979, Wheat rhizosphere pH and the biological control of take-all, in: *The Soil–Root Interface* (J. L. Harley and R. S. Russell, eds.), pp. 329–338, Academic Press, London.

Smith, M. S., and Tiedje, J. M., 1979, The effect of roots on soil denitrification, *Soil Sci. Soc. Am J.* **43**:951–955.

Smith, R. L., Bouton, J. H., Schank, S. C., Queensbury, K. H., Tyler, M. E., Milam, J. R., Gaskins, M. H., and Littell, R. C. 1976, Nitrogen fixation in grasses inoculated with *Spirillum lipoferum, Science* **193**:1003–1005.

Smith, S. E., and Daft, M. J., 1977, Interactions between growth, phosphate content and nitrogen fixation in mycorrhizal and non-mycorrhizal *Medicago sativa, Aust. J. Plant Physiol.* **4**:403–413.

Smith, S. E., Nicholas, D. J. D., and Smith, F. A., 1979, Effect of early mycorrhizal infection on nodulation and nitrogen fixation in *Trifolium subterraneum, Aust. J. Plant Physiol.* **6**:305–316.

Sneh, B., 1981, Use of rhizosphere chitinolytic bacteria for biological control of *Fusarium oxysporum* f. sp. *dianthi* in carnation, *Phytopathol. Z.* **100**:251–256.

Snellgrove, R. C., Splittstoesser, W. E., Stribley, D. P., and Tinker, P. B., 1982, The distribution of carbon and the demand of the fungal symbiont in leek plants with vesicular-arbuscular mycorrhizas, *New Phytol.* **92**:75–87.

Sparling, G. P., and Tinker, P. B., 1978, Mycorrhizal infection in Pennine grassland. II. Effects of mycorrhizal infection on the growth of some upland grasses on γ-irradiated soils, *J. Appl. Ecol.* **15**:951–958.

Sprent, J. I., 1979, *The Biology of Nitrogen-Fixing Organisms,* McGraw-Hill, London.

Stewart, W. D. P., Rowell, P., and Lockhart, C. M., 1979, Associations of nitrogen fixing prokaryotes with higher and lower plants, in: *Nitrogen Assimilation of Plants* (E. J. Hewitt and C. V. Cutting, eds.), pp. 45–66, Academic Press, London.

Stirling, G. R., and Wachtel, M. F., 1980, Mass production of *Bacillus penetrans* for the biological control of root-knot nematodes, *Nematologica* **26**:308–312.

Stirling, G. R., McKenry, M. V., and Mankau, R., 1979, Biological control of root-knot nematodes (*Meloidogyne* sp.) on peach, *Phytopathology* **69**:806–809.

Stotzky, G., and Burns, R. G., 1982, The soil environment: Clay–humus–microbe interaction, in: *Experimental Microbial Ecology* (R. G. Burns and J. H. Slater, eds.), pp. 105–133, Blackwell Scientific, Oxford.

Stribley, D. P., Tinker, P. B., and Rayner, J. H., 1980, Relation of internal phosphorus concentration and plant weight in plants infected by vesicular-arbuscular mycorrhizas, *New Phytol.* **86**:261–266.

Strobel, G. A., 1974, Phytotoxins produced by plant parasites, *Annu. Rev. Plant Physiol.* **25**:541–566.

Strullu, D. G., Harley, J. L., Gourret, J. P., and Garrec, J. P., 1982, Ultrastructure and microanalysis of the polyphosphate granules of the ectomycorrhizas of *Fagus sylvatica*, *New Phytol.* **92**:417–423.

Subba Rao, N. S. (ed.), 1982a, *Advances in Agricultural Microbiology*, Butterworth Scientific, London.

Subba Rao, N. S., 1982b, Biofertilizers, in: *Advances in Agricultural Microbiology* (N. S. Subba Rao, ed.), pp. 219–242, Butterworth Scientific, London.

Subba Rao, N. S., Bidwell, R. G. S., and Bailey, D. L., 1961, The effect of rhizoplane fungi on the uptake and metabolism of nutrients by tomato plants, *Can. J. Bot.* **39**:1759–1764.

Suslow, T. V., 1982, Role of root-colonizing bacteria in plant growth, in: *Phytopathogenic Prokaryotes*, Vol. 1 (M. S. Mount and G. H. Lacy, eds.), pp. 187–223, Academic Press, New York.

Swaby, R. L., 1942, Stimulation of plant growth by organic matter, *J. Aust. Inst. Agric. Sci.* **8**:136–163.

Szaniszlo, P. J., Powell, P. E., Reid, C. P. P., and Clive, G. R., 1981, Production of hydroxamate siderophore iron chelators by ectomycorrhizal fungi, *Mycologia* **73**:1158–1174.

Takai, Y., and Kamura, T., 1966, The mechanism of reduction in waterlogged paddy soil, *Folia Microbiol.* **11**:304–313.

Takijima, Y., 1964, Studies on organic acids in paddy field soils with reference to their inhibitory effects on the growth of rice plants, *Soil Sci. Plant Nutr.* **10**:14–21.

Teakle, D. S., and Hiruki, C., 1985, Soil-borne viruses of plants, *Curr. Top. Pathogen-Vector-Host Res.* (in press).

Thompson, L. K., and Hale, M. G., 1983, Effects of kinetin in the rooting medium on root exudation of free fatty acids and sterols from roots of *Arachis hypogaea* L. "Argentine" under axenic conditions, *Soil Biol. Biochem.* **15**:125–126.

Thornton, H. G., 1930, The influence of the host plant in inducing parasitism in lucerne and clover nodules, *Proc. R Soc. B* **106**:110–122.

Tien, T. M., Gaskins, M. H., and Hubbell, D. H., 1979, Plant growth substances produced by *Azospirillum brasiliense* and their effect of the growth of pearl millet (*Pennisetum americanum* L.), *Appl. Environ. Microbiol.* **37**:1016–1024.

Tirol, A. C., Roger, P. A., and Watanabe, I., 1982, Fate of nitrogen from a blue-green alga in a flooded rice soil, *Soil Sci. Plant Nutr.* **28**:559–569.

Tisdall, J. M., and Oades, J. M., 1979, Stabilization of soil aggregates by the root systems of ryegrass, *Aust. J. Soil Res.* **17**:429–441.

Trolldenier, G., 1972, L'influence de la nutrition potassique de haricots nams (*Phaseolus vulgaris* var. *nanus*) sur l'exsudation de substances organiques marguées au ^{14}C, le nombres de bactéries rhizosphériques et la respiration des racines, *Rev. Ecol. Biol. Sol* **9**:595–603.

Trolldenier, G., and Markwordt, U., 1962, Untersuchungen uber den Einfluss der Bodenmikroorganismen auf die Rubidium- und Calcium-Aufnahme in Nahrlosung Wachsender Pflanzen, *Arch. Mikrobiol.* **43**:148–151.

Tschudi, S., and Kern, H., 1979, Specific lysis of the mycelium of *Gaeumannomyces graminis* by enzymes of *Streptomyces lavendulae,* in: *Soil-Borne Plant Pathogens* (B. Schippers and W. Gams, eds.), pp. 611–615, Academic Press, London.

Turner, S. M., and Newman, E. I., 1984, Growth of bacteria on roots of grasses: Influence of mineral nutrient supply and interactions between species, *J. Gen. Microbiol.* **130**:505–512.

Uecker, F. A., Ayers, W. A., and Adams, P. B., 1978, A new hyphomycete on sclerotia of *Sclerotinia sclerotiorum, Mycotaxon* **7**:275–282.

Utkhede, R. S., and Rahe, J. E., 1983, Interactions of antagonist and pathogen in biological control of onion white rot, *Phytopathology* **73**:890–893.

Vaidehi, B. K., 1973, Effect of foliar application of urea on the behaviour of *Helminthosporium hawaiiensis* in the rhizosphere of rice, *Ind. J. Plant Pathol.* **3**:81–85.

Vamos, R., 1959, "Brusone" disease of rice in Hungary, *Plant Soil* **11**:65–77.

Vance, C. P., 1983, *Rhizobium* infection and nodulation: A beneficial plant disease, *Annu. Rev. Microbiol.* **37**:399–424.

Vancûra, V., 1967, Root exudates of plants III. Effect of temperature and "cold shock" on the exudation of various compounds from seeds and seedlings of maize and cucumber, *Plant Soil* **27**:319–328.

Vancûra, V., 1980, Fluorescent pseudomonads in the rhizosphere of plants and their relation to root exudates, *Folia Microbiol.* **25**:168–173.

Vancûra, V., and Kunc, F., 1977, The effect of streptomycin and actidione on respiration in the rhizosphere and non-rhizosphere soil, *Zentralbl. Bakteriol. Parasitenkd. Infektionskr.* **132**:472–478.

Vancûra, V., Přikryl, Z., Kalachovâ, L., and Wurst, M., 1977, Some quantitative aspects of root exudation, *Ecol. Bull. Stockholm* **25**:381–386.

Vandenbergh, P. A., Gonzalez, C. F., Wright, A. M., and Kunka, S., 1983, Iron-chelating compounds produced by soil pseudomonads: Correlation with fungal growth inhibition, *Appl. Environ. Microbiol.* **46**:128–132.

Van Egeraat, A. W. S. M., 1975, Exudation of ninhydrin-positive compounds by pea-seedling roots: A study of the sites of exudation and of the composition of the exudate, *Plant Soil* **42**:37–47.

Van Schreven, D. A., 1958, Some factors affecting the uptake of nitrogen by legumes, in: *Nutrition of the Legumes* (E. G. Hallsworth, ed.), pp. 137–163, London.

Van Vuurde, J. W. L., Kruyswyk, C. J., and Schippers, B., 1979, Bacterial colonization of wheat roots in a root–soil model system, in: *Soil-Borne Plant Pathogens* (B. Schippers and W. Gams, eds.), pp. 229–234, Academic Press, London.

Vesely, D., 1977, Potential biological control of damping-off pathogens in emerging sugar beet by *Pythium oligandrum, Phytopathol. Z.* **90**:113–115.

Vesely, D., 1979, Use of *Pythium oligandrum* to protect emerging sugar beet, in: *Soil-Borne Plant Pathogens* (B. Schippers and W. Gams, eds.), pp. 593–595, Academic Press, London.

Von Bulow, J. W. F., and Dobereiner, J., 1975, Potential for nitrogen fixation in maize genotypes in Brazil, *Proc. Natl. Acad. Sci. USA* **72**:2384–2393.

Vrany, J., 1974, Changes of microflora of wheat roots after foliar application of urea, *Folia Microbiol.* **19**:229–235.

Warembourg, F. R., Montange, D., and Bardin, R., 1982, The simultaneous use of $^{14}CO_2$

and $^{15}N_2$ labelling techniques to study the carbon and nitrogen economy of legumes grown under natural conditions, *Physiol. Plant* **56**:46–55.

Weinhold, A. R., Oswald, J. W., Bowman, T., Bishop, J., and Wright, D., 1964, Influence of green manures and crop rotation on common scab of potato, *Am. Potato J.* **41**:265–273.

Weller, D. M., 1983, Colonization of wheat roots by a fluorescent pseudomonad suppressive to take-all, *Phytopathology* **73**:1548–1553.

Weller, D. M., and Cook, R. J., 1983, Suppression of take-all of wheat by seed treatments with fluorescent pseudomonads, *Phytopathology* **73**:463–469.

Wells, H. D., Bell, D. K., and Jawenski, C. A., 1972, Efficacy of *Trichoderma harzianum* as a biocontrol for *Sclerotium rolfsii*, *Phytopathology* **62**:442–447.

Welte, E., and Trolldenier, G., 1962, Der Einfluss der Bodenmikroorganismen auf Trockensubstanzbildung und Aschegehalt in Nährlösung wachsender, *Pflanzen Arch. Mikrobiol.* **43**:138–147.

Wheeler, H., 1978, Disease alterations in permeability and membranes, in: *Plant Disease, An Advanced Treatise,* Vol. III. *How Plants Suffer from Disease* (J. G. Horsfall and E. B. Cowling, eds.), pp. 327–347, Academic Press, New York.

Wheeler, H., and Hanchey, P., 1968, Permeability phenomena in plant disease, *Annu. Rev. Phytopathol.* **6**:331–350.

Whipps, J. M., 1984, Environmental factors affecting the loss of carbon from the roots of wheat and barley seedlings, *J. Exp. Bot.* **35**:767–773.

Whipps, J. M., and Lewis, D. H., 1981, Patterns of translocation, storage and interconversion of carbohydrates, in: *Effects of Disease on the Physiology of the Growing Plant* (P. G. Ayres, ed.), pp. 47–83, Cambridge University Press, Cambridge.

Whipps, J. M., and Lynch, J. M., 1983, Substrate flow and utilization in the rhizosphere of cereals, *New Phytol.* **95**:605–623.

Wiedenroth, E., and Poskuta, J., 1981, The influence of oxygen deficiency in roots on CO_2 exchange rates of shoots and distribution of ^{14}C-photoassimilates of wheat seedlings, *Z. Pflanzenphysiol.* **103**:459–467.

Wilcox, H. E., 1983, Fungal parasitism of woody plant roots from mycorrhizal relationships to plant disease, *Annu. Rev. Phytopathol.* **21**:221–242.

Willey, C. R., 1970, Effect of short periods of anaerobic and near anaerobic conditions on water uptake by tobacco roots, *Agron. J.* **62**:224–229.

Williams, P. H., 1966, A cytochemical study of hypertrophy in clubroot of cabbage, *Phytopathology* **56**:521–524.

Williams, P. H., Keen, N. T., Strandberg, J. O., and McNabola, S. S., 1968, Metabolite syntheses and degradation during club root development in cabbage hypocotyls, *Phytopathology* **58**:921–928.

Williams, P. H., Aist, J. R., and Bhattacharya, P. K., 1973, Host–parasite relations in cabbage club root, in: *Fungal Pathogenicity and the Plant's Response* (R. J. W. Byrde and C. V. Cutting, eds.), pp. 141–155, Academic Press, London.

Williamson, F. A., and Wyn Jones, R. G., 1973, The influence of soil microorganisms on growth of cereal seedlings and on potassium uptake, *Soil Biol. Biochem.* **5**:569–575.

Woldendorp, J. W., 1963a, L'influence des plantes vivantes sur la dénitrification, *Ann. Inst. Pasteur* **105**:426–433.

Woldendorp, J. W., 1963b, The influence of living plants on denitrification, *Meded. Landbouwhogesch. Wageningen* **63**:1–100.

Woldendorp, J. W., 1981, Nutrients in the rhizosphere, in: *Agricultural Yield Potentials in Continental Climates,* pp. 89–115, International Potash Institute, Bern.

Wong, P. T. W., 1975, Cross protection against the wheat and oat take-all fungi by *Gaeumannomyces graminis* var. *graminis, Soil Biol. Biochem.* **7**:189–194.

Wong, P. T. W., 1981, Biological control by cross-protection, in: *Biology and Control of Take-All* (M. J. C. Asher and P. J. Shipton, eds.), pp. 417–431, Academic Press, London.

Wong, P. T. W., and Southwell, R. J., 1980, Field control of take-all by avirulent fungi, *Ann. Appl. Biol.* **94:**41–49.

Wood, R. K. S., Ballio, A., and Graniti, A. (eds.), 1972, *Phytotoxins in Plant Diseases,* Academic Press, London.

Wyse, D. L., Meggitt, W. F., and Penner, D., 1976, Factors affecting EPTC injury to navy bean, *Weed Sci.* **24:**1–4.

Zambolin, L., and Schenck, N. C., 1983, Reduction of the effects of pathogenic, root-infecting fungi on soybean by the mycorrhizal fungus, *Glomus mosseae, Phytopathology* **73:**1402–1405.

Zeyen, R. J., 1979, Viruses, in: *Ecology of Root Pathogens* (S. V. Krupa and Y. R. Dommergues, eds.), pp. 179–205, Elsevier, Amsterdam.

6

Measurement of Bacterial Growth Rates in Aquatic Systems from Rates of Nucleic Acid Synthesis

D. J. W. MORIARTY

1. Introduction

Marine microbiology has expanded rapidly as a scientific discipline in the last 10–20 years. A change in experimental approach, from isolation of individual organisms and pure culture studies to whole-community studies, has helped foster this expansion. New techniques, such as epifluorescence microscopy and the use of radioisotopes, have shown that bacteria are more numerous and active than had been generally accepted. Early work with radioisotopes showed that bacteria were actively metabolizing organic matter in the sea, but accurate measurements of growth rates and production were needed in order to quantify fully the role of bacteria in food chains and cycles of organic matter. Perhaps the ultimate expression of bacterial activity is cell division. If we can quantify this, then we can confidently make statements about other activities of bacteria.

It is the growth of heterotrophic bacteria that will be considered here. The growth of other microorganisms that fix carbon dioxide for their primary source of carbon, such as chemoautotrophic bacteria and cyanobacteria, can in principle be measured using $^{14}CO_2$. There has been no satisfactory technique until recently for measuring the growth of heterotrophic bacteria in aquatic environments. Laboratory methods for mea-

D. J. W. MORIARTY • CSIRO Marine Laboratories, Cleveland, Queensland 4163, Australia.

suring growth rates do not work in natural environments, due to the small size of heterotrophic bacteria, their aggregation on or in particles and in mixed species groups, and the effects of predation. Brock (1971) stressed that microbial ecologists need to study what bacteria are doing in nature. He discussed a number of techniques that could be used, but no generally useful methods were available then. With the development over the last few years of methods for calculating growth rates from the rates of incorporation of radioactive precursors into nucleic acids, considerable advances have been made. In his review on microbial growth rates, Brock (1971) discussed aspects of the labeling of DNA with radioactive thymidine, which showed that it could be applied as a technique to study growth rates. This review examines in detail methods for measuring bacterial growth rates in the natural environment using the rates of synthesis of nucleic acids, especially DNA. A general review of biomass and activity measures has recently been published (van Es and Meyer-Reil, 1982), so other techniques will not be discussed here.

Brock (1967) used tritiated thymidine and autoradiography to estimate the growth rate of *Leucothrix mucor* in various aquatic habitats. His method is applicable only to bacteria that can be grown in pure culture and can be recognized in their natural environment. When applied to the microbial populations in general, autoradiography can be used as an indication of metabolic activity or to estimate the number of growing cells, but not growth rates. For example, autoradiography has been used by Ramsay (1974) to quantify active bacteria on *Elodia canadensis* leaves and by Hoppe (1976) to count active bacteria in seawater. More recent work has centered on the measurement of radioactivity in extracted DNA as a measure of the growth rate of the whole community of aquatic bacteria.

Kunicka-Goldfinger (1976) used semicontinuous culture on membrane filters to show that there was a linear relationship between incorporation of tritiated thymidine into macromolecules insoluble in trichloroacetic acid (TCA) and bacterial growth. She stated that the measurement of DNA synthesis with tritiated thymidine would be a useful measure of bacterial growth in water. Tobin and Anthony (1978) demonstrated that DNA was labeled with tritiated thymidine by bacteria in lake sediments. The incorporation of tritiated uridine into RNA was used by La Rock et al. (1979) to show that there was a zone of active bacterial growth in a deep anoxic basin in the Gulf of Mexico.

At about the same time, Fuhrman and Azam (1980) and Moriarty and Pollard (1981) developed the use of tritiated thymidine for estimating bacterial growth rates in water and sediment, respectively. Fuhrman and Azam (1980, 1982) showed that growth rates of bacteria in coastal waters could be estimated from the rate of incorporation of tritiated thymidine into DNA in seawater samples. Moriarty and Pollard (1981,

1982) measured the growth rate of bacteria in marine sediments using tritiated thymidine incorporation into DNA. These methods hold considerable promise for advancing our knowledge of the productivity of bacteria and the flux of organic carbon in aquatic ecosystems and have been eagerly applied by other research workers. As it is only a short time since these methods were introduced, methodologies and interpretations of results need improvement.

As an alternative to using radioactive thymidine for labeling DNA, Karl (1981, 1982) has proposed the use of adenine. Adenine also labels RNA in addition to DNA, and rates of RNA synthesis have been used to estimate overall microbial growth rates in freshwater ponds and the sea (Karl, 1979; Karl et al., 1981). As will be discussed below, the microbial populations studied with adenine include microalgae and other groups in addition to heterotrophic bacteria.

Azam and Fuhrman (1984) have discussed the criteria necessary for the ideal method for measuring growth rates of bacteria in nature. These may be listed as follows: (1) The method should be specific for heterotrophic bacteria; (2) the method should not rely on balanced growth, or if it does, balanced growth should be shown to occur in the environment in question; (3) the growth rate should not be altered by any incubation or other procedures involved in the measurement of the growth rate; and (4) if the method requires a conversion factor, there must be a means of determining that factor with confidence for the sample.

The measurement of rates of DNA synthesis with tritiated thymidine complies with these criteria to a better extent than measurements using other precursors, provided certain conditions are met. There are, however, problems in obtaining accurate and ecologically meaningful results with any nucleic acid precursor that is used to measure growth rates. This is because methods that depend on measurement of nucleic acid synthesis do not in fact measure growth or cell division directly, but the rate of incorporation of a radioactive precursor into the macromolecule. The processes of uptake, or transport of the precursor into the cell, as well as metabolism within the cell, need to be taken into consideration in order to convert rates of precursor incorporation into a macromolecule to rates of cell division. Some aspects of the biochemistry of DNA and RNA synthesis that are relevant to the interpretation of results will be discussed below.

2. Bacterial Growth Processes

A good general introduction to the principles of bacterial growth and its measurement is available (Pirt, 1975). This text is aimed primarily at measurement of growth in pure cultures, but the basic principles discussed there are relevant to natural populations. The main phases in a classical bacterial growth cycle in batch culture are well known: lag, exponential, stationary, and, finally, death. These ideal stages can be identified with cultures in the laboratory, but in a natural environment the situation is more complex, with different bacteria in varying stages of growth at the same time. In marine environments many bacteria do not die, but develop what is known as a "starvation-survival" phase (Morita, 1982). When energy substrates are depleted, the bacteria divide rapidly and form many small cells. The final expression of growth of a bacterium is cell division and an increase in numbers and biomass of viable cells. In natural systems it is very difficult (or even impossible) to measure growth accurately by counting new cells formed. Before cells divide, there must be synthesis of new cellular components: walls, membranes, proteins, RNA, and, of course, DNA. Thus, growth may be defined as the increase in a particular component of the biomass of cells, which culminates in cell division.

The measurement of the rate of DNA synthesis is particularly useful, because it is related to cell division. Once DNA synthesis is initiated, it proceeds to completion, and triggers a cycle of cell division (Lark, 1969). Although metabolic turnover of DNA does not occur in a manner analogous to that of enzymes or messenger RNA, some turnover does occur due to processes such as excision and repair of incorrect bases. The rate of thymidine incorporation due to such processes would be insignificant compared to normal replication. Variations in the rate of synthesis of other cellular components, such as RNA or protein, may not reflect rates of cell division. A direct correlation between rates of cell division and increase in mass of most cellular components, such as RNA and protein, occurs only when growth is balanced (Campbell, 1957). According to Campbell, growth is balanced over a time interval if, during that interval, every extensive property (e.g., protein, RNA, DNA, cell number) increases by the same factor. This state of growth applies during the exponential phase of an ideal growth curve. In other words, during balanced growth the relationship between growth rate and the rate of increase of biomass or any cellular substance (x) is an exponential one. The rate of formation of any component x is proportional to the amount of x at any given time, and may be expressed as follows (Pirt, 1975):

$$dx/dt = \mu x$$

where μ is the specific growth rate; it has dimensions of time^{-1}. The value of μ will be the same for every extensive property during balanced growth. The DNA method for measuring microbial growth rates gives an estimate of the number of bacteria dividing per unit time. Knowing the actual number present, one can estimate the doubling time for the population. Some workers use specific growth rate; this is related to doubling time or generation time g as follows:

$$g = (\ln 2)/\mu$$

During balanced growth, the rate of synthesis of any component; e.g., RNA, may be chosen to measure growth rate. Conditions for balanced growth have been defined for particular enteric bacteria. Detailed studies of the physiological states and chemical composition of *Salmonella typhimurium* have been made by Schaechter *et al.* (1958) and Kjeldgaard *et al.* (1958). They pointed out that if the growth rate was increased, by supplying better quality nutrients for example, the rates of RNA and protein synthesis immediately increased, whereas DNA synthesis and cell division rates continued at the old rate for some time and then abruptly increased. Conversely, when growth rate was lowered by withdrawing nutrients, RNA and protein synthesis slowed or stopped, but DNA synthesis and cell division continued at the fast rate for some time, and then rapidly slowed to the new rate. Rapidly growing cells were larger and contained more RNA, protein, and DNA than slowly growing cells.

Brunschede *et al.* (1977) found a similar situation in *Escherichia coli*. After an increase in growth rate, RNA and protein synthesis increased immediately, but DNA synthesis was stable for 30 min and then increased. The rates of synthesis of all three macromolecules took a period of 2 hr or longer to stabilize.

DNA synthesis is regulated primarily at the site of initiation. Faster growth rates are achieved by having more than one replication fork proceeding along the chromosome. The rate of travel of a replication fork is constant (Lark, 1969). Thus, provided the rate of thymidine incorporation is measured over a shorter time interval than is required for one cycle of DNA synthesis, small changes in the external environment due to sampling should not affect the measurement. There are difficulties, of course, in extrapolating from the laboratory to the field, and obviously major changes, such as aerating an anaerobic sediment, will affect energy metabolism and thus directly affect the rate of thymidine incorporation. For this and other reasons, studies on bacterial growth processes in sediments are particularly difficult, but not impossible. The thymidine technique is a promising one, although present results need to be accepted with caution.

The ratios of rates of stable RNA (transfer and ribosomal RNA) to DNA synthesis in *E. coli* were directly related to growth rate during balanced growth and at fast growth rates (Dennis and Bremer, 1974). At slow growth rates (less than 0.67 hr^{-1}), the relationship between rates of stable

RNA and DNA syntheses did not hold in their experiments. This growth rate, although slow for a culture, is much faster than is observed in many natural systems. There may be problems, therefore, in using techniques for measuring growth that depend on the ratio of RNA synthesis to DNA synthesis unless it is shown that balanced growth occurs (e.g., Karl, 1981). This illustrates one of the problems confronting microbial ecologists. Much of the laboratory work from which conclusions are drawn and extrapolated to the natural environment has been undertaken on organisms such as *E. coli*. The need to move between the laboratory and nature and carry out complementary experiments has been emphasized by Brock (1971).

Caution must be observed when applying the principles of the growth processes described above to all bacteria, especially in the natural environment, where nutrient supply is usually small and variable. Even in a chemostat, an organism growing at a constant rate may not be in a state of balanced growth. For example, the fermentative bacterium *Zymomonas* exhibits unbalanced or uncoupled growth (Swings and de Ley, 1977). The rate of substrate dissimilation per unit weight of organism is independent of growth rate. The amounts of glucose metabolized and ATP produced do not vary with growth rate in minimal media where a growth factor is limiting. Swings and de Ley (1977) suggest that the uncoupling occurs because *Zymomonas* lacks adequate mechanisms to control energy charge levels and the link between ATP production and its consumption during biosynthesis. In the well-studied *E. coli,* however, such mechanisms do exist. This illustrates the danger in generalizing to all bacteria from studies with *E. coli.* If some marine or freshwater bacteria are more akin to *Zymomonas* in their lack of fine cellular control mechanisms, the measurement of processes such as protein or RNA synthesis or oxygen uptake may not be related to growth rate, even if there is a constant supply of nutrient. Thus, relating growth rate to most extensive properties (such as RNA content) can only be done under well-defined conditions and such studies cannot be extrapolated to all marine bacteria. It is unlikely, therefore, that methods for measurement of growth rate based on RNA synthesis meet the second criterion listed in Section 1. DNA synthesis, on the other hand, is more closely correlated to cell division and growth rates. Because there are more than one or two replication forks present in rapidly growing cells, multiple copies of the genome will be present; thus, the DNA content per cell is prone to vary with growth rate. The DNA content per unit biomass, however, does not vary to the same degree, because rapidly growing cells are larger than slowly growing ones (Maaloe and Kjeldgaard, 1966).

Where bacteria live in conditions of very low nutrient supply, such as in the ocean, the assimilated nutrients may be needed more for cell maintenance (e.g., protein turnover and osmotic balance) than for growth. Thus, at very low growth rates, the extra energy required for maintenance will alter the relationship between growth rate and energy production, which could lead to uncoupled growth (Pirt, 1975). In dealing with growth processes in natural systems, the principles gained from the study of pure cultures may not be applicable. Christian *et al.* (1982) have demonstrated this need for caution in their studies with mixed batch cultures from natural populations. They found that growth rate constants calculated from a variety of different measures of growth were variable. Referring to the criteria listed above, we see from the discussion in this section that the measurement of DNA synthesis appears to meet the second and third criteria better than the measurement of rates of synthesis of other macromolecules.

3. Biochemistry of Nucleic Acid Synthesis

3.1. Introduction

In principle, the measurement of growth rates using nucleic acid synthesis is simple, involving the measurement of the rate of incorporation of radioactive precursor into a macromolecule. Several conditions need to be observed: (1) The specific radioactivity of the precursor immediately before incorporation into the macromolecule must be known; (2) the radioactivity measured at the end of the experiment must only be in the macromolecule under consideration; and (3) the added radioactive molecule should be incorporated into the macromolecule by only one biosynthetic pathway, to avoid complications due to differing kinetics or degrees of participation. The processes of uptake of thymidine, assimilation into cellular constituents, and incorporation into DNA are all distinct, and the rates of occurrence of each vary between types of microorganism.

It is the final process, the immediate incorporation of thymidine into DNA through the action of thymidine kinase, that enables us to distinguish bacterial activity from that of other microbes, and thus makes thymidine so useful as a measure of heterotrophic growth rates. Thymine nucleotides, unlike those of other nucleic acid bases, have only one function in cells, participation in DNA synthesis (O'Donovan and Neuhard, 1970). In this respect, thymidine meets the first criterion listed in Section 1, in contrast with adenine or other precursors.

These points will be elaborated in the section below, where aspects of the biochemistry of DNA and RNA synthesis relevant to use of nucleic acid bases and nucleosides in natural aquatic systems are discussed. Most knowledge of these processes has been gained from studies on enteric bacteria, particularly *Escherichia coli*. As discussed above (Section 2), conclusions drawn from such work may suggest what happens in a natural community, but are not necessarily directly applicable. Furthermore, the application of rates of nucleic acid synthesis to growth rates in the sea involves studying the net activity of a whole community of different bacteria. By reporting a single growth rate we are making the simplifying assumption that the community behaves as a single species or population of a species.

A detailed description of DNA synthesis and pathways of nucleotide biosynthesis is given by Kornberg (1980). Some of the following is based on Kornberg's book, which may not always be cited specifically in the text.

3.2. Measurement of DNA Synthesis with Thymidine

3.2.1. Biochemistry of Thymidine Incorporation into DNA

Thymidine meets the criteria for pulse labeling DNA in bacteria reasonably well. It is rapidly and efficiently taken up by bacteria, is stable during uptake, is converted rapidly into nucleotides, and labels DNA with little or no dilution by intracellular pools (Kornberg, 1980). There are, however, pitfalls in its use, which must be considered. For environmental studies, the main problems are (1) dilution of the labeled thymine moiety during incorporation into DNA by other sources of thymine nucleotides; (2) degradation of thymidine within cells and subsequent random distribution of label; (3) the possible uptake of thymidine by microorganisms other than bacteria and, conversely, the lack of uptake by some bacteria; and (4) effects on the rate of DNA synthesis caused by disturbing the interactions between different microorganisms and their environment during experimental manipulation. These points are considered below.

There are two principal pathways for nucleotide biosynthesis in cells: (1) the *de novo* route, in which the nucleotides are synthesized from basic cellular components; and (2) the salvage pathway, in which free bases and nucleosides arising from breakdown of excess nucleotides or nucleic acids are converted back to nucleotide triphosphates.

Thymidine itself does not occur in the *de novo* pathway. Thymidine monophosphate (TMP)* is synthesized directly from deoxyuridine

*A note on nomenclature is necessary here. Deoxyribonucleotides are generally abbreviated as, e.g., dAMP for deoxyadenine monophosphate, and ribonucleotides as, e.g., AMP. Thymidine was originally known only as the deoxy form, as it is not (or rarely) found in RNA, and so the terms thymidine, deoxythymidine, or thymidine deoxyribose are used interchangeably, as are TMP, dTMP, etc. The following abbreviations are used in this chapter: A, adenine; Ad, adenosine; dAd, deoxyadenosine; dR-1-P, deoxyribose-1-phosphate; R-1-P, ribose-1-phosphate; I, inosine; IMP, inosine monophosphate; PRPP, phosphoribosylpyrophosphate; Tdr, thymidine; dTMP, dTDP, and dTTP, thymidine mono-, di-, and triphosphate, respectively; dUMP, dUDP, and dUTP, deoxyuridine mono-, di-, and triphosphate, respectively; dC, deoxycytidine; dCMP, dCDP, and dCTP, deoxycytidine mono-, di-, and triphosphate, respectively.

monophosphate by the enzyme thymidylate synthetase (Fig. 1). In the salvage pathway, thymidine is phosphorylated to form TMP by the enzyme thymidine kinase (Figs. 1 and 2). The base (thymine) is converted into the nucleoside (thymidine) by the action of the enzyme thymidine phosphorylase (Fig. 2). Deoxyribose-1-phosphate is needed, which is often supplied by another nucleoside; e.g., deoxyadenosine. Unless there is an adequate supply of deoxyribose-1-phosphate, this reaction does not occur, and in fact the reverse reaction may predominate. This is why thymidine and not thymine is used as a precursor to measure rates of DNA synthesis.

The relevance of these pathways to the use of labeled thymidine in aquatic environments is twofold. Firstly, much of the label may be lost by catabolism to thymine and then by loss of the labeled methyl group during further degradation of thymine (Fig. 2) (Fink and Fink, 1962; Vogels and van der Drift, 1976). There are no direct routes for tritium label to be incorporated into RNA or DNA after degradation of thymine. The tritiated methyl group enters the general pool of metabolites in the cell and eventually tritium may be distributed into all compounds, including RNA and DNA, but it will be considerably diluted. The amount of isotope incorporated into nucleic acids in such a way would be small compared to direct incorporation of thymidine into DNA, and would take some time (see Section 3.2.3). In short-term experiments (generally 10–30 min), such nonspecific labeling is insignificant (Pollard and Moriarty, 1984; Moriarty *et al.*, 1985a; Riemann, 1984). If the bacteria are not growing, not only DNA, but other macromolecules also are not labeled significantly above background levels (D. J. W. Moriarty, unpublished results). The effects of catabolism on the kinetics of labeling will be discussed below (Section 3.2.5).

Secondly, the main mechanism for the dilution of isotope in DNA is the action of thymidylate synthetase, in which TMP is derived from dUMP and is mixed with TMP formed by thymidine kinase from thymidine. In order to calculate the rate of DNA synthesis using radioactive thymidine, the specific radioactivity of the thymidine triphosphate pool must be known. Other sources of the thymidine inside and outside the cell may dilute the isotope, but on present experience this is not usual in bacteria from aquatic habitats. During conversion to thymidine triphosphate, the isotope may be diluted by synthesis of nucleotides from sources other than the added precursor. As the salvage and *de novo* pathways converge at the synthesis of dTMP (Fig. 1), it is here that the major dilution of isotope is likely to occur in growing cells. As dTDP and dTTP are synthesized only from dTMP, isotope in dTDP and dTTP would not be diluted further. There are pathways by which cytidine nucleotides can be converted to thymidine nucleotides via dUTP or dUMP (Fig. 1). Thus, dUTP and dUMP are key intermediates in the biosynthesis of thymidine nucleotides.

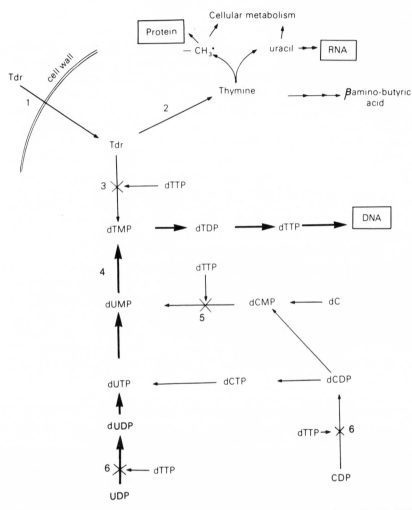

Figure 1. Pathways of thymine nucleotide metabolism. Tdr, Thymidine; dTMP, dTDP, and dTTP, thymidine mono-, di-, and triphosphate, respectively; dUMP, dUDP, and dUTP, deoxyuridine mono-, di-, and triphosphate, respectively; dC, deoxycytidine; dCMP, dCDP, and dCTP, deoxycytidine mono-, di-, and triphosphate, respectively. Enzymes: (1) active transport mechanism; (2) thymidine phosphorylase; (3) thymidine kinase; (4) thymidylate synthetase; (5) deoxycytidylate deaminase; (6) ribonucleoside reductase. The sites of feedback inhibition by dTTP are shown. Salvage and degradative pathways are shown with thin arrows; *de novo* pathways are shown with bold arrows.

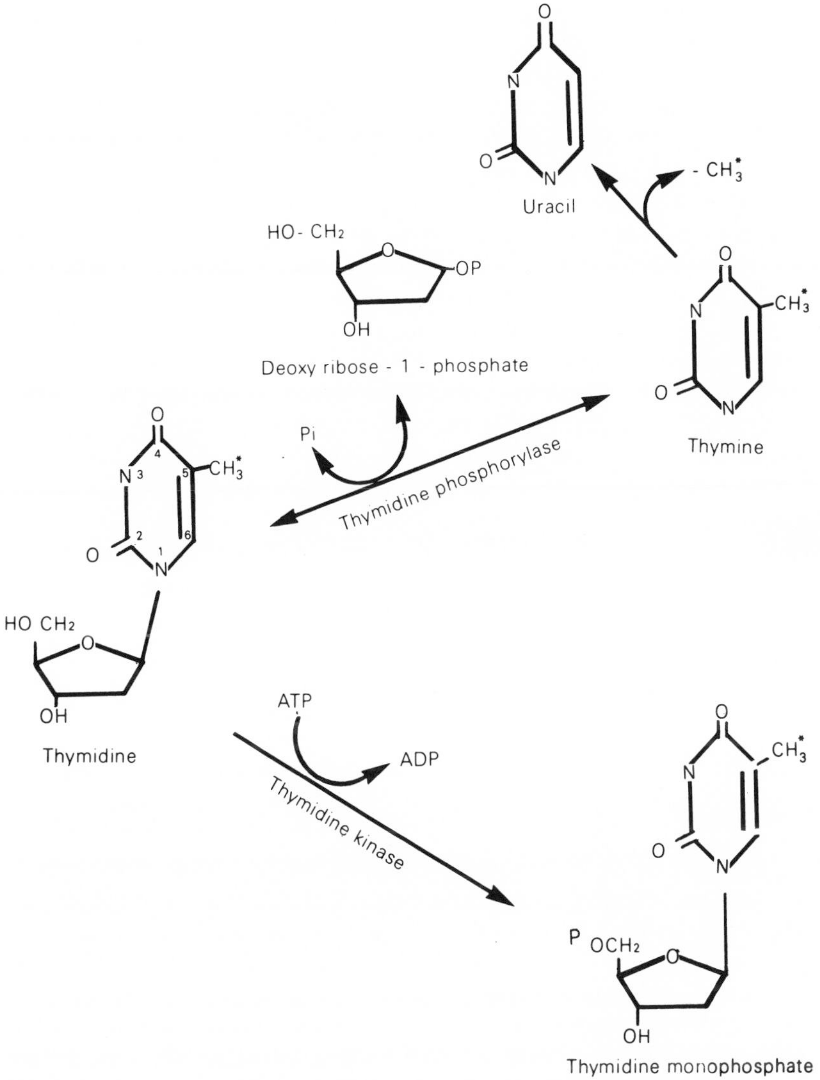

Figure 2. Reactions and structures of some compounds involving thymidine. The location of the tritium label is shown as an asterisk.

There is evidence that DNA polymerase is closely associated with a number of nucleotide kinases and that its activity is very much greater when the enzymes are organized together (Mathews *et al.,* 1979). This means that there are functional compartments of nucleotide precursors inside the multienzyme complexes in which high concentrations of precursors are available at the site where they are used; i.e., the replication fork on the DNA. These pools would be small and turning over rapidly, in contrast to larger pools of the same nucleotides that would be dispersed more generally in the cytoplasm and be available for repair of DNA and perhaps regulation of nucleotide biosynthesis, but not normal DNA synthesis. There must, therefore, be more than one pool of dTTP (and other trinucleotides) in cells. Evidence for such pools of dTTP in bacteria was provided by Werner (1971), who showed that the rate of labeling of DNA by thymine or thymidine was much faster than the rate of labeling of the extractable dTTP pool. Thus, the specific activity of the dTTP pool that is used for DNA synthesis cannot be measured by extracting dTTP; it can, however, be measured using an isotope-dilution analysis (see Section 3.2.6).

Probably all organisms regulate nucleotide biosynthesis to prevent unnecessary buildup of nucleotide triphosphates. There are complex interactions between many nucleotides involving activation or inhibition of certain key biosynthetic enzymes. Thymidine triphosphate is an important regulator in pyrimidine nucleotide biosynthesis and this means that the pool size of dTTP must be closely regulated in order for it to act in this way (Maley and Maley, 1972). There are two routes for the *de novo* synthesis of thymidine nucleotides, one via cytidine nucleotides and one via uridine nucleotides (Fig. 2). Both are regulated by dTTP, which acts as an inhibitor of ribonucleoside reductase and deoxycytidylate deaminase. An increase in the dTTP pool size will slow down or turn off the supply of dTMP by *de novo* synthesis, depending on the size of the pool (Kuebbing and Werner, 1975). When measurements are made of bacterial growth rates in natural systems, sufficient thymidine can be added to inhibit *de novo* synthesis completely (Pollard and Moriarty, 1984). Thus, isotope dilution can be prevented. This makes ecological work simpler, because the specific activity of dTTP used for DNA synthesis is not altered from that of the tritiated thymidine supplied, and only one measurement is needed. As noted below (Section 3.2.6), this conclusion is based on laboratory cultures with copiotrophs. It may not apply to oligotrophs and needs to be checked in oligotrophic environments. [Oligotrophs are bacteria that are adapted to growth at very low concentrations of nutrients, whereas copiotrophs are bacteria that grow only when copious quantities of nutrients are supplied, for example, in the usual laboratory media; see Poindexter (1981).]

3.2.2. Transport of Thymidine

Before DNA can be labeled with radioactive thymidine, the thymidine needs to be rapidly and efficiently taken up by the cells. Furthermore, the enzyme thymidine kinase must be present, as shown in Section

3.2.1 and discussed in detail in Section 3.2.3. Very little information is available on pyrimidine nucleoside transport into most bacteria. Thymidine, uridine, cytidine, and deoxycytidine are all transported chemically intact into *E. coli* by an energy-dependent process. The purines adenosine and guanosine are transported similarly. At least two different transport mechanisms have been identified (Munch-Petersen *et al.*, 1979). In normal wild-type cells, the processes of transport and catabolism are closely linked but separate. The catabolic enzymes, although located close to the cell surface, were inside the membrane. A similar separation of transport and catabolism could be expected to occur in other bacteria, but no information is available.

There are interactions between nucleosides during uptake, indicating that at least one mechanism is shared by different nucleosides. Thymidine competed with uridine during transport (Roy-Burman and Visser, 1981). Mygind and Munch-Petersen (1975) also showed that nucleosides competed for the same transport mechanism, or at least part of the transport mechanism. Thus, the extent of labeling of DNA with tritiated thymidine will be influenced by the presence of other nucleotides in the environment outside the cells. The rate of tritiated thymidine incorporation into DNA in sediment bacteria was depressed by uridine (a pyrimidine), but not by hypoxanthine (a purine) (D. J. W. Moriarty and P. C. Pollard, unpublished results). The most likely explanation for this is that uridine competed for uptake sites with thymidine in the population of sediment bacteria.

Probably not all bacteria have mechanisms for efficient transport of nucleosides or bases. Some species of marine *Pseudomonas* do not incorporate labeled thymidine into DNA because they are unable to take up thymidine into the cell (Pollard and Moriarty, 1984). Bacteria with very limited nutrient requirements, such as chemolithotrophic bacteria, may also lack such transport systems. Sulfate-reducing bacteria have limited nutrient requirements, and it has been found that tritiated thymidine was very poorly incorporated into their DNA (D. J. W. Moriarty and G. W. Skyring, unpublished results). Bacteria that cannot transport nucleosides may not be able to transport free bases either.

3.2.3. Thymidine Kinase

After transport into the cell, the next stage in incorporation of thymidine into DNA is its conversion to TMP by the salvage pathway enzyme thymidine kinase (Fig. 2). Although most organisms were once thought to have this salvage pathway (Kornberg, 1980), some groups of microorganisms are now known to lack it. Thymidine kinase is not present in the fungi *Neurospora crassa, Aspergillus nidulans,* and *Saccharo-*

myces cerevisiae, the alga *Euglena gracilis* (Grivell and Jackson, 1968), and the cyanobacteria *Anacystis* and *Synechocystis* (Glaser *et al.*, 1973). The nuclei of the eucaryotic algae *Bryopsis, Chlamydomonas, Dictyota, Euglena, Padina,* and *Spirogyra* lack thymidine kinase; although a small amount of label was incorporated, it required hours or days of incubation to be shown by autoradiography (Stocking and Gifford, 1959; Sagan, 1965; Steffensen and Sheridan, 1965; Swinton and Hanawalt, 1972). According to Sagan (1965), this low rate of labeling was probably due to degradation of thymidine and incorporation into RNA and protein. We have examined species from three genera of microalgae (*Thalassiosira, Isochrysis,* and *Platymonas*) and a marine *Synechococcus* and have found no significant incorporation of tritiated thymidine into their DNA (Pollard and Moriarty, 1984). Similarly, Bern (1985) has shown that bacteria took up tritiated thymidine from lake water, whereas a variety of blue-green and eucaryotic algae did not.

It was the lack of tritium incorporation into DNA in some eucaryotic fungi and algae and the nonspecific labeling of DNA and RNA by [^{14}C]thymidine that led Grivell and Jackson (1968) to show that these organisms did not have thymidine kinase. Since all cyanobacteria, eucaryotic algal nuclei, and fungi that have been investigated do not contain thymidine kinase, it seems reasonable to conclude that that is a general phenomenon in these groups and thus that they do not incorporate thymidine into DNA. In support of this generalization, Fuhrman and Azam (1980) reported that over 90% of total incorporation of thymidine into TCA-insoluble matter was into particles less than 1 μm in size. There are reports of thymidine incorporation into DNA in algae and protozoa. Some of these (e.g., Grivell and Jackson, 1968) used [2-^{14}C]thymidine, not [^{3}H-methyl]thymidine. Much longer time periods and higher concentrations of thymidine than are used to measure bacterial growth rates were needed to demonstrate incorporation of radioactivity into DNA in these eucaryotes.

Even if some algae do take up thymidine, they cannot incorporate the tritium label on the thymidine into DNA immediately. The thymidine is catabolized, and after some time the labeled methyl group is distributed among various cellular components, including protein and RNA. Protozoa, however, probably do contain thymidine kinase (Plaut and Sagan, 1958; Stone and Prescott, 1964). They do not feed primarily on dissolved organic compounds, and thus are unlikely to have efficient transport mechanisms that can take up the nanomolar concentrations of thymidine supplied during experimental manipulation. Cycloheximide, an inhibitor of DNA synthesis in eucaryotes, has no effect on the incorporation of thymidine into DNA in marine sediments and seawater (Moriarty and Pollard, 1982; and unpublished observations). Thus, although bacteria may not be the only organisms that can utilize thymidine for DNA synthesis, they are the only ones that do so over a short period.

The experimental evidence discussed in this review shows clearly that thymidine is useful as a measure of bacterial growth rates, but there are pitfalls. In particular, experiments should be for short time intervals at nanomolar concentrations. Under these circumstances, the rate of labeling of DNA in eucaryotes will be slow compared to that in bacteria.

3.2.4. Uptake of Thymidine by Different Organisms

Many microorganisms other than bacteria may be able to take up thymidine, but unless they also have the enzyme thymidine kinase, DNA will not be rapidly labeled. As discussed in Section 3.2.3 on thymidine kinase, this means that labeling of DNA by thymidine is specific for bacteria in short-term experiments. The use of short-term experiments needs to be stressed here. Over a long period (2–24 hr) label may be incorporated into DNA of eucaryotes, including protozoa. In at least one seawater sample examined, it seems that only a small proportion of the active bacteria cannot utilize thymidine. This conclusion was reached by Fuhrman and Azam (1982) after a careful autoradiographic study. Ramsay (1974) found that fewer bacteria were labeled with tritiated thymidine than with glucose in a freshwater habitat. It is possible that some of the difference may have been due to a different amount of isotope taken up, as discussed by Fuhrman and Azam (1982), but this cannot be the full explanation. *Pseudomonas fluorescens* and a pseudomonad strain isolated from a freshwater lake could not be labeled with [³H]thymidine (Ramsay, 1974). A few species of aquatic pseudomonads have been found not to incorporate tritiated thymidine into DNA (Pollard and Moriarty, 1984). Thus, not all marine or freshwater heterotrophic bacteria can utilize thymidine, probably due to a deficiency in transport of thymidine. Those bacteria that cannot utilize thymidine probably constitute only a small proportion of the total number of bacteria in aquatic systems. Measurements of growth rates with tritiated thymidine agree well with measurements made using direct counts of bacterial numbers in containers (Bell *et al.,* 1983; Fuhrman and Azam, 1982; Kirchman, *et al.,* 1982). The work of Kirchman *et al.* (1982) indicates that most bacteria in the system they studied utilized thymidine.

3.2.5. Degradation of Thymidine

Thymidine is rapidly degraded within cells, initially by the inducible enzyme thymidine phosphorylase. The tritium on the methyl group is transferred to other compounds, including water, and eventually will be incorporated by *de novo* pathways into RNA, DNA, and protein. It will be considerably diluted during these processes, but these processes would become apparent during any long-term experiments. Thus, other compounds insoluble in cold trichloroacetic acid (TCA), in addition to DNA, would be labeled in mixed populations of microorganisms, but in the short term (usually less than 1 hr) specific labeling of bacterial DNA predominates. Karl (1982) disputes this statement, but unfortunately, he used long-term incubations and used acid in the initial extraction procedure. Deoxyribonucleic acid is very labile to acid, and readily fragments on subsequent treatment with alkali, thus appearing in RNA fractions (Munro and Fleck, 1966). Riemann (1984) has studied the distribution of tritium from thymidine into various macromolecules and showed that, in general, most was incorporated into DNA.

In marine sediments, thymidine was incorporated into DNA at a linear rate for an initial period of 8 min at warm (31°C) temperatures and up to 20 min at colder (18°C) temperatures (Fig. 3). Slower rates of incor-

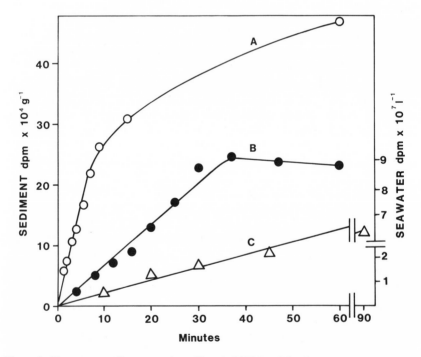

Figure 3. Time course of incorporation of [methyl-^{3}H]thymidine into DNA. (A) Sediment, temperature 31°C; (B) sediment, temperature 18°C; (C) seawater, temperature 18°C. [Redrawn from Moriarty and Pollard (1981, 1982).]

poration often followed the initial period. These kinetic studies indicate that the supply of labeled thymidine became limiting after a short period, due to adsorption of thymidine in the sediment (D. J. W. Moriarty and P. C. Pollard, unpublished results). Thus, in working with sediments, higher concentrations of thymidine and short-term analyses are needed (over a period of 5–20 min). Only results based on the initial linear period of incorporation can be used to calculate growth rates.

Linear rates of incorporation occurred for a longer time in seawater, probably because the concentration of thymidine was higher than in sediment interstitial water (Fig. 3). Effects due to labeling of other macromolecules may become apparent after a short time. In one experiment (Fig. 4) the rate of incorporation of label into extracted DNA differed from the rate of incorporation of label into TCA-insoluble compounds after about 30–40 min. In the first 30 min, the two rates agreed well, indicating that all label was being incorporated into DNA (i.e., bacterial DNA) initially. Hollibaugh *et al.* (1980) found that 82% of the tritium in macromolecules was in DNA in the first hour of incubation of a seawater sample. As Riemann (1984) has shown, the chemical methods for separating DNA from other macromolecules do not give clear-cut results. The relative differences between the amounts of label in DNA and protein and the rates of label incorporation into these macromolecules differ between water bodies, depending on factors such as temperature, microbial composition, and nutrient availability.

Correct ecological interpretations cannot be made without accurate values for the rate of incorporation of thymidine into DNA. If the rate of incorporation of tritium into any other macromolecules or organic matter is measured, growth rates cannot be calculated with confidence because, unless growth is balanced, only DNA synthesis is directly cor-

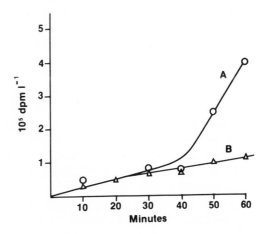

Figure 4. Time course for incorporation of tritiated thymidine into all macromolecules insoluble in (A) trichloroacetic acid and (B) purified DNA in coral reef waters. [From Moriarty *et al.* (1985b).]

related with cell growth (see Section 2). In the water column it is sufficient to use incorporation into TCA-insoluble matter over a short time interval, because mostly DNA is labeled initially (Moriarty *et al.,* 1985a; Pollard and Moriarty, 1984). A correction factor for tritium in other macromolecules may give incorrect results; this factor can change with time. The validity of the technique for each new environment is best established using a time course experiment.

3.2.6. Isotope Dilution

It is not possible to measure the specific activity of any nucleotide at the site of DNA replication simply by extracting the nucleotides and measuring it directly, because there are functionally separate pools of nucleotides in the multienzyme complexes (see Section 3.2.1).

The specific radioactivity of precursors at the site of macromolecule synthesis can be determined using an isotope dilution analysis (Forsdyke, 1968, 1971; Sjostrom and Forsdyke, 1974; Scott and Forsdyke, 1976, 1980). This procedure has been used to measure rates of DNA synthesis in marine environments (Moriarty and Pollard, 1981; 1982).

The principle of the technique is as follows: a series of samples are incubated with a constant amount of radioactive thymidine to which increasing amounts of unlabeled thymidine are added. The DNA is extracted and the reciprocals of the amounts of radioactivity in DNA are plotted against the amounts of thymidine present (Fig. 5). If there is no dilution of the isotope incorporated into DNA by any sources other than the unlabeled thymidine that was added, the plot will pass through zero (e.g., Fig. 5A). A negative intercept on the ordinate is an estimate of the amount of dilution of isotope by other sources of thymine in DNA (e.g., Fig. 5B). It is not strictly a pool of thymidine, but represents the sum of all pools that dilute the tritiated thymidine prior to incorporation into DNA.

The isotope dilution method measures the dilution of labeled thymine in dTTP, the final precursor to DNA, by all sources of thymine, because the effect of added thymidine on incorporation of radioactivity into DNA itself is measured. A necessary condition is that the rate-limiting step for incorporation of thymidine be the final one, i.e., DNA polymerase. Provided the concentration of thymidine is sufficiently high, this condition is met in bacteria with normal regulatory mechanisms (Pollard and Moriarty, 1984). Oligotrophic bacteria may not, however, regulate DNA synthesis in the normal way, and dilution could still occur (see this section below).

Isotope dilution experiments need not be carried out on every sample if a sufficiently high concentration of labeled thymidine is used (Pol-

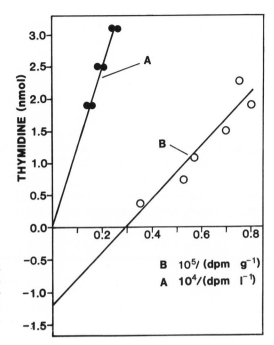

Figure 5. Isotope dilution plots of incorporation of tritiated thymidine into DNA in (A) seawater and (B) epiphytic flocculent sediment on seagrass. [From Moriarty and Pollard (1982).]

lard and Moriarty, 1984). Isotope dilution was more apparent in sediments, especially when growth rates were high, because thymidine was adsorbed to the sediment and thus the effective concentration was much less than that expected from the level of addition (Moriarty and Pollard, 1981, 1982). Large amounts of thymidine, compared to the amounts used in water column work, need to be added to sediment to block *de novo* synthesis and thus prevent isotope dilution. Alternatively, small amounts of sediment can be used.

An alternative procedure for checking whether isotope is being diluted as a result of *de novo* synthesis is to add increasing amounts of tritiated thymidine at the same specific activity. The concentration of thymidine at which no further radioactivity is incorporated should be used. *De novo* synthesis should be fully inhibited at this concentration; typical values for this are 5 nM in lake water (Bell *et al.,* 1983), and 5–10 nM in seawater (Fuhrman and Azam, 1982; Kirchman *et al.,* 1982). From personal observations and discussions with H. W. Ducklow and J. A. Fuhrman, I would recommend that a minimum of 20 nM thymidine be used. Where there are no large pools of thymidine (i.e., >1 nM), this experiment measures the effect of *de novo* synthesis. If, however, pools of thymidine are present, a double reciprocal plot would be needed to measure the pool size (Hunter and Francke, 1974), but this is not accurate, because a very large pool will have a value close to zero when plotted as a reciprocal. In my experience, pools of thymidine are rarely, if at all, large enough to have any effect and thus the alternative procedure does work.

Hagstrom (1984) has suggested that a better approach is to extract pure DNA from natural samples and measure the dilution of tritiated thymidine directly. It may be possible to purify DNA from water, but it is very difficult to do so from sediment, because humic compounds adsorb strongly to DNA (Torsvik, 1980). There is, however, a problem in deciding what proportion of purified DNA is from growing bacteria and what is from other organisms or inactive bacteria. Unless this proportion is known, a growth rate cannot be estimated.

The best method for measuring dilution of radioactive thymidine is being debated in the literature. This debate is unresolved because no method that has been proposed thus far is free of methodological problems when applied to field samples. It is better to attempt to minimize or measure dilution than to assume an arbitrary value that has been published for other environments or populations.

The process of tritiated thymidine incorporation into DNA is distinct from the process of uptake of an isotopically labeled substrate into cells. The velocity of uptake is dependent on substrate concentration, but, as stressed above, the synthesis of DNA is independent of uptake and biochemical conversions of precursors, i.e., the rate of DNA synthesis is independent of added precursor concentration. Bacteria take up thymidine very much faster than they incorporate it into DNA and thus the rate of incorporation of label into DNA is not influenced by radioactive thymidine concentration while zeroth-order kinetics applies. Laws (1983) has incorrectly criticized the work of Forsdyke and his colleagues (cited above) by concluding that the mathematics was faulty and thus that the isotope dilution analysis could not be used as proposed here. The mathematics is correct, but it is only a tool and cannot prove or disprove the hypothesis concerning measurement of isotope dilution. It is the biological premise on which the mathematics is based that must be disproved, and, as Laws stated, that rests on the assumption that the velocity of uptake is completely independent of the concentration of the radioactive precursor. In the case of DNA synthesis the premise is true for bacteria with normal regulatory mechanisms.

The isotope dilution methodology has been validated by experiments with the marine bacterium *Alteromonas undina* grown in a chemostat (Pollard and Moriarty, 1984). Growth rates measured directly compared well with rates calculated using the isotope dilution technique with tritiated thymidine (Table I). Two different levels of isotope dilution were

Table I. Comparison of Two Methods for Measuring
Bacterial Growth Rates[a]

Percentage dilution	Specific growth rate (hr^{-1})	
	Direct microscopy	DNA synthesis
0	0.3 ± 0.04	0.28 ± 0.03
44	0.22 ± 0.04	0.29 ± 0.07

[a]From Pollard and Moriarty (1984). *Alteromonas undina* was grown in a
chemostat and growth rates were measured by direct microscopy of acridine orange-stained preparations and by the rate of incorporation of tritiated thymidine into DNA, using the isotope dilution methodology.

used. Therefore, in those experiments the isotope dilution method did
give an adequate estimate of the specific activity of dTTP, the final precursor of thymine in DNA. Such experiments, where growth rates measured using direct microscopy agree with those estimated from the rate
of thymidine incorporation into DNA, show that uptake and thymidine
kinase were not rate-limiting steps. The rate-limiting step for tritiated
thymidine incorporation must have been at the level of DNA polymerase, otherwise the thymidine method would have underestimated growth
rates.

Isotope dilution is not invariant in particular populations of bacteria, but is dependent on a number of factors, particularly the concentration of tritiated thymidine around the bacteria. To summarize the discussion above on the biochemistry of DNA synthesis, we see that bacteria
regulate the concentration of dTTP. If thymidine is supplied at a sufficiently high concentration, it is used preferentially for DNA synthesis
(provided it can be taken up), and *de novo* synthesis of thymidine nucleotides is inhibited. A note of caution is necessary here, however. Further
work is needed to check the validity of the thymidine method in the field.
For example, if a growing population is dominated by bacteria that cannot easily take up thymidine, transport may be the rate-limiting step, and
thus isotope dilution would still occur, but not be measurable. Oligotrophic bacteria may not regulate DNA synthesis in the same way as copiotrophic bacteria, such as *Alteromonas undina,* and thus would not
respond in the same way to added thymidine. Ideally, metabolism in oligotrophic bacteria is regulated by the concentration of substrates (Poindexter, 1981). Thus, studies of growth with thymidine (or adenine and
other precursors of nucleic acids) in environments where oligotrophs
might predominate, such as the open ocean, need to be interpreted with
caution. The good correlations between values for bacterial production in
the North Sea estimated with the thymidine method and two other methods indicate that isotope dilution did not occur (Lancelot and Billen,

1984). These and other studies (see Section 3.2.7.), which show a conversion factor of about 2×10^{18} cells dividing/mole thymidine incorporated, support the argument that isotope dilution is prevented by a sufficiently high concentration of thymidine.

Fuhrman and Azam (1982) measured DNA synthesis in bacteria from coastal waters and open ocean using the thymidine technique and a [^{32}P]phosphate technique. They concluded that the thymidine was being diluted, particularly in the open ocean bacteria. Although it is difficult to measure rates of DNA synthesis accurately with ^{32}P (Fuhrman and Azam, 1982; Moriarty, 1984) and thus decide how great the thymidine dilution is, the difference between coastal and open ocean bacteria is difficult to explain. Ducklow and Hill (1985) have reported that rates of thymidine incorporation could not be reconciled with changes in the numbers of bacteria growing in seawater cultures in the open ocean. It may be that oligotrophic bacteria are common in open ocean water and differ significantly in their biochemistry from "ordinary" bacteria.

Some workers have reported nonlinear isotope dilution plots and hence uncertainty in interpreting results (Riemann and Sondergaard, 1984; Riemann et al., 1984). Such problems may be methodological, because there is continual improvement in the methods used for measuring thymidine incorporation into DNA (Pollard and Moriarty, 1984). It is also possible that there are populations of bacteria that have different biochemical regulatory mechanisms, as suggested above, and thus do not respond as predicted to changes in thymidine concentration. More studies are necessary on these problems in the use of tritiated thymidine.

3.2.7. Calculation of Growth Rates

In order to calculate growth rates of organisms from the rate of synthesis of a macromolecule, we need to know the amount of that macromolecule per cell; the amount should also remain constant per unit biomass, or nearly so. For this reason, DNA synthesis is much better than RNA synthesis as a measure of growth. As pointed out in Section 2, RNA content may be very variable and its rate of synthesis cannot be used as a measure of growth in natural systems. The genome size of most bacteria is within the range $(1-3.6) \times 10^9$ daltons (Gillis et al., 1970; Wallace and Morowitz, 1973). The average genome size is 2.5×10^9 daltons, or 4×10^{-15} g DNA cell^{-1}. Fuhrman and Azam (1982) measured the DNA content of a mixed population of marine bacteria and obtained a value of 2.6×10^{-15} g cell^{-1}.

Thymine bases in DNA are assumed to be 25% of the total number of bases. For the average of a mixed population this is a reasonable value. Most GC ratios (guanine + cytosine) of marine bacteria lie in the range

35–70%. A factor of 1.3×10^{18} to convert moles of thymidine incorporated into number of DNA molecules synthesized (i.e., number of bacteria dividing) was derived by Moriarty and Pollard (1981, 1982). Fuhrman and Azam (1980) used a range of 2.0×10^{17} to 1.3×10^{18} and later increased this to an average of 1.7×10^{18} for nearshore waters and 2.4×10^{18} for offshore waters, which led to better agreement with other measurements (Fuhrman and Azam, 1982). Using the average amount of DNA cell^{-1} determined by Fuhrman and Azam (1982), I conclude that a conversion factor of 2×10^{18} is a good round figure in agreement with information currently available, at least for nonoligotrophic environments.

Some recent studies by Bell et al. (1983) show that this conversion factor is applicable to freshwater bacterial growth as well. They measured bacterial growth in lakewater cultures using direct microscopy and tritiated thymidine incorporation and calculated conversion factors that ranged from 1.9×10^{18} to 2.2×10^{18} cells dividing mole^{-1} thymidine incorporated. Further work is needed to improve the accuracy of conversion factors and to determine whether they differ significantly between environments. Ducklow and Hill (1985) have found that a conversion factor of 4.0×10^{18} best fitted data obtained from seawater cultures of natural populations in warm core Gulf Stream rings, which are oligotrophic.

The value of 2.0×10^{18} is based solely on the size of the DNA molecule and its thymine content. It is assumed either that the radioactivity measured in the experiments is in DNA only, and that there is no dilution of the radioactivity, or that adequate corrections have been made for these sources of error. Where very different conversion factors are found, it may indicate that one of the above assumptions is not valid. For example, isotope dilution could occur, but not be measurable as discussed in the previous section (3.2.6).

As Fuhrman and Azam (1982) point out, errors arise in the calculation of conversion factors because assumptions and extrapolations are needed. They attempted to calculate a factor directly by measuring the incorporation of thymidine into DNA and the increase in number of bacteria in two experiments with samples of water treated to remove predators on bacteria. In one experiment, water was passed through a 3-μm filter and allowed to incubate; in the other, water sterilized by filtration was inoculated with a natural population and incubated. Their results gave a conversion factor of 1.3×10^{18} cells produced/mole thymidine incorporated.

A somewhat different approach was used by Kirchman et al. (1982). They diluted seawater tenfold with filtered seawater and compared the rate of increase in cell numbers with incorporation of tritium from thymidine into TCA-insoluble matter. They found conversion factors ranging from 1.9×10^{18} to 6.8×10^{18} cells produced mole^{-1} thymidine incorporated into TCA-insoluble matter. They developed a mathematical model to relate growth and cell division to the rate of incorporation of a radioactive precursor into a macromolecule. An assumption was made that the bacterial population growing in nature was the same as that growing in the seawater culture. Two conditions need to be met for the application of their method: (1) the time period for assay in culture and the environment should be short enough to ensure that DNA is the only macromolecule labeled; and (2) isotope dilution should be measured or, better still, the concentration of thymidine should be high enough to eliminate it. These conditions are necessary because growth rates and the rates of synthesis of different macromolecules may be different in the seawater culture than in the natural environment. An advantage of their procedure for estimating a conversion factor for a particular environment is that it takes into account bacteria that cannot utilize thymidine, as well as avoiding the need to make assumptions about the genome size and proportion of thymine in DNA.

In calculating conversion factors from data obtained with culture studies, further problems arise in determining the type of growth curve to be analyzed. Growth may be linear or exponential and, in diluted water from natural aquatic systems, it may be difficult to decide which form the growth curve takes and therefore the time period over which the conversion factor should be estimated. In addition, a lag phase may occur (Christian *et al.*, 1982).

3.3. Biochemistry of Adenine Incorporation into DNA and RNA

In contrast to pathways for thymidine involvement in DNA synthesis, those of adenine and its nucleotides are very complex. Adenine nucleotides have many functions in cellular metabolism, particularly in regulation of biosynthesis and in energy transfer and storage. For the sake of clarity only some of the pathways of nucleic acid synthesis are shown in Fig. 6. These may differ in detail for different microorganisms. Salvage pathways are very important in supplying nucleotides for nucleic acid synthesis. There is constant degradation of nucleic acids, mainly various forms of RNA and especially messenger RNA, which has a short half-life. Some of the salvage pathways involving adenine, and interconversions between nucleosides and nucleotides are shown in Fig. 6. The *de novo* route of synthesis is via inosine monophosphate to AMP for ribonucleotides. Deoxyribonucleotides are synthesized by reduction of the corresponding ribonucleotide diphosphate in most organisms. Some bacteria

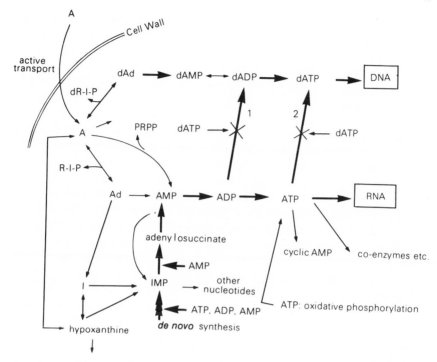

Figure 6. Pathways of adenine nucleotide metabolism. The principal pathways for *de novo* synthesis are shown by heavy arrows. Unlabeled arrows indicate other reactions not shown here. A, Adenine; Ad, adenosine; dAd, deoxyadenosine; I, inosine; IMP, inosine phosphate; R-1-P, ribose-1-phosphate; dR-1-P, deoxyribose-1-phosphate; PRPP, phosphoribosylpyrophosphate. Enzymes: (1) ribonucleoside reductase, usual type; (2) ribonucleoside reductase, less common. Sites of feedback inhibition by dATP and other nucleotides are shown.

have an enzyme that reduces the ribonucleotide triphosphate (Thelander and Reichard, 1979).

Adenine is transported into the cell intact by active transport (i.e., it is energy dependent), which is facilitated by rapid conversion to adenosine or adenosine monophosphate (Roy-Burman and Visser, 1975; Burton, 1977). The label from the adenine would be distributed among many adenine compounds (Fig. 6). Tritiated adenine (labeled in the [2-H] position) would not give rise to labeled guanosine nucleotides, because the 2 position is substituted differently.

The complexity of the pathways for distribution of labeled adenine and compartmentalization of different processes, as well as individual differences between different microorganisms, cannot be adequately depicted here (Fig. 6). For example, the synthesis of ATP is shown here mediated by the enzyme adenylate diphosphate kinase as a precursor to RNA or DNA synthesis. Elsewhere in the cell, however, there are other mechanisms for ATP synthesis, particularly substrate-level phosphorylation and oxidative phosphorylation in heterotrophs and photophosphorylation in algae. Labeled adenine would undoubtedly enter the pools of adenine nucleotides used in energy metabolism.

Because there are separate sites for DNA synthesis, RNA synthesis, and various energy supply processes and the reaction kinetics vary, particularly where multienzyme complexes are involved, rapid equilibration of labeled adenine in various pools would not be expected to occur. Indeed, Karl et al. (1981) show that the specific activity of ATP pools in a complex freshwater community and in a pure culture of a marine bacterium do not equilibrate for up to 4 hr. They found that pools of AMP and ATP were separated in different compartments. Winn and Karl (1984a) have discussed the problems in determining rates of nucleic acid synthesis with tritiated adenine when the specific activity of ATP changes during the incubation. They showed that the specific activity did stabilize if adenine was always present to excess in the medium and a sufficiently long incubation time (10% of the generation time) was used. If, therefore, tritiated adenine is used as a short-term pulse label, results will be difficult to interpret. There is, however, a possibility that the confinement of microorganisms in a bottle for a long time may alter growth rates, and this effect may be difficult to recognize.

The specific activity of the ATP pool that is used for RNA synthesis cannot be measured directly by extracting ATP from cells, because the different pools of ATP cannot be extracted separately (Fuhrman and Azam, 1980). Karl (1981) has attempted to circumvent the problems of actually measuring the specific activity of ATP in natural environments by relating the ratio, (rate of RNA synthesis/rate of DNA synthesis), to specific growth rate. The disadvantage of this technique is that the incorporation of adenine into nucleic acids is not specific to bacteria and if growth is not balanced, the rate of RNA synthesis may not be proportional to DNA synthesis or growth, particularly in mixed communities of microorganisms.

In separate, but not mixed, cultures of rapidly growing bacteria and algae, the adenine procedure did give good estimates of the true growth rates (Winn and Karl, 1984a). In order to explain their results, the authors had to demonstrate that bacterial activity was minimal in the algal culture. More work is needed to show not only whether the rates of nucleic acid synthesis measured in the field can be reliably related to growth, but also the microorganisms to which the rates apply. At this stage, it is my conclusion that the difficulties with interpretation are too great for adenine to be a useful precursor for measuring microbial or bacterial growth rates in natural environments.

There are a number of reasons why adenine cannot be used to measure accurately growth rates of microbes in the natural environment; some of these have been mentioned above and are summarized here. Problems in the measurement of specific activity of ATP, the final precursor to RNA synthesis, and dATP, the precursor to DNA synthesis, by extraction of ATP are further compounded in the natural environment. Even if there were no compartments within a cell, in a community of organisms there would be many that were not growing or that had very different pool sizes of ATP and differing growth rates.

Bacteria, with their efficient transport mechanisms, are likely to take up adenine more rapidly than are algae or protozoa, and as their ATP pools are smaller, the specific radioactivity of ATP would be higher. The rates of turnover and synthesis of new nucleic acids would differ between eucaryotes and procaryotes, which would complicate interpretation of results. Techniques are needed that allow measurement of processes within smaller ecosystem compartments and thus let us refine models for ecosystem function. The microbial world includes primary producers using light or chemical energy, herbivores, decomposers, and carnivores; in other words, many different trophic levels are included. The adenine method lumps all these together.

There may be particular environments where one group of microorganisms (e.g., bacteria) predominates, and thus results of the adenine method could be meaningfully interpreted. If, however, bacteria predominate, the thymidine method is preferable. A direct comparison between the two methods in such environments would be interesting.

Adenine, or precursors other than thymidine, may be useful in restricted environments where information is needed on RNA synthesis, or perhaps DNA synthesis in eucaryotes (e.g., fungi). The isotope dilution procedure of Forsdyke (see Section 3.2.6) would be preferable for determining specific radioactivity. Selective inhibitors could also be useful.

4. Ecological Significance of Bacterial Growth Rate Measurements

4.1. Introduction

Many research workers are now using the tritiated thymidine method to determine bacterial growth rates in aquatic systems. Most workers agree that the thymidine method gives a good measure of bacterial growth, although there is some discussion about how close the measured values are to the real values. Continual improvements in methodology and interpretation in ecological terms are occurring. It should be noted, however, that conversion factors are not necessarily universal, and may vary between different aquatic systems or microbial communities. Independent, corroborative data on bacterial growth in nature are necessary, but are difficult to obtain (Ducklow and Hill, 1985). In their original paper, Fuhrman and Azam (1980) commented on the need for caution in calculating growth rates from the thymidine data; this comment still applies. Nevertheless, the thymidine method is very promising and deserves further development. It has not only provided information that supports estimates of bacterial growth rates from other methods, but has also enabled microbial ecologists to study bacterial growth in environments (e.g., sediments) where other techniques do not work. Some of the applications and results obtained from the use of the tritiated thymidine and adenine methods are discussed below.

4.2. Comparison of Thymidine and Other Methods

Values for bacterial growth rates in natural systems determined with the thymidine method have been compared directly with those made using other techniques. Fuhrman and Azam (1980, 1982) have compared growth rates measured with tritiated thymidine and direct counts of bacteria in enclosed water samples and found good agreement. Newell and Fallon (1982) also tried to compare the two techniques, but did not obtain consistent growth with enclosed samples. In order to measure growth rates with direct counts in culture, predators of bacteria have to be removed (usually by filtration with 3-μm filters). Algae will also be removed, and because there is a coupling between bacterial growth and supply of nutrients from algae, filtration is likely to disrupt bacterial growth rates. Furthermore, many apparently free bacteria may be utilizing nutrients on surfaces, and may be separated from surfaces by shear forces during filtration (Hermansson and Marshall, 1985). This problem illustrates one of the advantages of using tritiated thymidine. Experiments may be carried out in a time short enough to avoid the effects of predators or change in supply of substrates. As pointed out in Section 2, DNA synthesis is unlikely to be immediately affected by a change in nutrient status, but should continue without change until replication is completed. Toxic effects (e.g., contaminants in sample bottles or O_2 in a very anaerobic environment) are likely to have an immediate effect, however. Oligotrophic bacteria may respond immediately to changes in substrate concentration. Thus, changes in growth rates may be induced by the enclosure of water in bottles and such effects need to be considered.

Ducklow and Hill (1985) showed that short-term (15–45 min) assays of thymidine incorporation agreed well with measured increases in cell numbers. They concluded that the thymidine method was a valid and useful technique for determining rates of bacterial growth in the sea. They also pointed out that there were some discrepancies between changes in direct counts and thymidine incorporation rates in oligotrophic waters. Reasonable agreement between the thymidine method and direct microscopy has also been reported by Bell *et al.* (1983) and Bell and Kuparinen (1984) and for some experiments by Riemann *et al.* (1984). In other experiments, the thymidine technique underestimated the growth rates (Riemann *et al.*, 1984).

A number of comparisons have been made between the thymidine and frequency-of-dividing-cells methods (Hagstrom, 1984; Newell and Fallon, 1982; Riemann *et al.*, 1984; Riemann and Sondergaard, 1984). Good correlations between values for growth rates in the water column were found in most instances, indicating that both methods were measuring growth. Discrepancies were found in many cases, however, in the absolute values of growth rate. Newell and Fallon (1982) found that values were two to seven times lower with the thymidine method. Values calculated from the frequency-of-dividing-cells method were unrealistically high (5–50 g C m^{-2} d^{-1}) compared to measured values for oxygen utilization. It is very difficult in sediments to distinguish dividing cells from cells that have divided but remain attached to each other in a filament. Riemann and Sondergaard (1984) also reported that the thymidine method gave lower values for growth rate than the frequency-of-dividing-cells method. Reasonably close correspondence has been reported by Hagstrom (1984) and Riemann *et al.* (1984).

Very good agreement has been found between the thymidine method and two other methods for estimating heterotrophic bacterial production or utilization of carbon (Lancelot and Billen, 1984). The other two methods involved determination of the uptake of dissolved sugars, amino acids, and carboxylic acids by bacteria and an estimate from exoproteolytic enzyme activities. The thymidine method measures cell division rates with reasonable accuracy, but not production, because it is difficult to measure bacterial cell sizes accurately to determine their carbon content and thus estimate production. Furthermore, not all bacteria may be growing, so an average cell size may over- or underestimate production. The work of Lancelot and Billen (1984) is valuable, therefore, in helping decide whether production estimates from thymidine incorporation rates are accurate. Information is also needed on conversion efficiencies before a complete assessment can be made. More studies on conversion efficiencies in natural systems are necessary.

4.3. Water Column

The growth rates of bacteria that have been measured in various seas are generally all within two orders of magnitude (Table II). These growth rates agree quite well with rates measured by other techniques, some examples of which are given in Table II. Doubling times of 7–37 days have been reported for a freshwater lake (Riemann *et al.* (1982). In a eutrophic lake in Sweden, bacterial doubling times ranged from 0.2 to 2.9 days in summer (Bell *et al.*, 1983).

The growth rates that have been measured are composite ones for the whole community. The thymidine method does not distinguish between different populations with different growth rates. They ought to be a true average because the relationship between thymidine incorporation and growth is linear. Although the proportion of active bacteria in a community may be assessed by autoradiography, it is difficult to determine the proportion that are actually growing. Kirchman *et al.* (1982) have suggested a way to determine the proportion of growing bacteria. By analyzing mathematically the relationship between the change in bacterial abundance in culture with time, they estimated that at least 50% of bacteria present were dividing.

Several factors contribute to the variability in growth rates shown in Table II. The most important ones are temperature and nutrient supply to bacteria. Growth rates are faster in warm water than in cold water, and so seasonal temperature differences will be directly correlated with growth rates. Heterotrophic bacteria require a supply of organic matter, which in the open ocean comes originally from phytoplankton. Thus, high bacterial growth rates are found in seasons and regions where phytoplankton are present. Fuhrman *et al.* (1980) found that bacterial growth rates off the coast of California were correlated more with abundance of phytoplankton than with primary production. They suggested that in this case, bacterial growth probably depended on organic matter released from algal cells as a result of zooplankton feeding. Evidence to show that bacterial growth rates were faster in the presence of zooplankton has been reported (Eppley *et al.*, 1981). They suggested that this was due to the release of organic matter from the phytoplankton by the zooplankton. Bacterial growth in the sea is also dependent to some extent on excretion of organic matter from phytoplankton during photosynthesis (Smith *et al.*, 1977; Williams and Yentsch, 1976). Using the tritiated thymidine method, Bell *et al.* (1983) showed that algal exudates supported between 10 and 80% of bacterial growth in a freshwater lake. During a spring bloom in a Swedish freshwater lake, bacterial growth (including respiration) accounted for about 20% of gross primary production. The bacterial growth was supported by algal excretory products (Bell and Kuparinen,

1984). Ducklow and Kirchman (1983) found that there was a coupling between bacteria and phytoplankton density in shelf waters in the New York Bight.

A very tight coupling occurred between bacterial growth and the spring phytoplankton bloom in the North Sea (Lancelot and Billen, 1984). Bacteria utilized 44–68% of total primary production, up to 100% of dissolved organic matter production, and some particulate organic production. Rapid recycling of nitrogen via bacterial activity was necessary to support the primary production.

Not all the variation in bacterial growth rates is due to seasonal changes in temperature or phytoplankton density. Diel variation may be quite marked in some environments. Doubling times varied from 1.4 days at noon to 12 days at night in water over a seagrass bed (Table III) (Moriarty and Pollard, 1982). A similar variation of one order of magnitude was found in the water column near coral reefs (Table III). Diel variation of bacterial growth in planktonic communities may be influenced by excretion of organic matter from algae during photosynthesis as well as by the release of organic matter during feeding of zooplankton. Thus, various factors may influence bacterial growth rates in planktonic communities, and if these do not act in concert, simple diel cycles linked to photosynthetic production may not occur. Small diel cycles in bacterial growth rates in a freshwater lake were found with the use of tritiated thymidine (Riemann *et al.*, 1982). The thymidine method has been used to show diel variations in growth rates in coastal seawater and freshwaters by Hagstrom (1984), Riemann and Sondergaard (1984), and Riemann *et al.* (1984).

More detailed studies on small-scale temporal and spatial variation in bacterial growth rates are needed because bacteria respond quickly to changes in nutrient concentration. The tritiated thymidine technique is ideally suited for conducting such studies because incubation times need only be short and small samples are sufficient.

Values for bacterial productivity ranging from 5 to 45% of primary productivity have been obtained (Table II). If the efficiency of utilization of organic matter is taken to be 50% (Payne, 1970), these values show that from 10 to 90% of primary production is needed to support the bacterial production. Ducklow and Kirchman (1983) suggested that their high value of 35% for bacterial production as a proportion of primary production was due to utilization of allochthonous organic matter in a river plume. The proportion of primary production that is utilized by bacteria in the water column probably depends on the growth state of the phytoplankton. In the early stages of a bloom, excreted dissolved organic matter is likely to be the main source of carbon for bacteria, and this is not a high proportion of primary production. In a senescing bloom, where a

Table II. Growth Rates of Bacteria in Marine Water Columns[a]

Location	Season	Production (ng C/liter per hr)	Specific growth rate (day^{-1})	Bacterial production as percent primary production	Reference
Thymidine technique					
North Sea	Spring	80–240	—	5–20	Lancelot and Billen (1984)
Antarctic	Summer	0.2–121	0.007–0.46	10–12	Fuhrman and Azam (1980)
New York Bight	Spring (3°C)	294–556	0.09	23–35	Ducklow et al. (1982)
California coast	Spring	196–2210	0.5–2.3	—	Fuhrman and Azam (1980)
California coast	Spring	80–800	0.2–1.0	5–25	Fuhrman and Azam (1982)
Chesapeake Bay	Spring	292–3130	1.1–6.9	—	Ducklow (1982)
Georgia coast	Summer	208–2080	0.07–0.3	—	Newell and Fallon (1982)
Moreton Bay, East Australia	Autumn, day	300	1.3	5–10	Moriarty and Pollard (1982)
Moreton Bay, East Australia	Autumn, night	90	0.3	—	Moriarty and Pollard (1982)
Great Barrier Reef	Summer	375–1830	0.5–6.9	—	Moriarty et al. (1985b)
Great Barrier Reef	Winter	10–151	0.01–0.3	—	Moriarty et al. (1985b)
Harelin Pool, West	Spring	130–320	0.02–0.1	—	Moriarty (1983)

Frequency of dividing cells					
Antarctic	Summer	0.02–1.4	0.8–2.3	15–45	Hanson et al. (1983)
Baltic Sea	Spring, Summer, Autumn	175	0.2–1.7	25	Hagstrom et al. (1979)
Georgia coast	Summer	1700–5600	0.5–0.8	—	Newell and Fallon (1982)
Growth in enclosed chambers					
California coast	Spring	416–1416	0.4–1.4	—	Fuhrman and Azam (1980)
California coast	Spring	—	0.7–3.5	—	Carlucci and Shimp (1974)
Growth in diffusion cultures					
North Atlantic	—	—	2.3–3.5	—	Sieburth et al. (1977)
Baltic	Summer	416–2375	0.2–1.4	29	Meyer-Reil (1977)
North Atlantic	—	25–1817	—	—	Delattre et al. (1979)
Growth in continuous culture (single species)					
North Atlantic	—	—	0.1–0.3	—	Jannasch (1969)

[a]Where conversions were needed the following factors were used: biomass, 20 fg C/cell nearshore and 10 fg C/cell offshore; growth rate, 2×10^{-18} cells/mole thymidine incorporated.

Table III. Daily Fluctuations in Bacterial Growth Rates in Seawater Over a Seagrass Bed in March and Over Three Zones of a Coral Reef in July[a]

Time of day	Seagrass bed		Coral reef[b]					
	Specific growth rate (day^{-1})	Number of bacteria $(10^9$ liter$^{-1})$	Specific growth rate (day^{-1})			Number of bacteria $(10^8$ liter$^{-1})$		
			Outside	Crest	Flat	Outside	Crest	Flat
0600	0.07	6.8	—	—	—	—	—	—
0900	0.14	4.2	0.6	0.2	0.5	6.0	5.0	3.8
1100–1200	0.35	2.8	0.5	0.2	0.4	7.1	7.7	3.7
1400	0.06	3.6	—	—	—	—	—	—
1600	—	—	0.1	0.4	2.8	5.6	6.4	2.9
1900	0.03	4.5	—	—	—	—	—	—

[a]Data from Moriarty and Pollard (1982) and Moriarty et al. (1985b).
[b]Zones studied are 500 m outside the reef, the reef crest, and the reef flat.

large amount of organic matter is available to bacteria as a result of cell death, lysis, and feeding by zooplankton, bacterial production may be high in proportion to primary production. Lancelot and Billen (1984) found that bacteria utilized 44–68% of primary production during a spring bloom in the North Sea. They estimated from comparisons of thymidine incorporation rates and uptake of dissolved organic carbon that the bacterial growth efficiency was 10–30%.

There is enough comparable information from different sources and different methodologies to show that bacterial growth in seawater generally utilizes at least 25% of primary production. Growth rates are faster inshore than offshore (Fuhrman et al., 1980; Newell and Fallon, 1982). The reasons for these differences are complex and require further study. Runoff from land supplies nutrients directly for bacterial growth and for phytoplankton growth. More particulate matter from rivers and disturbance of sediments in shallow coastal regions would provide greater surface area for concentrating nutrients and thus providing a site for bacterial growth (Kjelleberg et al., 1982).

Particulate matter in the water column may be an important site for bacterial growth, not only as a surface for concentrating nutrients, but also as a source of organic nutrients. The thymidine method was used by Ducklow et al. (1982) to show that the sedimentation of particles with attached bacteria in the Hudson River plume removed from 3 to 67% of daily bacterial production. Bacterial growth on particles that are probably derived from mucus accounted for at least 50% of bacterial production over coral reefs. About 50% of tritiated thymidine incorporation occurred on particles in water from a coral reef lagoon and over a seagrass bed on a reef flat. About 30% of tritiated thymidine incorporation occurred on particles in water outside the reef area (Moriarty *et al.,* 1985b). In a freshwater pond, only a small proportion (2.8%) of thymidine incorporation into DNA occurred on particles. Thus, the estimated production of particle-bound bacteria and the mineralization of organic matter by attached bacteria was low compared to the activities of free bacteria (Kirchman, 1983). The thymidine method has proved to be useful for studying the effects of particles on bacterial growth rates because it can be used over a short enough time interval to minimize changes due to sampling.

If the sampling procedure includes filtration, it is likely that bacterial activity at particle surfaces will be underestimated because shear forces will easily remove reversibly attached bacteria (Hermansson and Marshall, 1985). By labeling bacteria with tritiated thymidine and substrate (stearic acid) with ^{14}C, Hermansson and Marshall (1985) showed that many apparently free bacteria were able to utilize substrates bound to particle surfaces.

The adenine method has been used to measure microbial growth rates in the tropical north Pacific Ocean by Winn and Karl (1984b). They estimated production in the photic zone (0–150 m) to be 400 mg C m^{-2} day^{-1}, which was four times higher than previous estimates of primary production in that area (Bienfang and Gundersen, 1977). Because phytoplankton and bacterial production cannot be discriminated with the adenine technique, such values cannot be compared with those from the thymidine technique. In deeper waters (150–900 m), however, bacteria were probably the main microbial producers and in this depth zone, Winn and Karl (1984b) estimated production to be 790 mg C m^{-2} day^{-1}. If bacteria are 30% efficient in utilizing carbon in the sea (Lancelot and Billen, 1984), the bacteria in the deep zone would require an input of 2.6 g C m^{-2} day^{-1}, which is far greater than any measured or estimated values [e.g., 3–24 mg C m^{-2} hr^{-1} (Beinfang and Gundersen, 1977)], particularly when much of the primary production is probably utilized in the upper 150 m. These results support the conclusion (Section 3.3) that tritiated adenine is not a useful precursor to use for determining microbial or bacterial growth rates in natural environments.

4.4. Sediments

The measurement of bacterial biomass and growth rates in sediments is more difficult than in water. Perhaps for this reason few quantitative studies have been carried out, and yet in shallow environments benthic productivity is important in food chains and in nutrient cycles. Macrophytes are significant primary producers in many coastal waters, and much of their production has to be cycled through bacteria before it becomes available to animals.

Thymidine is strongly adsorbed to particulate matter, so concentrations of added thymidine in sediments need to be 100 to 1000 times greater than those used for experimental work in the water column. Likewise, if there is a large amount of particulate matter in the water column, higher concentrations of thymidine will be needed. If thymidine concentrations are not high enough, isotope dilution will be substantial, making results difficult to interpret correctly (Pollard and Moriarty, 1984).

Bacterial production measured with the thymidine method compares well with estimates from an independent technique using [32P]phosphate incorporation into phospholipid (Table IV) (Moriarty *et al.*, 1985d). Close agreement was not expected because phosphate is used by all microbes, although phospholipids were mainly labeled in bacteria (Moriarty *et al.*, 1985d). Furthermore, phosphate does not disperse as readily in sediment as does thymidine. Support for the use of phospholipid synthesis as an alternative measure of growth is shown by the close agreement with the thymidine method where aerobic bacterial growth on seagrass leaf detritus suspended in the water column was measured (Table V). Because the results of the two methods agreed well for sediment also, it may be concluded that the thymidine method measures the growth of most bacteria in the sediment, including anaerobic bacteria. These experiments were carried out with care to ensure as little disturbance as possible. Rates of DNA synthesis did not respond immediately to disturbance of sediment, which makes the technique a useful one for sediments. Rates of phospholipid synthesis, however, increased rapidly

Table IV. Comparison of Measurements of Bacterial Production in Sediments and Water Made Using the Thymidine Technique and Rates of [32P]Phosphate Incorporation into Phospholipid[a]

Type of sample	Depth (cm)	Thymidine[b] (mg C m^{-2} hr^{-1})	[32P]lipid[b] (mg C m^{-2} hr^{-1})
Sand bank	0–1	9.0 ± 1.4	12.3 ± 1.1
Halodule sediment	0–2	7.6 ± 0.7	8.4 ± 2.3
Halodule sediment	2–4		1.2 ± 0.1
Halodule sediment	8–10	0.5 ± 0.14	0.3 ± 0.03
Water column, seagrass leaf detritus	—	6.0 ± 1.2[c]	4.7 ± 0.3[c]

[a]Moriarty, *et al.* (1985d, and unpublished work).
[b]Standard errors are shown for 6–12 measurements each.
[c]Value is g C/liter per hr.

in a well-mixed slurry (Moriarty *et al.*, 1985d). Further work is needed to determine the responses of DNA synthesis in sediments to handling.

Most bacterial production occurs at or near the surface of sediments, as seen in the example in Table IV. Bacterial production was most rapid in the upper 2 cm of sediment in a seagrass bed and was more than an order of magnitude lower at 8–10 cm.

The productivity of bacteria in sediment from a number of environments is shown in Table V. Results have been normalized to 10 mm depth for comparison, which may have over- or underestimated some values. Seagrasses are very productive, and in turn they support an active bacterial community.

High rates of bacterial productivity have been found in sediments of seagrass *(Zostera capricorni)* beds, with a marked diel variation (Fig. 7). Growth rates at noon were about an order of magnitude faster than at night (Moriarty and Pollard, 1982). Growth rates were low, and showed no diel variation in sediments without seagrass. It was estimated that bacterial growth in the water and sediments utilized between about 10 and 20% of net primary production of the seagrass system.

Thymidine can be used in anoxic sediments to estimate the growth rates of many anaerobic bacteria. This has been demonstrated in an experiment using an enrichment culture of a mixed community of bacteria from a seagrass bed sediment growing on glucose and yeast extract.

Table V. Bacterial Productivity in the Top 10 mm of Sediment from Different Environments

Site	Production (mg C m^{-2} day^{-1})	Actual depth[a] (cm)	Reference
Halodule bed, Gulf of Mexico, U.S.	85	2	D. J. W. Moriarty (unpublished results)
Spartina creek bank, Georgia	10.5	20	Fallon *et al.* (1983)
Nearshore, Georgia	40.5	25	Fallon *et al.* (1983)
Six km offshore, Georgia	16.5	25	Fallon *et al.* (1983)
Zostera bed, Moreton Bay, Australia	60	0.5	Moriarty *et al.* (1985a)
Aquaculture ponds, Malaysia	150–500	1	Moriarty (1986)
Mangrove creek bank, Malaysia	230	1	D. J. W. Moriarty (unpublished results)

[a]Because different workers studied different depths of sediments, results were normalized to the top 10 mm for comparison; actual values can be obtained using these depth values.

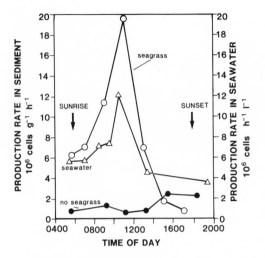

Figure 7. Diurnal variation in bacterial cell production rates in sediments and water column associated with seagrass beds. [From Moriarty and Pollard (1982).]

The growth rate of the culture measured by direct microscopy agreed well with that estimated using the thymidine method (Pollard and Moriarty, 1984). Fermentative bacteria, an important group in sediments, were probably predominant in the culture, and thus it is likely that the thymidine method measures their growth. Experiments with pure cultures are needed to check this. Not all anaerobic bacteria can take up thymidine. Tritiated thymidine does not label DNA in *Desulfovibrio* (G. Skyring, personal communication) or acetate-utilizing, sulfate-reducing bacteria (D. J. W. Moriarty, unpublished results). Because these bacteria can use only a very limited range of substrates, they may lack pyrimidine transport enzymes. The thymidine method thus gives minimum values for growth rates in anoxic sediments.

Bacterial productivity in coral reef sediments is high and shows seasonal and diel variations (Moriarty *et al.,* 1985c). Holothurians feed on bacteria in reef sediments (Moriarty, 1982), and it was estimated that they ate 10–40% of daily bacterial production in summer (Moriarty *et al.,* 1985c). Heavy grazing by them depresses bacterial production (Fig. 8). Benthic microalgae probably contribute some nutrients to bacteria, but mucus and slime settling from the water column are also likely to be significant sources of nutrients for bacteria (Ducklow and Mitchell, 1979). Bacterial productivity is too high in proportion to primary production in reef sediments to be linked closely to the growth of microalgae. Other sources of organic matter, such as mucus from animals (especially corals) are needed to support some of the high rates of bacterial production (Table VI).

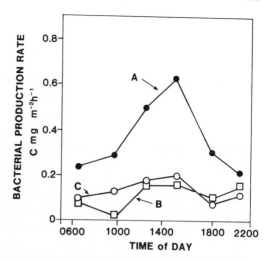

Figure 8. Bacterial cell production rates in coral reef sand, measured using the rate of tritiated thymidine incorporation into DNA. (A) Sediment from which holothurians were excluded. (B) Sediment on which holothurians were confined by a cage. (C) control area, no cage. [From Moriarty *et al.* (1985c).]

Table VI. Comparison of Bacterial and Primary Productivities in the Top 10 mm of Coral Reef Sediments[a]

	Primary productivity (mg C /m² per day)	Bacterial productivity (mg C /m² per day)	Respiration rate (mg C/ m² per day)
Normal area	350	190	630
Cage without holothurians	670	370	870
Cage with holothurians	230	190	600

[a]From Moriarty et al. (1985c). Bacterial productivity was measured using tritiated thymidine and microalgal productivity with $^{14}CO_2$ incorporation or O_2 changes in chambers.

Microbial production in coral reef sediments in Kaneohe Bay, Hawaii has been estimated with the adenine method to be 1.1–5.8 g C m^{-2} hr^{-1} (Burns *et al.*, 1984). This production would require an input of about 30–250 g C m^{-2} day^{-1}, which seems very unlikely. Sediment to a depth of 5 cm was collected and homogenized, so it is likely that only bacterial DNA synthesis was measured with the adenine. By comparison with bacterial DNA synthesis measured with thymidine at Lizard Island (see this section above), these rates are about two orders of magnitude too high. These rates are also one to two orders of magnitude higher than those measured on the Kaneohe Bay reef by Sorokin (1978), who used a dark ^{14}C-fixation method. Thus, these results of Burns *et al.* (1984) also indicate that adenine is not useful as a measure of microbial or bacterial growth rates.

4.5. Growth State

The growth state of bacteria is relevant to the methodology of growth rate measurement. For balanced growth (see Section 2), a constant supply of nutrients is needed. Bacteria in seawater and sediments are generally not in an environment where there is a copious and continual supply of nutrients. Many bacteria exist as small, starved cells and are adapted to respond quickly to change in the environment (Kjelleberg et al., 1982). If they are starved, nutrients may be used for maintenance rather than growth and division. This conclusion is supported by studies on bacterial growth rates in the Antarctic. More tritium from thymidine was incorporated into protein than into DNA in bacteria in deep Antarctic water compared to bacteria in surface waters, where a lower proportion of tritium was used for protein synthesis (Hanson and Lowery, 1983).

The diel variation that occurs in growth rates shows that bacteria do respond rapidly to changes in nutrient concentration. It is unlikely, therefore, that they will be in a state of balanced growth as would occur for bacteria in culture with a constant supply of nutrients. Hanson and Lowery (1983) attempted to determine whether bacteria in Antarctic waters were in a state of balanced growth by comparing rates of [^{3}H]adenine incorporation into RNA and DNA and [^{3}H]thymidine incorporation into DNA. Rate ratios of RNA/DNA synthesis are difficult to interpret when two different precursors are used, because adenine may be utilized by most microbes [i.e., algae, protozoa, and bacteria (Karl et al. 1981)], whereas thymidine is incorporated into DNA of bacteria only. It does seem likely, however, that oceanic microplankton are in various growth states. This conclusion is supported by the wide variation in doubling times for bacteria in both water and sediment from one environment (see Tables III and VII) (Moriarty and Pollard, 1982). Short doubling times (0.1–0.3 day) indicate that probably most bacteria in a community are growing, whereas long times suggest that many may be dormant.

4.6. Food Chain Dynamics

Tritiated thymidine has been used to label bacteria and study their utilization by animals (Hollibaugh et al., 1980). The advantage of this technique is that a short-term pulse label can be employed during which bacterial DNA is labeled specifically (see Section 3.2.3). Furthermore, natural assemblages of bacteria may be labeled, thus avoiding the need to work with cultures of possibly unnatural species. Hollibaugh et al. (1980) used this technique to study grazing rates of microzooplankton on bacteria. Bacteria have been shown to constitute only a low proportion of the carbon requirements of an isopod that fed on leaf detritus (Findlay *et al.*, 1984).

The generally large proportion of primary production that is utilized by bacteria implies that bacteria should have an important role in aquatic food chains. The biomass or number of bacteria in sediments without seagrass is not much lower than those with seagrass, yet the growth rates of bacteria around seagrass may be considerably higher (Moriarty, 1980; Moriarty and Pollard, 1982). This suggests that grazers on bacteria are more active in the seagrass zone than in bare sediments, and that the bacteria are being grazed at a rate approximately equal to the growth rate. Bacterial numbers or biomass in sediments from a wide diversity of environments are similar (e.g., Dale, 1974; Moriarty, 1980; 1982; Newell and Fallon, 1982). A possible explanation for this is that these values for bacterial biomass are near the minimum for effective grazing by bacteriovores. The expenditure of energy in searching for food may be greater than that gained at this level of bacterial biomass. A correlation was observed between bacterial biomass and doubling time and meiofauna numbers in an aquaculture pond (Moriarty, 1986). Three pens within a pond were fertilized daily with chicken manure for periods of 1, 2, and 3 weeks, respectively. After 1 week, bacterial biomass was high, but after 2 weeks, bacterial biomass fell twofold and after 3 weeks, it had fallen fivefold (Table VII). Conversely, bacterial doubling times increased fourfold over the 3-week period. These changes were correlated with increased numbers of meiofauna. Although meiofauna clearly have a role both in controlling bacterial numbers and, in this case, stimulating growth rates, protozoa are probably also important in sediments.

A similar situation occurs in the water column where microflagellates limit bacterial populations (Fenchel, 1982). Detailed studies of the interactions between bacterial production and grazers can now be studied

Table VII. Effect of Meiofauna on Bacterial Biomass and Growth Rates in the Upper 1 cm of Sediment in a Tropical Aquaculture Pond[a]

	Bacteria		Meiofauna
Pond treatment	Biomass $(g\ C\ m^{-2})$	Specific growth rate (day^{-1})	number (per $10\ cm^2)$
Untreated	3.4	6	140
Manure, 1 week	4.3	8	300
Manure, 2 weeks	1.9	6	1500
Manure, 3 weeks	0.8	2	1500

[a]From Moriarty (1986). Bacterial biomass was measured with muramic acid and growth rates with the tritiated thymidine technique.

because the thymidine technique permits short-term measurements of growth rates under almost natural conditions with algae and bacteriovores present. The results of Ducklow and Kirchman (1983) for bacterial growth rates in the Hudson River plume suggest that bacterial grazers play a significant role in limiting bacterial populations. During a spring bloom in a Swedish lake, the actual numbers of bacteria that accumulated were about 50% less than the numbers predicted by the thymidine method. A likely explanation for this effect was considered to be grazing by protozoans (Bell and Kuparinen, 1984). Hagstrom (1984) reached a similar conclusion when he compared rates of bacterial growth, measured by both the frequency-of-dividing-cells and thymidine methods, with the actual increase in bacterial numbers in coastal seawater samples. Now that we can measure bacterial growth rates in natural systems, another major problem that can be investigated is the quantitative contribution of protozoa and other larger animals that feed on bacteria to the food chain and nutrient cycles. Ducklow (1983) and Williams (1984) have both reviewed this problem, and suggest that bacteria contribute significantly to the nutrition of protozoa and microzooplankton. Whether bacteria are an important food source for higher organisms in the water column is still open to question, but seems doubtful. In the sediments, however, bacteria are an important food source for large animals (e.g., Moriarty, 1982; Moriarty *et al.*, 1985c).

ACKNOWLEDGMENTS. I am grateful to many colleagues for helpful discussions. I wish to thank Hugh W. Ducklow and Peter C. Pollard for comments on an early draft of the manuscript, and especially Peter J. leB. Williams for much help and detailed criticism.

References

Azam, F., and Fuhrman, J. A., 1984, Measurement of bacterioplankton growth in the sea and its regulation by environmental conditions, in: *Heterotrophic Activity in the Sea* (J. Hobbie and P. J. leB. Williams, eds.), pp. 179–196, Plenum Press, New York.

Bell, R. T., and Kuparinen, J., 1984, Assessing phytoplankton and bacterioplankton production during early spring in Lake Erken, Sweden, *Appl. Environ. Microbiol.* **48:**1221–1230.

Bell, R. T., Ahlgren, G. M., and Ahlgren, I., 1983, Estimating bacterioplankton production by measuring [³H]thymidine incorporation in a eutrophic Swedish lake, *Appl. Environ. Microbiol.* **45:**1709–1721.

Bern, L., 1985, Autoradiographic studies of [*methyl*-³H]thymidine incorporation in a cyanobacterium *(Microcystis wesenbergii)*–bacterium association and in selected algae and bacteria, *Appl. Environ. Microbiol.* **49:**232–233.

Bienfang, P., and Gundersen, K., 1977, Light effects on nutrient-limited, oceanic primary production, *Mar. Biol.* **43**:187–199.

Brock, T. D., 1967, Bacterial growth rate in the sea: Direct analysis by thymidine autoradiography, *Science* **155**:81–83.

Brock, T. D., 1971, Microbial growth rates in nature, *Bacteriol. Rev.* **35**:39–58.

Brunschede, H., Dove, T. L., and Bremer, H., 1977, Establishment of exponential growth after a nutritional shift-up in *Escherichia coli* B/r: Accumulation of deoxyribonucleic acid, ribonucleic acid, and protein, *J. Bacteriol.* **129**:1020–1033.

Burns, D., Andrews, C., Craven, D., Orrett, K., Pierce, B., and Karl, D.; 1984, Microbial biomass, rates of DNA synthesis and estimated carbon production in Kaneohe Bay, Hawaii, *Bull. Mar. Sci.* **34**:346–357.

Burton, K., 1977, Transport of adenine, hypoxanthine and uracil into *Escherichia coli*, *Biochem. J.* **168**:195–204.

Campbell, A., 1957, Synchronization of cell division, *Bacteriol. Rev.* **21**:263–272.

Carlucci, A. F., and Shimp, S. L., 1974, Isolation and growth of a marine bacterium in low concentrations of substrate, in: *Effect of the Ocean Environment on Microbial Activities* (R. R. Colwell and R. Y. Morita, eds.), pp. 363–367, University Park Press, Baltimore.

Christian, R. R., Hanson, R. B., and Newell, S. Y., 1982, Comparison of methods for measurement of bacterial growth rates in mixed batch cultures, *Appl. Environ. Microbiol.* **43**:1160–1165.

Dale, N. G., 1974, Bacteria in intertidal sediments: Factors related to their distribution, *Limnol. Oceanogr.* **19**:509–518.

Delattre, J. M., Delesmont, R., Clabaux, M., Oger, C., and Leclerc, H., 1979, Bacterial biomass, production and heterotrophic activity of the coastal seawater at Gravelines (France), *Oceanol. Acta* **2**:317–324.

Dennis, P. P., and Bremer, H., 1974, Macromolecular composition during steady-state growth of *Escherichia coli* B/r, *J. Bacteriol.* **119**:270–281.

Ducklow, H. W., 1982, Chesapeake Bay nutrient and plankton dynamics. 1. Bacterial biomass and production during spring tidal destratification in the York River, Virginia, estuary, *Limnol. Oceanogr.* **27**:651–659.

Ducklow, H. W., 1983, The production and fate of bacteria in the oceans, *Bioscience* **33**:494–501.

Ducklow, H. W., and Hill, S. M., 1985, Tritiated thymidine incorporation and the growth of heterotrophic bacteria in warm core rings, *Limnol. Oceanogr.* **30**:260–272.

Ducklow, H. W., and Kirchman, D. L., 1983, Bacterial dynamics and distribution during a spring diatom bloom in the Hudson River plume, U.S.A., *J. Plankton Res.* **5**:333–355.

Ducklow, H. W., and Mitchell, R., 1979, Bacterial populations and adaptations in the mucus layers on living corals, *Limnol. Oceanogr.* **24**:715–725.

Ducklow, H. W., Kirchman, D. L., and Rowe, G. T., 1982, Production and vertical flux of attached bacteria in the Hudson River plume of the New York Bight as studied with floating sediment traps, *Appl. Environ. Microbiol.* **43**:769–776.

Eppley, R. W., Horrigan, S. G., Fuhrman, J. A., Brooks, E. R., Price, C. C., and Sellner, K., 1981, Origins of dissolved organic matter in Southern California coastal waters, experiments on the role of zooplankton, *Mar. Ecol. Prog. Ser.* **6**:149–159.

Fallon, R. D., Newell, S. Y., and Hopkinson, C. S., 1983, Bacterial production in marine sediments: Will cell-specific measures agree with whole-system metabolism? *Mar. Ecol. Prog. Ser.* **11**:119–127.

Fenchel, T., 1982, Ecology of heterotrophic microflagellates. IV. Quantitative occurrence and importance as bacterial consumers, *Mar. Ecol. Prog. Ser.* **9**:35–42.

Findlay, S., Meyer, J. L., and Smith, P. J., 1984, Significance of bacterial biomass in the nutrition of a freshwater isopod (*Lirceus* sp.), *Oecologia (Berlin)* **63**:38–42.

Fink, R. M., and Fink, K., 1962, Relative retention of H^3 and C^{14} labels of nucleosides incorporated into nucleic acids of *Neurospora, J. Biol. Chem.* **237**:2889–2891.

Forsdyke, D. R., 1968, Studies of the incorporation of [5-^{3}H] uridine during activation and transformation of lymphocytes induced by phytohaemagglutinin, *Biochem J.* **107**:197–205.

Forsdyke, D. R., 1971, Application of the isotope-dilution principle to the analysis of factors affecting the incorporation of [^{3}H]uridine and [^{3}H]cytidine into cultured lymphocytes, *Bioch. J.* **125**:721–732.

Fuhrman, J. A., and Azam, F., 1980, Bacterioplankton secondary production estimates for coastal waters of British Columbia, Antarctica, and California, *Appl. Environ. Microbiol.* **39**:1085–1095.

Fuhrman, J. A., and Azam, F., 1982, Thymidine incorporation as a measure of heterotrophic bacterioplankton production in marine surface waters: Evaluation and field results, *Mar. Biol.* **66**:109–120.

Fuhrman, J. A., Ammerman, J. W., and Azam, F., 1980, Bacterioplankton in the coastal euphotic zone: Distribution, activity and possible relationships with phytoplankton, *Mar. Biol.* **60**:201–207.

Gillis, M., De Ley, J., and De Cleene, M., 1970, The determination of molecular weight of bacterial genome DNA from renaturation rates, *Eur. J. Biochem.* **12**:143–153.

Glaser, V. M., Al-Nui, M. A., Groshev, V. V., and Shestakov, S. V., 1973, The labelling of nucleic acids by radioactive precursors in the blue-green algae, *Arch. Mikrobiol.* **92**:217–226.

Grivell, A. R., and Jackson, J. F., 1968, Thymidine kinase: Evidence for its absence from *Neurospora crassa* and some other micro-organisms, and the relevance of this to the specific labelling of deoxyribonucleic acid, *J. Gen. Microbiol.* **54**:307–317.

Hagstrom, A., 1984, Aquatic bacteria: Measurements and significance of growth, in: *Current Perspectives in Microbial Ecology* (M. J. Klug and C. A. Reddy, eds.), pp. 95–501, American Society for Microbiology, Washington, D.C.

Hagstrom, A., Larsson, U., Horstedt, P., and Normark, S., 1979, Frequency of dividing cells, a new approach to the determination of bacterial growth rates in aquatic environments, *Appl. Environ. Microbiol.* **37**:805–812.

Hanson, R. B., and Lowery, H. K., 1983, Nucleic acid synthesis in oceanic microplankton from the Drake Passage, Antarctica: Evaluation of steady-state growth, *Mar. Biol.* **73**:79–89.

Hanson, R. B., Lowery, H. K., Shafer, D., Sorocco, R., and Pope, D. H., 1983, Microheterotrophs in the Antarctic Ocean (Drake Passage): Nutrient uptake, productivity estimates, and biomass, *Appl. Environ. Microbiol.* **45**:1622–1632.

Hermansson, M., and Marshall, K. C., 1985, Utilization of surface localized substrate by non-adhesive marine bacteria, *Microb. Ecol.* **11**:91–105.

Hollibaugh, J. T., Fuhrman, J. A., and Azam, F., 1980, Radioactively labeling natural assemblages of bacterioplankton for use in trophic studies, *Limnol. Oceanogr.* **25**:171–181.

Hoppe, H.-G., 1976, Determination and properties of actively metabolizing heterotrophic bacteria in the sea, investigated by means of micro-autoradiography, *Mar. Biol.* **36**:291–302.

Hunter, T., and Francke, B., 1974, *In vitro* polyoma DNA synthesis characterization of a system from infected 3T3 cells, *J. Virol.* **13**:125–139.

Jannasch, H. W., 1969, Estimations of bacterial growth rates in natural waters, *J. Bacteriol.* **99**:156–160.

Karl, D. M., 1979, Measurement of microbial activity and growth in the ocean by rates of stable ribonucleic acid synthesis, *Appl. Environ. Microbiol.* **38**:850–860.

Karl, D. M., 1981, Simultaneous rates of RNA and DNA syntheses for estimating growth and cell division of aquatic microbial communities, *Appl. Environ. Microbiol.* **42:**802–810.

Karl, D. M., 1982, Selected nucleic acid precursors in studies of aquatic microbial ecology, *Appl. Environ. Microbiol.* **44:**891–902.

Karl, D. M., Winn, C. D., and Wong, D. C. L., 1981, RNA synthesis as a measure of microbial growth in aquatic environments. I. Evaluation, verification and optimization of methods, *Mar. Biol.* **64:**1–12.

Kirchman, D., 1983, The production of bacteria attached to particles suspended in a freshwater pond, *Limnol. Oceanogr.* **28:**858–872.

Kirchman, D., Ducklow, H. W., and Mitchell, R., 1982, Estimates of bacterial growth from changes in uptake rates and biomass, *Appl. Environ. Microbiol.* **44:**1296–1307.

Kjeldgaard, N. O., Maaloe, O., and Schaechter, M., 1958, The transition between different physiological states during balanced growth of *Salmonella typhimurium, J. Gen. Microbiol.* **19:**607–616.

Kjelleberg, S., Humphrey, B. A., and Marshall, K. C., 1982, Effect of interfaces on small, starved marine bacteria, *Appl. Environ. Microbiol.* **43:**1166–1172.

Kornberg, A., 1980, *DNA Replication,* Freeman, San Francisco.

Kuebbing, D., and Werner, R., 1975, A model for compartmentation of *de novo* and salvage thymidine nucleotide pools in mammalian cells, *Proc. Natl. Acad. Sci. USA* **72:**3333–3336.

Kunicka-Goldfinger, W., 1976, Determination of growth of aquatic bacteria by measurements of incorporation of tritiated thymidine, *Acta Microbiol. Polon.* **25:**279–286.

Lancelot, C., and Billen, G., 1984, Activity of heterotrophic bacteria and its coupling to primary production during the spring phytoplankton bloom in the southern bight of the North Sea, *Limnol. Oceanogr.* **29:**721–730.

Lark, K. G., 1969, Initiation and control of DNA synthesis. *Annu. Rev. Biochem.* **38:**569–604.

LaRock, P. A., Lauer, R. D., Schwarz, J. R., Watanabe, K. K., and Wiesenburg, D. A., 1979, Microbial biomass and activity distribution in an anoxic, hypersaline basin, *Appl. Environ. Microbiol.* **37:**466–470.

Laws, E. A., 1983, Plots of turnover times versus added substrate concentrations provide only upper bounds to ambient substrate concentrations, *J. Theoret. Biol.* **101:**147–150.

Maaloe, O., and Kjeldgaard, N. O., 1966, *Control of Macromolecular Synthesis. A Study of DNA, RNA, and Protein Synthesis in Bacteria,* Benjamin, New York.

Maley, G. F., and Maley, F., 1972, The regulatory influence of allosteric effectors on deoxycytidylate deaminase, *Curr. Top. Cell. Regul.* **17:**177–228.

Mathews, C. K., North, T. W., and Reddy, G. P. V., 1979, Multienzyme complexes in DNA precursor biosynthesis, in: *Advances in Enzyme Regulation,* Vol. 17 (G. Weber, ed.), pp. 133–156. Pergamon Press, Oxford.

Meyer-Reil, L.-A., 1977, Bacterial growth rates and biomass production, in: *Microbial Ecology of a Brackish Water Environment* (G. Rheinheimer, ed.), pp. 223–236, Springer-Verlag, Berlin.

Moriarty, D. J. W., 1980, Measurement of bacterial biomass in sandy sediments, in: *Biogeochemistry of Ancient and Modern Environments* (P. A. Trudinger, M. R. Walter, and B. J. Ralph, eds.), pp. 131–138, Australian Academy of Science, Canberra and Springer-Verlag, Berlin.

Moriarty, D. J. W., 1982, Feeding of *Holothuria atra* and *Stichopus chloronotus* on bacteria, organic carbon and organic nitrogen in sediments of the Great Barrier Reef, *Aust. J. Mar. Freshwater Res.* **33:**255–263.

Moriarty, D. J. W., 1983, Bacterial biomass and productivity in sediments, stromatolites and water of Hamelin Pool, Shark Bay, W. A., *Geomicrobiol. J.* 3:121–133.

Moriarty, D. J. W., 1984, Measurements of bacterial growth rates in some marine systems using the incorporation of tritiated thymidine into DNA, in: *Heterotrophic Activity in the Sea* (J. E. Hobbie and P. J. leB. Williams, eds.), pp. 217–231, Plenum Press, New York.

Moriarty, D. J. W., 1986, Bacterial productivity in ponds used for culture of penaeid prawns, Gelang Patah, Malaysia, *Microb. Ecol.* 12 (in press).

Moriarty, D. J. W., and Pollard, P. C., 1981, DNA synthesis as a measure of bacterial productivity in seagrass sediments, *Mar. Ecol. Prog. Ser.* 5:151–156.

Moriarty, D. J. W., and Pollard, P. C., 1982, Diel variation of bacterial productivity in seagrass *(Zostera capricorni)* beds measured by rate of thymidine incorporation into DNA, *Mar. Biol.* 72:165–173.

Moriarty, D. J. W., Boon, P., Hansen, J., Hunt, W. G., Poiner, I. R., Pollard, P. C., Skyring, G. W., and White, D. C., 1985a, Microbial biomass and productivity in seagrass beds, *Geomicrobiol. J.* 4:21–51.

Moriarty, D. W., Pollard, P. C., and Hunt, W. G., 1985b, Temporal and spatial variation in bacterial productivity in coral reef waters, measured by rate of thymidine incorporation into DNA, *Mar. Biol.* 85:285–292.

Moriarty, D. J. W., Pollard, P. C., Hunt, W. G., Moriarty, C. M., and Wassenberg, T. J., 1985c, Productivity of bacteria and microalgae and the effect of grazing by holothurians on a coral reef flat, *Mar. Biol.* 85:293–300.

Moriarty, D. J. W., White, D. C., and Wassenberg, T. J., 1985d, A convenient method for measuring rates of phospholipid synthesis and their relevance to the determination of bacterial productivity, *J. Microbiol. Meth.* 3:321–330.

Morita, R. Y., 1982, Starvation-survival of heterotrophs in the marine environment, in: *Advances in Microbial Ecology,* Vol. 6 (K. C. Marshall, ed.), pp. 171–198, Plenum Press, New York.

Munch-Petersen, A., Mygind, B., Nicolaisen, A., and Pihl, N. J., 1979, Nucleoside transport in cells and membrane vesicles from *Escherichia coli* K12, *J. Biol. Chem.* 254:3730–3737.

Munro, H. N., and Fleck, A., 1966, The determination of nucleic acids, in: *Methods of Biochemical Analysis* (D. Glick, ed.), pp. 113–176, Interscience, New York.

Mygind, B., and Munch-Petersen, A., 1975, Transport of pyrimidine nucleosides in cells of *Escherichia coli* K12, *Eur. J. Biochem.* 59:365–372.

Newell, S. Y., and Fallon, R. D., 1982, Bacterial productivity in the water column and sediments of the Georgia (USA) coastal zone: Estimates via direct counting and parallel measurement of thymidine incorporation, *Microb. Ecol.* 8:33–46.

O'Donovan, G. A., and Neuhard, J., 1970, Pyrimidine metabolism in microorganisms, *Bacteriol. Rev.* 34:278–343.

Payne, W. J., 1970, Energy yields and growth of heterotrophs, *Annu. Rev. Microbiol.* 24:17–52.

Pirt, S. J., 1975, *Principles of Microbe and Cell Cultivation,* Blackwell Scientific, Oxford.

Plaut, W., and Sagan, A., 1958, Incorporation of thymidine in the cytoplasm of *Amoeba proteus, J. Biophys. Biochem. Cytol.* 4:843–847.

Poindexter, J. S., 1981, Oligotrophy: Fast and famine existence, in: *Advances in Microbial Ecology,* Vol. 5 (M. Alexander, ed.), pp. 63–89, Plenum Press, New York.

Pollard, P. C., and Moriarty, D. J. W., 1984, Validity of the tritiated thymidine method for estimating bacterial growth rates: The measurement of isotope dilution during DNA synthesis, *Appl. Environ. Microbiol.* 48:1076–1083.

Ramsay, A. J., 1974, The use of autoradiography to determine the proportion of bacteria metabolizing in an aquatic habitat, *J. Gen. Microbiol.* **80**:363–373.

Riemann, B., 1984, Determining growth rates of natural assemblages of freshwater bacteria by means of ^{3}H-thymidine incorporation into DNA: Comments on methodology, *Arch. Hydrobiol. Beih.* **19**:67–80.

Riemann, B., and Sondergaard, M., 1984, Measurements of diel rates of bacterial secondary production in aquatic environments, *Appl. Environ. Microbiol.* **47**:632–638.

Riemann, B., Fuhrman, J., and Azam, F., 1982, Bacterial secondary production in freshwater measured by ^{3}H-thymidine incorporation method, *Microb. Ecol.* **8**:101–114.

Riemann, B., Nielsen, P., Jeppesen, M., Marcussen, B., and Fuhrman, J. A., 1984, Diel changes in bacterial biomass and growth rates in coastal environments, determined by means of thymidine incorporation into DNA, frequency of dividing cells (FDC), and microautoradiography, *Mar. Ecol. Prog. Ser.* **17**:227–235.

Roy-Burman, S., and Visser, D. W., 1975, Transport of purines and deoxyadenosine in *Escherichia coli, J. Biol. Chem.* **250**:9270–9275.

Roy-Burman, S., and Visser, D. W., 1981, Uridine and uracil transport in *Escherichia coli* and transport-deficient mutants, *Biochim. Biophys. Acta* **646**:309–319.

Sagan, L., 1965, An unusual pattern of tritiated thymidine incorporation in *Euglena, J. Protozool.* **12**:105–109.

Schaechter, M., Maaloe, O., and Kjeldgaard, N. O., 1958, Dependency on medium and temperature of cell size and chemical composition during balanced growth of *Salmonella typhimurium, J. Gen. Microbiol.* **19**:592–606.

Scott, F. W., and Forsdyke, D. R., 1976, Isotope-dilution studies of the effects of 5-fluorodeoxyuridine and hydroxyurea on the incorporation of deoxycytidine and thymidine by cultured thymus cells, *Can. J. Biochem.* **54**:238–248.

Scott, F. W., and Forsdyke, D. R., 1980, Isotope-dilution analysis of the effects of deoxyguanosine and deoxyadenosine on the incorporation of thymidine and deoxycytidine by hydroxyurea-treated thymus cells, *Biochem. J.* **190**:721–730.

Sieburth, J. McN., Johnson, K. M., Burney, C. M., and Lavoue, D. M., 1977, Estimation of *in situ* rates of heterotrophy using diurnal changes in dissolved organic matter and growth rates of picoplankton in diffusion culture, *Helgol. Wiss. Meeresunters.* **30**:565–574.

Sjostrom, D. A., and Forsdyke, D. R., 1974, Isotope-dilution analysis of rate-limiting steps and pools affecting the incorporation of thymidine and deoxycytidine into cultured thymus cells, *Biochem. J.* **138**:253–262.

Smith, W. O., Barber, R. T., and Huntsman, S. A., 1977, Primary production off the coast of Northwest Africa: Excretion of dissolved organic matter and its heterotrophic uptake, *Deep Sea Res.* **24**:35–47.

Sorokin, Y. I., 1978, Microbial production in the coral reef community, *Arch. Hydrobiol.* **83**:281–323.

Steffensen, D. M., and Sheridan, W. F., 1965, Incorporation of ^{3}H-thymidine into chloroplast DNA of marine algae, *J. Cell Biol.* **25**:619–626.

Stocking, C. R., and Gifford, E. M., 1959, Incorporation of thymidine into chloroplasts of *Spirogyra, Biochem. Biophys. Res. Commun.* **1**:159–164.

Stone, G. E., and Prescott, D. M., 1964, Cell division and DNA synthesis in *Tetrahymena pyriformis* deprived of essential amino acids, *J. Cell Biol.* **21**:275–281.

Swings, J., and de Ley, J., 1977, The biology of *Zymomonas, Bacteriol. Rev.* **41**:1–46.

Swinton, D. C., and Hanawalt, P. C., 1972, *In vivo* specific labeling of *Chlamydomonas* chloroplast DNA, *J. Cell Biol.* **54**:592–597.

Thelander, L., and Reichard, P., 1979, Reduction of ribonucleotides, *Annu. Rev. Biochem.* **48**:133–158.

Tobin, R. S., and Anthony, D. H. J., 1978, Tritiated thymidine incorporation as a measure of microbial activity in lake sediments, *Limnol. Oceanogr.* **23**:161–165.

Torsvik, V. L., 1980, Isolation of bacterial DNA from soil, *Soil Biol. Biochem.* **12**:15–21.

Van Es, F. B., and Meyer-Reil, L.-A., 1982, Biomass and metabolic activity of heterotrophic marine bacteria, in: *Advances in Microbial Ecology*, Vol. 6 (K. C. Marshall, ed.), pp. 111–170, Plenum Press, New York.

Vogels, G. D., and van der Drift, C., 1976, Degradation of purines and pyrimidines by microorganisms, *Bacteriol. Rev.* **40**:403–468.

Wallace, D. C., and Morowitz, H. J., 1973, Genome size and evolution, *Chromosoma* **40**:121–126.

Werner, R., 1971, Nature of DNA precursors, *Nature New Biol.* **233**:99–103.

Williams, P. J. leB., 1984, Bacterial production in the marine food chain: The emperor's suit of clothes?, in: *Flow of Energy and Materials in Marine Ecosystems: Theory and Practice* (M. J. Fasham, ed.), pp. 271–299, Plenum Press, New York.

Williams, P. J. leB., and Yentsch, C. S., 1976, An examination of photosynthetic production, excretion of photosynthetic products, and heterotrophic utilization of dissolved organic compounds with reference to results from a coastal subtropical sea, *Mar. Biol.* **35**:31–40.

Winn, C. D., and Karl, D. M., 1984a, Laboratory calibrations of the [^{3}H]adenine technique for measuring rates of RNA and DNA synthesis in marine microorganisms, *Appl. Environ. Microbiol.* **47**:835–842.

Winn, C. D., and Karl, D. M., 1984b, Microbial productivity and community growth rate estimates in the tropical North Pacific Ocean, *Biol. Oceanogr.* **3**:123–145.

7

Microelectrodes: Their Use in Microbial Ecology

**NIELS PETER REVSBECH and
BO BARKER JØRGENSEN**

1. Introduction

Among the fundamental goals of microbial ecology is the development of methods that will enable the identification and counting of the important microorganisms in nature, the determination of their physical and chemical microenvironment, and the analysis of their metabolic processes and interactions. Due to the small size of the organisms, much effort has been devoted to the development of high-resolution techniques for the observation and understanding of the world of bacteria on a microscale. Scanning and transmission electron microscopy and fluorescent staining, immunofluorescence and other techniques for light microscopy have been the most successful in terms of reaching a high spatial resolution. With respect to our understanding of the microbial microenvironments and of the nature of the microorganisms that carry out the measured metabolic activities, there is still a long way to go. Most chemical and radiotracer techniques in use today operate on a centimeter or at best on a millimeter scale and in most cases their results cannot be directly related to the relevant microorganisms. One notable exception to

NIELS PETER REVSBECH and BO BARKER JØRGENSEN • Institute of Ecology and Genetics, University of Aarhus, DK-8000 Aarhus C, Denmark.

this is the combined use of autoradiography and fluorescence microscopy on microbial communities.

High-resolution electrochemical techniques have been available, however, for many years without catching the attention of microbial ecologists. Simple, intracellular capillary electrodes with tip diameters of 1 μm or less have been used in neurophysiology since the 1950s for the study of action potentials and of ion transport over biological membranes. Ion-sensitive microelectrodes were later developed for the analysis of H^+, Na^+, K^+, and Ca^{2+}. These were either solid-state electrodes based on ion-selective glasses or liquid-membrane electrodes based on organic ion-exchangers. Again, the main use of the microelectrodes was in cell physiology, where they were applied mostly for intracellular analysis. In the 1960s polarographic microelectrodes with sensing tips of only a few micrometers appeared, which allowed measurements of oxygen in whole tissues and in cell cultures.

It may seem surprising that this important development of new analytical microtechniques, of obvious applicability to many problems in microbial ecology, had not reached this field many years ago. This is a classic example of the communication barrier between the different biological and technological disciplines, which has repeatedly delayed the introduction of new approaches.

In 1978, we started to use oxygen microelectrodes in our studies of the ecology and biogeochemistry of marine sediments and other microbial environments. The first aim was simply to analyze the microscale distribution of oxygen in order to identify the boundaries of aerobic and anaerobic metabolism in nature. We later developed microelectrode techniques to measure photosynthesis and respiration at high spatial and temporal resolution, and we also included pH and sulfide microelectrodes in our research.

Because so little work of relevance to microbial ecology had been done previously with these techniques, the present chapter necessarily is based heavily on our own work. The reader should therefore not anticipate a review of of a well-established research field, but rather a summary of our own limited experience with a new and fascinating methodology. Our motivation to write the review at this early stage has mostly been the interest we have experienced from many colleagues, who wish to introduce microelectrodes in their own research. We hope this chapter will be a help and inspiration to such new research. In the presentation we have included both a brief description of how to make and use microelectrodes as well as examples of different applications, which should give an impression of the technical problems involved and the many potential applications.

2. Microelectrodes and Measuring Equipment

2.1. Types of Microelectrodes

The four different types of microelectrodes used in microbial ecology are shown in Fig. 1. They comprise two types of oxygen microelectrodes, a sulfide microelectrode, and a pH microelectrode. Some publications describe measurements with "microelectrodes" that were several millimeters in diameter. We will not try to define an absolute size limit of a microelectrode, but an electrode having a diameter of more than 0.2 mm does not qualify as "micro." The term "electrode" is also used here for a combined sensor, such as the one shown in Figs. 1B and 2, which actually contains two electrodes, a sensing electrode and a reference electrode, both situated behind a polymer membrane. Two of the electrodes are commercially available: the cathode-type oxygen microelectrode (Diamond Electro-Tech, P.O. Box 2387, Ann Arbor, MI 48106) and the pH microelectrode (Microelectrodes, P.O. Box 365 Oak Hill Park, Londonderry, NH 03053). We have, however, always made our own microelectrodes, as described in the following sections.

2.1.1. Cathode-Type Oxygen Microelectrode

The type of oxygen microelectrode (Baumgärtl and Lübbers, 1983; Revsbech 1983) shown in Fig. 1A has been used in most published studies, but a slightly different design (Whalen et al., 1967) also has been applied. Whalen's electrode is built around a core of Wood's metal instead of a platinum wire; Wood's metal is an alloy that melts at very low temperatures. The tip of Whalen's electrode is coated with gold, so that, electrochemically speaking, it is a tiny gold cathode, like the electrode shown in Fig. 1A. Both electrodes can be made with tip diameters of less than 1 μm.

Oxygen is sensed only at the very tip of the cathode-type oxygen microelectrode, where the membrane-covered gold surface is exposed to the medium. The microelectrode is charged at about -0.75 V versus a standard calomel reference electrode, and the current originating from the reduction of oxygen at the gold surface is proportional to the oxygen partial pressure in the surrounding medium. Because of the small size of the electrode tip, diffusion is fast relative to convective transport (Purcell, 1977; Crank, 1983) in supplying oxygen to the reducing metal surface. Convection by stirring therefore gives only a 1–5% increase in current compared to the current in stagnant medium. Because of the small size and short diffusion path, the 90% response time can be less than 0.2 sec.

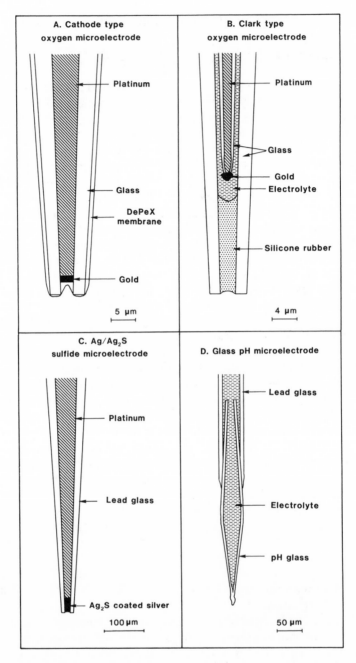

Figure 1. Sensing tips of the four types of microelectrodes that have been used in microbial ecology. [Redrawn from Revsbech (1983).]

The response time can be measured by quickly introducing the electrode from air into deoxygenated water. Electrical contact between the cathode and the reference electrode while the microelectrode is in water-saturated air can be maintained via the wet microelectrode surface.

The oxygen microelectrode can be made to withstand high hydrostatic pressures by filling the air space within the shaft with oil (C. E. Reimers, unpublished data). These pressure resistant electrodes are now being used for *in situ* measurement of oxygen microprofiles in pelagic sediments at 4000 m water depth.

A coaxial design of the electrode shown in Fig. 1A has been described by Baumgärtl and Lübbers (1983). The glass shaft of their electrode is coated with multiple thin layers, the outermost consisting of silver. The silver layer shields the platinum wire inside the microelectrode from electrical noise, and at the same time the chlorinated silver layer serves as a reference electrode. Such an external reference electrode may, however, be affected by some of the ions (e.g., sulfide) in the medium, and this will prevent its application in many environments. The greatest problems in the construction and use of the oxygen microelectrodes described above are associated with the membrane. The membrane must allow oxygen to pass through, and at the same time it must be electrically conducting; i.e., it must be permeable to at least some ions. Both Mg^{2+} and Ca^{2+} poison the electrodes, most probably by precipitation of hydroxides and carbonates on the gold surface (Baumgärtl and Lübbers, 1983), which becomes alkaline during the reduction of oxygen to OH^-. Dissolved organic matter may also poison the electrode. The membrane should therefore not be permeable to these poisoning substances. It is difficult to make a sufficiently thin membrane that at the same time will prevent large, double-charged cations to pass through and still be permeable to oxygen and small ions such as K^+ and Na^+. The membrane often has to be applied several times before a satisfactory result is obtained (Revsbech 1983). Because of these problems with the membrane, the signal from this type of oxygen microelectrode is seldom very stable when used in natural media. The electrodes described by Whalen *et al.* (1967) can be made with a very deep recess at the tip. The deep recess improves the stability of the electrode when used in animal or human tissue (P. Nair, personal communication), but it is not known whether this will be the case also in other media. According to P. Nair, a membrane should not even be necessary when the recess is sufficiently deep.

It is our experience that a usable response of the oxygen microelectrodes described above cannot be obtained in media that are both high in Mg^{2+} or Ca^{2+} and low in Na^+ and K^+. The instability is expressed both as a slow drift and as an increased sensitivity to oxygen after periods at low oxygen tensions. The increase in sensitivity after periods at low oxygen is probably due to dissolution of precipitates at the gold surface during these less OH⁻-producing conditions.

The oxygen microelectrodes are poisoned by sulfide (Revsbech *et al.*, 1980b), but after the first exposure, the effect of sulfide is small. Electrodes for use in media where they may encounter hydrogen sulfide should therefore be poisoned before use. A "conditioning" of the electrode in sediment before using it to record oxygen profiles within the sediment was recommended by Reimers *et al.* (1984). The conditioning also improved the stability when no hydrogen sulfide was present.

2.1.2. Clark Microelectrode

The problems with the membrane of the cathode-type oxygen microelectrodes described above were solved (Revsbech and Ward 1983) by developing a combined microsensor (Figs. 1B and 2) that is a small version of the conventional Clark electrode (Clark *et al.*, 1953). In this microsensor, the cathode is situated behind an electrically insulating membrane of silicone rubber (Fig. 1B), which is extremely permeable to

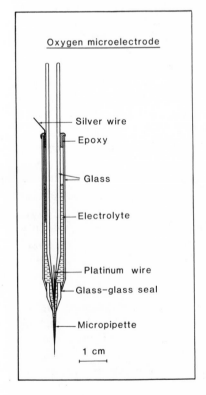

Oxygen microelectrode

— Silver wire

— Epoxy

— Glass

— Electrolyte

— Platinum wire

— Glass–glass seal

— Micropipette

1 cm

Figure 2. Oxygen microelectrode of the Clark type. The electrode tip is shown in Fig. 1B. [From Revsbech and Ward (1983).]

oxygen (Shiver, 1969). The cathode is bathed in an electrolyte solution of 1 M KCl into which a Ag/AgCl reference electrode (Fig. 2) is immersed. The electrolyte solution also serves as an electrical shielding of the cathode. This electrode has a response time comparable to the cathode-type oxygen microelectrode shown in Fig. 1A, and it is also insensitive to stirring. The signal from the electrode is much more stable, however. The current drift is often less than 2%/hr even when used in natural media containing various salts and dissolved organic compounds. The linearity of the signal versus oxygen tension is 100% for almost all electrodes. The tip diameter has been made smaller than 2 μm, but it is also possible to make thicker and much sturdier electrodes for use in systems like soils (Sexstone *et al.*, 1985) and coarse sediments, which require mechanical strength.

2.1.3. Sulfide Microelectrode

The sulfide microelectrode (Revsbech *et al.*, 1983) shown in Fig. 1C is a small Ag/Ag$_2$S electrode. The problems associated with the use of sulfide electrodes are discussed by Berner (1962) and Guterman and Ben-Yaakov (1983). The chemical species actually being sensed by the electrode is S^{2-} ion, the concentration of which is dependent on both the total concentration of dissolved sulfide and pH. Knowledge about the exact pH at the point being analyzed is therefore essential for the interpretation of the readings obtained using the sulfide microelectrode. Theoretical calculations of the total amount of dissolved sulfide from electrode potential and pH are not sufficiently accurate, and in practice it is always necessary to calibrate the electrode. There are, however, often problems with drift of the potential, so that recalibration may be necessary. Due to the high pK of the HS$^-$–S^{2-} pair (11.96), the detection limit for total sulfide is not very low, about 10 μM at pH 7–8, and the calibration curve of electrode potential versus log(S^{2-}) is nonlinear at concentrations below 100 μM total sulfide.

2.1.4. pH Microelectrode

The pH microelectrode shown in Fig. 1D is a small version of the large, commercial glass pH electrodes. The electrode shown in Fig. 1D is relatively sturdy and it is also easy to make. Various other designs of pH microelectrodes are possible (Thomas, 1978), and tip diameters of less than 1 μm can be obtained. Some types of pH microelectrodes have the pH- sensitive glass cone within a recess at the electrode tip, whereas others are made with a liquid pH-sensitive membrane (Ammann *et al.*, 1981). The electrodes with a liquid membrane have inferior specificity to hydrogen ions as compared to the glass electrodes, and they also have a shorter lifetime. They are, however, relatively easy to construct and can be made with tip diameters of less than 1 μm.

Calibration of the pH microelectrodes should be carried out in buffers of similar ionic composition as the media in which they are used, as considerable interference by other ions, especially Na$^+$, may occur. Old and "overhydrated" electrodes were found to give a potential almost one pH unit off the actual value when first calibrated in standard pH buffers and afterward used in a seawater salt brine of 80% salinity (N. P. Revsbech, unpublished data). Correctly hydrated electrodes (Section 2.2.4) gave a negligible error. Good pH microelectrodes with correctly hydrated pH-glass have linear calibration curves in the range from pH 4 to 10 with slopes of 57–59 mV/pH unit when calibrated at room temperature. Excessively hydrated electrodes may have slopes of less than 50 mV/pH unit.

2.2. Construction of Microelectrodes

A very good introduction to working with glass in small dimensions and to the construction of various ion-selective microelectrodes is given by Thomas (1978). It is not possible to outline all the details for the construction of microelectrodes here, but we will give a short description of how the electrodes shown in Fig. 1 are made.

The equipment necessary for making the electrodes is quite simple, the most expensive parts being two micromanipulators ($400–600 each), a light microscope with the ability to give a good resolution at 100× and 400× magnification, and a dissection microscope with magnifications adjustable from about 10× to 50×. Physiologists often use electrode pullers to make the very delicate microelectrodes for intracellular use, but we have never found such equipment necessary or advantageous for the construction of our electrodes.

2.2.1. Construction of Cathode-Type Oxygen Microelectrode

The oxygen microelectrode shown in Fig. 1A is made around a core of tapered platinum wire (99.99% purity, annealed/slightly hard-drawn condition, 0.1-mm thermocouple wire from Engelhard Industries, Surrey, U.K.). The platinum wire is finely tapered by etching in a 90% saturated KCN solution while 2–7 V ac is applied. The other electrode immersed into the KCN solution to complete the circuit consists of a graphite rod from a pencil. A voltage of 7 V is applied while 1 cm of the wire is immersed in the KCN solution. The current is reduced to 2 V when the 1-cm tip has almost disappeared, and the etching is continued for a few seconds. The initially high voltage during etching creates a smooth surface on the platinum, but the tip becomes blunt. Etching at low voltage tends to create cavities in the platinum, and it should therefore only be applied for a short period, but it is essential for the creation of a thin tip.

The etched platinum wires are cleaned in water, sulfuric acid, and alcohol, and are then inserted into precleaned soda-lime capillaries (AR-glas, Glaswerk Wertheim, West Germany). These capillaries should be thin-walled, but the internal diameter should be at least 0.25 mm. The platinum is coated with glass by hanging the capillary vertically through an electrically heated nichrome loop (Fig. 3). Melting of the glass should start around the nonetched platinum wire, about 1 cm from the etched part. When the glass melts, gravity pulls the platinum wire through the melting zone, and a thin glass coating is produced. It is often necessary to attach a 2-g weight to the electrode shaft to obtain a sufficient pull by gravity.

The tip of the platinum is totally covered with glass after a successful melting as described above. The very tip of the platinum must therefore be exposed by grinding off excess glass on a rotating disc covered with $0.25\text{-}\mu m$ grain size diamond paste. For electrodes of the type shown in

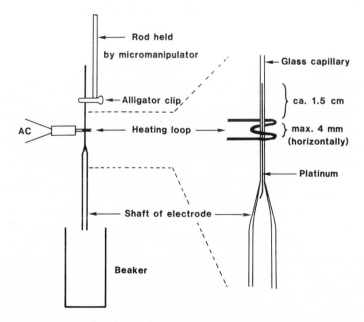

Figure 3. Left: Coating of platinum wire with glass. When the loop is heated, the glass capillary inside the loop melts, and the glass shaft falls down into the beaker. Right: Position of platinum wire relative to heating loop before heat is applied.

Fig. 1A, it is advisory to proceed with the grinding until a tip diameter of 5–9 μm is obtained. After grinding, the platinum at the tip should be etched briefly in KCN so that a 5–10 μm-long recess is formed.

The platinum in the recess is coated with gold by electroplating as described by Baumgärtl and Lübbers (1983). The gold-plating solution [5% KAu(CN)$_2$ in water] may be contained in a thin capillary so that the process can be observed under the microscope. The electrode should soak in water for 1 day after the electroplating. After this, the tip of the electrode is coated with a polymer membrane as described by Revsbech (1983). The electrode is ready for a test of performance after hydration of the membrane for 1 day.

2.2.2. Construction of Clark Microelectrode

The cathode of the Clark microelectrode shown in Figs. 1B and 2 is made largely as described above for the cathode-type oxygen microelectrode. The platinum wire must, however, be etched to a longer and thinner tip, since the cathode is then easier to fit into an outer casing. Furthermore, a recess is not etched at the cathode tip before gold plating, and a membrane is not applied on the cathode.

The tip of the outer casing is shaped in several steps by gravitational pull in electrically heated metal loops as shown in Fig. 3. A small loop made of 25-μm-thick platinum wire (also used for making pH microelectrodes as shown in Fig. 4) is used for the final step. The membrane of silicone rubber is applied by inserting the tip into uncured silicone rubber (Silastic, Medical Adhesive type A, Dow Corning, Midland, MI 48640). The membrane enters the tip by capillary suction, and the tip should be withdrawn from the silicone rubber when 10–20 μm of the capillary is filled. The cathode is inserted into the outer casing using a micromanipulator, and a partial seal of epoxy is applied between cathode and casing. After the epoxy and silicone have cured for 1 day, the electrolyte consisting of 1 M KCl saturated with thymole is injected, a chlorinated silver wire is inserted, and the electrode is ready for use.

2.2.3. Construction of Sulfide Microelectrode

The Ag/Ag$_2$S sulfide microelectrode is constructed by procedures rather similar to the ones described above for the cathode-type oxygen microelectrode. The soda-lime glass is, however, substituted with the electrically more insulating lead glass (e.g., 28% PbO Bleiglas, no. 8095, Schott Glaswerke, Mainz, West Germany). The silver layer is applied by electroplating (Revsbech et al., 1983), and the electrode is immersed into 0.1 M ammonium sulfide solution to form the Ag$_2$S coating.

2.2.4. Construction of pH Microelectrode

The pH microelectrode shown in Fig. 1D is made by a simpler procedure than that outlined by Thomas (1978). It is constructed from pH-
sensitive glass [Corning 0150, suppliers listed by Thomas (1978)] and
lead glass. The pH-glass is drawn in a soft flame to 0.2- to 0.3-mm-thick
capillaries, which are then pulled by gravity in an electrically heated
metal loop as described above and shown in Fig. 3. The weight attached
to the capillary should be large ~10 g, and the heat applied should be
low. The capillary will then form a conical tip when it falls down. The
extreme tip is broken off so that a 5- to 10-μm-wide opening is formed.

The lead glass used for the shaft of the electrode is pulled in a flame
so that a capillary with a diameter of 0.5–1.0 mm is formed. This capillary is pulled by gravity as described above for the pH-sensitive glass, but
here the temperature of the heating loop should be high and the weight
attached to the capillary only 2 g. The lead glass will then form a very
long, 20- to 50-μm-wide capillary. The thin glass capillary is broken about
1 cm from the shoulder to the thicker capillary. A clean cut is preferable,
and this can be obtained by scraping the capillary with silicon carbide
crystals before breaking.

The lead glass capillary is positioned, tip upward, on a small adjust-
able stand. The pH-glass cone is now introduced into the lead glass capillary using a micromanipulator. After making a glass–glass seal (Fig. 4),
the small stand is lowered so that the lead glass is hanging in the pH-glass
capillary. The pH-sensitive cone is now made by heating the pH-glass just
above the fusion zone. Finally, the tip of the electrode is closed by positioning the very hot platinum heating loop just above the tip. The filling
of the electrode with electrolyte and the hydration at high temperatures
are described by Thomas (1978). Our electrodes are normally allowed to
hydrate for 3 or 4 days at 70°C.

2.3. Measuring Equipment

The impedances in the microelectrode measuring circuits are about
10^{-10} Ω for the oxygen microelectrodes and between 10^{-7} and 10^{-11} Ω for
the pH microelectrodes, depending on the size of the individual elec-
trode, the degree of hydration, etc. Because of the high impedances, the
signals from the electrodes are very sensitive to electromagnetic interference. The measuring circuits used should thus be designed to minimize
such electrical noise, and they should also be able to amplify the small
signals so that accurate readings can be obtained. Requirements of the
measuring circuit other than stability and high sensitivity may, however,

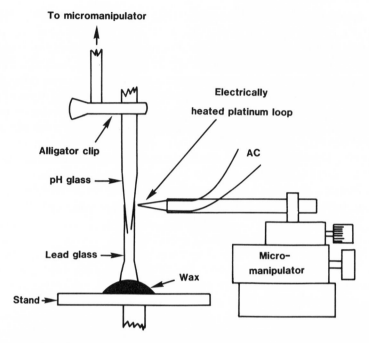

Figure 4. Fusion of pH and lead glasses for a pH microelectrode. The sizes of capillaries, lead glass shaft, etc., are not drawn to the same scale.

be important for optimal use of the microelectrodes in different ecological applications.

2.3.1. Measuring Circuits

A setup for measuring oxygen in a sediment core is shown schematically in Fig. 5. The microelectrode is held by a micromanipulator, which can be used to introduce the microelectrode tip into the substratum with a depth resolution of better than 10 μm. A micromanipulator with a motordrive (e.g., Mertzheuser, Steindorf/Wetzlar, West Germany) can be recommended, since it is difficult to operate the micromanipulator without causing uncontrolled vibration of the microelectrode tip. All cables conducting the primary signal must be of "low-noise" quality (with additional graphite shielding inside the copper shield) to minimize electrical noise associated with movement of the cables. The cables should be mounted on the microelectrodes in such a way that the contact between the inner conductor of the cable and the signal conducting metal wire of the microelectrode can be kept completely dry. The cable can be mounted

(Fig. 6) on the combined oxygen microelectrode as shown in Fig. 1B. An electrode holder made of Plexiglas (Revsbech, 1983) is, however, more convenient for laboratory use. Desiccant in the electrode holder is a necessity, since even moderate air humidity may cause erratic signals originating from currents creeping along the humid glass surfaces. The current in the circuit must be measured by a very sensitive ammeter with a range down to 1 pA (10^{-12} A). The amplified signal can then be recorded on a strip-chart recorder. The tracing shown in Fig. 5 is the actual tracing from the measurement of an oxygen profile in a dark-incubated, sandy sediment. The time it took to obtain a stable reading when the electrode tip was moved to a new layer (Fig. 5) was caused by the response time of the electrode, disturbance of the oxygen gradients when sand grains moved during the advancement of the electrode, and the time it actually took to move the electrode. The oxygen profile calculated from the tracing is shown to the right in Fig. 5.

In the measuring circuits for sulfide and pH, the picoammeter and voltage source in the oxygen microelectrode circuit are substituted by a high-impedance voltmeter. The cable between the voltmeter and electrode should be kept as short as possible.

Under conditions where electrical noise originating from extensive

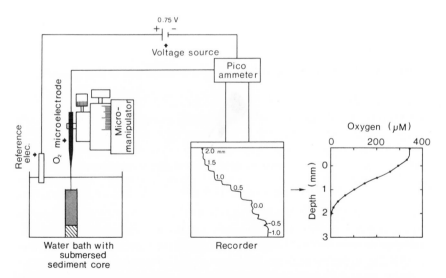

Figure 5. Simple circuit for measurement with oxygen microelectrodes. The microelectrode tip was first positioned in the water 1 mm above the sediment. It was then advanced downward into the sediment at increments of 0.25 mm. The current in the circuit at various depths was recorded on a strip-chart recorder. The oxygen profile to the right was calculated from this tracing.

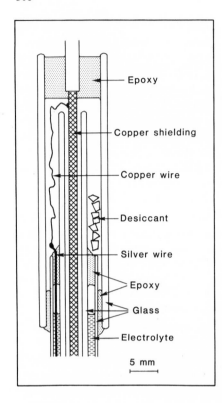

Figure 6. Connection of shielded cable with oxygen microelectrode of the type shown in Fig. 2. Electrical contact with the silver wire serving as reference electrode was made through the copper shielding of the cable. Electrical connection between the platinum wire in the tip region of the electrode and the inner conductor of the cable was made by simple contact. [From Revsbech and Ward (1983).]

movement of the cables becomes unacceptable, e.g., on board vibrating ships or under windy conditions in the field, it may be an advantage to build an amplifier into the electrode holder itself. Such devices have been described for both oxygen microelectrodes (Reimers *et al.*, 1984) and pH microelectrodes (Thomas 1978). Reimers and coworkers found that the electrical noise level on board ships could be reduced ten-fold in this way. Also, the positioning of the electrode within the substratum may be poorly defined when the vibration level is high. Especially when measuring with very small pH microelectrodes, the slow response caused by the capacity of an even moderate length of shielded cable may be unacceptable, and this problem is also solved by placing an amplifier in the electrode holder.

All the equipment for experimental work with microelectrodes can be made portable, and it is thus possible to measure microprofiles of oxygen, sulfide, and pH in the field. There are no serious problems in doing this when the substratum is mechnically stable, when the sun is shining, and when the air humidity is low. Useful data have been obtained by field work in Solar Lake, Sinai (Jørgensen *et al.*, 1979) and in hot springs of

Yellowstone National Park (Revsbech and Ward, 1983, 1984a,b). It has even been possible to do *in situ* experiments at a water depth of several meters, where the micromanipulator had to be operated by a diver (Lindeboom and Sandee, 1984), and it is expected that measurements will soon be done in the deep sea with the microelectrodes mounted on a free-falling vehicle or operated from a submersible. Field work in coastal sediments with the delicate electrodes may, however, also be a frustrating experience, especially due to the many snails, crabs, fish, etc., which always seem determined to approach and break the microelectrodes.

2.3.2. Computerized Data Collection

The circuit shown in Fig. 5 is able to record the data for later analysis, but for many applications it is advantageous to feed the signal from the picoammeter directly into a computer. We have often found that simple data handling and calculation took much more time than the actual measurements. An example of a custom-built, computer-based data collection system we have developed is shown in Fig. 7. The signal from the picoammeter is fed into a 12-bit analog-to-digital converter, which transforms the signal into a figure between 0 and 4095. Light intensity and other analog signals also may be fed into the converter. The digital signals from the converter are fed into a data collection unit, and the pulses from a stepper motor on the micromanipulator are also counted by this unit. In this way, coupled data points of electrode current and depth are received. (A micromanipulator equipped with stepper motor is sold by Fairlight Scientific & Industrial Equipment, Postbus 4055, 3006 AB Rotterdam, The Netherlands.) The data collection unit is in fact a small, independent computer, which sorts out the essential sets of information from the large amounts of data entering through the input channels. The

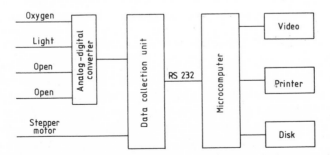

Figure 7. Schematic diagram of computerized data collection from microelectrode circuit. The positioning of the motor-driven micromanipulator and other parameters, such as light, can also be read into the computer.

data, which are transmitted further to the microcomputer via a serial RS 232 interface, may for example, only be transmitted for every 0.1-mm depth interval or for every 0.1 sec. The programs based upon which the data collection is performed are contained in an ROM (read only memory) chip built into the data collection unit. The microcomputer performs calculations on the data simultaneously with the collection of new data. Both data and results of calculations are stored on "floppy disks." The results of the calculations can be displayed on a video monitor and printed out on a dot-matrix printer. Another data collection system based on a programmable calculator (HP41C from Hewlett Packard) has been developed for the measurement of *in situ* oxygen microprofiles in deep-sea sediments (C. E. Reimers, unpublished results), and in connection with a datalogger (3421A from Hewlett Packard), for photosynthesis measurements (B. B. Jørgensen, unpublished results).

3. Microelectrode Applications

Microelectrode studies have been done in a wide variety of environments and microbial communities, mostly in sediments, and microbial mats. Because of the large amount of data available for sedimentary ecosystems, examples of microelectrode measurements in sediments will be treated first, and data obtained in sediments will be used to illustrate the general principles, applications, and problems associated with the use of microelectrodes in microbial ecology.

3.1. Oxygen Distribution in the Microenvironment

3.1.1. Oxic versus Oxidized Conditions

It is of fundamental importance for the understanding of microbial processes to know the environmental conditions at the site where the processes occur. It is especially important to know whether oxygen is present or not. Before the introduction of oxygen microelectrodes in ecological research, the oxygenated surface layer in sediments was often assumed to be identical to the brown, oxidized surface layer. The brown surface layer is virtually identical with the layer having a positive redox potential as read with a redox electrode (Fenchel 1969). Profiles of oxygen and redox typical of coastal, sandy sediments (Revsbech *et al.*, 1980a,b) are shown in Fig. 8A. The oxic zone shown in Fig. 8A was only 2 mm thick, whereas a positive redox potential was found down to a depth of 3.5 cm. The oxic–anoxic interface could not be distinguished on the redox profile, and substances other than molecular oxygen must consequently be able to

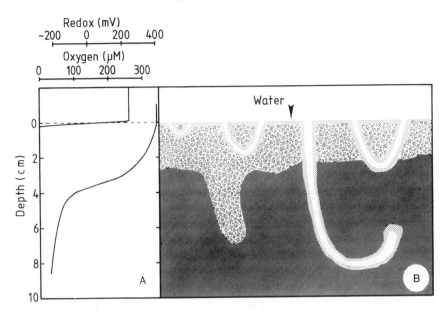

Figure 8. Oxic versus oxidized conditions. (A) Oxygen and redox profiles from the sediment shown in (B) (measured at point marked with arrow). Oxygen was found only in the uppermost 2 mm of the sediment and in the immediate vicinity of animal burrows (light dotted areas). The periodic introduction of oxygen into parts of the sediment caused by the activity of the infauna kept an ~3-cm-thick surface layer oxidized ("grainy" area) and even deep in the generally reduced sediment (dark area), deeply burrowing infauna caused local oxic and/or oxidizing conditions.

give the sediment a high, positive redox potential. This is not surprising, as the sediments often contain considerable amounts of iron and manganese, which in their oxidized forms can be responsible for such positive redox potentials. It is more difficult to understand, however, how such predominantly insoluble compounds are kept oxidized without oxygen being present. Oxidized manganese and iron compounds are continuously being reduced by the metabolic activity of some bacteria (Jones *et al.,* 1983), and they are also reduced by hydrogen sulfide formed by sulfate-reducing bacteria. It is likely (Revsbech *et al.,* 1980a) that the major part of the reoxidation at any given sediment location does not occur continuously, but occurs by local or periodic events. These events may be reworking by infauna, or introduction of oxygen due to impact of currents and waves during storms. The reworking of the sediment by infauna ["bioturbation" (Aller, 1977)] is often able to keep a relatively thick surface layer oxidized (Edwards, 1958; Davis, 1974). Localized strongly reducing conditions may be found close to the surface in such sediments if by chance for an extended period of time no animal has oxygenated and oxidized that particular area. The concept of bioturbation as an electron carrier in sediments is illustrated in Fig. 8B. The oxidized sediment layer may constitute an ecologically important redox buffer (Board, 1976; Jørgensen, 1977) in the case of periodic oxygen deficiency of the overlying water.

3.1.2. Factors Affecting the Oxygen Distribution in Sediment

The very superficial penetration of oxygen into the sediments shown in Figs. 5 and 8 is typical for most shallow water sediments (Sørensen et al., 1979; Revsbech et al., 1980a; 1980b; Lindeboom and Sandee, 1984). The penetration of oxygen is governed by the rate of diffusional supply of oxygen to the sediment and by the rate of net oxygen consumption per unit volume of sediment. Assuming a uniform depth distribution of oxygen consumption, one can calculate the depth of penetration h of oxygen into a sediment from (Revsbech et al., 1980a)

$$h = 2D_sC_0\phi/J \qquad (1)$$

where D_s is the apparent diffusion coefficient of oxygen in the sediment, ϕ is the porosity, C_0 is the oxygen concentration at the sediment surface, and J is the oxygen consumption per unit area of the sediment surface. The inverse relationship between oxygen penetration and oxygen consumption described by Eq. (1) is in accordance with the thick oxic surface layers in deep sea sediments where the oxygen consumption is low (Smith and Baldwin, 1984; Smith and Hinga 1983). Deep oxygen penetrations have been measured by microelectrodes (Reimers et al., 1984) and by gas chromatography [>60 cm measured by Murray and Grundmanis (1980); >100 cm measured by Sørensen (1984)] in cores collected in the deep sea.

Any impact causing a change of the variables on the right side of Eq. (1) would result in a different oxygen penetration. Three oxygen profiles measured in the same sediment core collected from a shallow, sandy area are illustrated in Fig. 9. The profile shown in Fig. 9A was recorded while a water current of 5 cm/sec was maintained above the sediment surface. The diffusive boundary layer (Jørgensen and Revsbech, 1985) above the sediment was 0.2 mm thick and caused a decrease in the concentration of oxygen from 252 μM in the overlying water to 228 μM at the sediment surface. The stirring of the water had been stopped for 1.5 hr when the profile shown in Fig. 9B was measured. The lack of stirring caused a thickening of the diffusive boundary layer above the sediment to about 1 mm. The thicker diffusional barrier caused a reduction of the oxygen con-

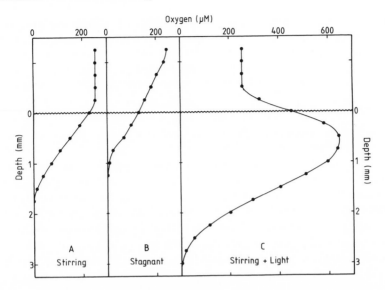

Figure 9. Oxygen profiles in a shallow, sandy sediment under various environmental conditions. (A) Current rate of ~5 cm/sec above the sediment, (B) stagnant overlying water, (C) as in (A), but with photosynthetic oxygen production by the benthic microbiota caused by illumination at 200 μEinst/m^2 per sec. Temperature at 22°C.

centration at the sediment surface to 130 μM and a reduction of the oxygen penetration from 1.7 to 1.2 mm.

The oxygen profile after an illumination period of 1.5 hr is shown in Fig. 9C. The photosynthesis in the surface layers caused both very high oxygen concentrations and an increase in the thickness of the oxic zone to 3.0 mm.

3.1.3. Diffusive Boundary Layers

The diffusive boundary layers described in Section 3.1.2 exist around all surfaces covered by liquids. The diffusive boundary layers are created by viscous forces, which cause a thin film of water to "stick" to the surface. The water in these films does not participate in the general circulation. Within the diffusive boundary layer, the transport of solutes occurs by molecular diffusion, whereas eddy diffusion is insignifcant as a transport mechanism (Santschi *et al.,* 1983). The diffusive boundary layer thus constitutes a diffusional barrier, which may limit the exchange of dissolved molecules between the surface and the bulk of the liquid. The boundary layers are especially important for the exchange of nutrients between microbial communities and their aquatic environment. Oxygen microelectrodes are ideal tools for the study of diffusive boundary layers (Jørgensen and Revsbech, 1985). The electrodes give information about both the thickness of the boundary layer and the extent to which it is a stable diffusive layer, i.e., whether the oxygen profile through the boundary layer can be explained by steady-state molecular diffusion. We use the term *true diffusive boundary layer* for the layer in which the linear oxygen profile can be explained by pure molecular diffusion, and the term *effective diffusive boundary layer* as explained in Fig. 10 (cf. Jørgensen and Revsbech, 1985).

The diffusive boundary layer covers the sea bottom as a thin blanket as shown in Fig. 11. The bulk water flow above the sediment was 5–10 cm/sec when we measured the oxygen profiles. The diffusive boundary layer above relatively smooth surfaces is about 0.1–0.2 mm thick at such current rates, whereas it is about 1 mm thick in stagnant water. Such thick diffusive boundary layers must cover the entire deep sea floor (Baudreau and Guinasso, 1982; Wimbush, 1976), where the currents are very slow. Small protrusions from the sediment surface are able to penetrate through most of the boundary layer, as its thickness is not only dependent on the current velocity above the substratum, but also on the size of the

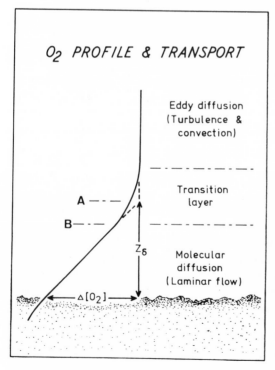

Figure 10. The diffusive boundary layer at the sediment–water interface as determined by the oxygen microgradients. Here A and B define the outer limits of the effective diffusive boundary layer and the true diffusive boundary layer, respectively. The thickness of the effective diffusive boundary layer is marked Z_δ. [From Jørgensen and Revsbech (1985).]

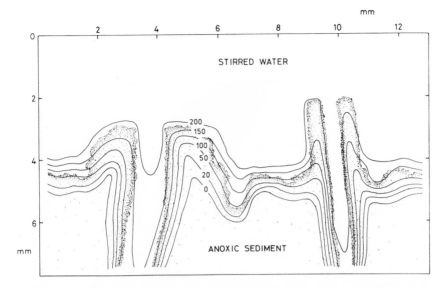

Figure 11. Isopleths of oxygen in a shallow sediment inhabited by two small polychaete worms. The isopleths show how the diffusive boundary layer created an oxygen gradient just above the sediment surface, and how this oxygen gradient followed the topography of the sediment. [From Jørgensen and Revsbech (1985).]

object. Small objects that protrude like the worm tubes shown in Fig. 11 thus reach into more oxic water than found at the normal sediment surface. This is especially important during periods when the water becomes stagnant and oxygen availability may be limited. Other organisms have similar strategies. Several ciliate species, e.g., *Vorticella* sp., have stalks that allow them to stretch out through the diffusive boundary layer of the substratum to which they are attached.

3.2. Microbial Photosynthesis

3.2.1. Effect of Photosynthesis on Oxygen Microprofiles

As shown in Fig. 9C, microbial photosynthesis can cause high concentrations of oxygen in the photic layers of sediments. The buildup of high oxygen concentrations within the upper layers of a sandy sediment during the first 30 min of a light period is shown in Fig. 12, while the disappearance of the oxygen peak within the first 30 min of the following dark period is shown in the lower portion of the figure. Most of the oxygen peak had actually formed after only 7 min of light incubation, and the steady-state dark profile was approached after only 7 min in the dark.

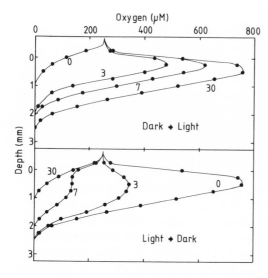

Figure 12. Oxygen profiles in a shallow, sandy sediment from Randers Fjord, Denmark, during a light–dark cycle. The time in minutes after start of illumination at 160 μEinst/m^2 per sec (dark → light) or darkening (light → dark) is written at each profile. Temperature 22°C.

The sediment analyzed in Fig. 12 had a relatively sparse population of diatoms and was only moderately photosynthetically active. Considerably faster changes and higher oxygen concentrations can be measured in sediments with denser microalgae (Revsbech et al., 1980b), cyanobacterial mats from saline environments (Jørgensen et al., 1979, 1983; Revsbech et al., 1983), or in algal mats from hot springs (Revsbech and Ward, 1983, 1984a,b; Ward et al., 1984).

Changes in oxygen concentration occurring within a sediment during light–dark cycles can be analyzed as in Fig. 12, where each profile was measured by quickly advancing the electrode from the sediment surface and downward, stopping for only a few seconds at 0.25-mm intervals. The disadvantage of this method is that all the data points used to construct the profiles are not measured simultaneously. It took about 1 min to measure a total oxygen profile from top to bottom, and the times written on each profile in Fig. 12 indicate only the start of the measurement. It is also possible, however, to fix the electrode tip at some depth in the substratum and then record the oxygen concentration continuously at that particular depth. A copy of the recorder output from such an experiment is shown in Fig. 13. The sediment was the same as analyzed in Fig. 12, but the light intensity was higher. The electrode was fixed at 0.75 mm depth within the dark-incubated sediment, and the sediment was then illuminated for 16 min. A high steady-state oxygen concentration of 1460 μM O$_2$ was approached at the end of the light incubation. When the light was turned off, the oxygen concentration decreased rapidly, and after 16 min in the dark, the former steady-state oxygen concentration in the dark was approached.

3.2.2. Photosynthetic Rates Measured with Microelectrodes

3.2.2a. Principle of Method. The stable oxygen concentration approached when the sediment had been exposed to light for 16 min (Fig. 13) must be the result of an equilibrium between oxygen-consuming and oxygen-producing processes at each depth. The processes removing oxygen from any particular layer are (1) biological and chemical consumption of oxygen within this layer and (2) molecular diffusion away from the layer. The only process that produces oxygen is oxygenic photosynthesis as carried out by microalgae and cyanobacteria. When photosynthesis was stopped instaneously by darkening (Fig. 13), consumption and diffusional transport of oxygen continued at initially unchanged rates. There was, however, no longer a production of oxygen to counterbalance the losses of oxygen, and the oxygen concentration therefore started to decrease at a rate equal to the former photosynthetic rate. Measurement of the initial rate of decrease in oxygen concentration after darkening can thus be used to quantify the photosynthesis of the oxygenic phototrophs (Revsbech *et al.,* 1981) in systems where these organisms are concentrated, e.g., in shallow water sediments and in epiphytic communities.

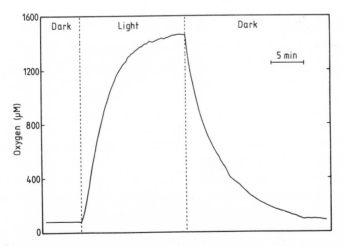

Figure 13. Continuous recording of oxygen concentrations at 0.75mm depth in a sandy sediment during a light–dark cycle. The sediment was taken from 0.5 m water depth in Randers Fjord, Denmark. The light intensity was 700 μEinst/m^2 per sec; temperature, 22°C.

The arguments above can also be expressed mathematically using Fick's second law of diffusion (Crank, 1983). In the light, the change in oxygen concentration $C(x,t)$ at depth x over time t is given by

$$\frac{\delta C(x,t)}{\delta t} = D_s \frac{\delta^2 C(x,t)}{\delta x^2} - R(x,t) + P(x,t) \tag{2}$$

where $R(x,t)$ is the rate of oxygen consumption and $P(x,t)$ is the rate of oxygen production. D_s is the apparent diffusion coefficient of oxygen in the substratum. At steady state in the light, $\delta C(x,t)/\delta t = 0$, and Eq. (2) can be written

$$P(x,t) = -\left[D_s \frac{\delta^2 C(x,t)}{\delta x^2} - R(x,t) \right] \tag{3}$$

In the dark, $P(x,t) = 0$, and Eq. (2) can be written

$$\frac{\delta C(x,t)}{\delta t} = D_s \frac{\delta^2 C(x,t)}{\delta x^2} - R(x,t) \tag{4}$$

Immediately after darkening, the oxygen gradients have not changed significantly, and the value of the diffusion term in Eqs. (3) and (4) is therefore the same. The rate of oxygen consumption $R(x,t)$ in the light and just after turning off the light can also be assumed to be the same [discussed by Revsbech and Jørgensen (1983)]. Therefore, by combining Eqs. (3) and (4) one obtains

$$P(x,t) = -\frac{\delta C(x,t)}{\delta t} \tag{5}$$

Equation (5) expressed that the initial rate of decrease in oxygen concentration after darkening equals the former gross photosynthetic rate.

3.2.2b. Capabilities and Limitations of the Method; Comparison with Other Methods. Comparison of the oxygen microelectrode method wtih conventional methods for measuring photosynthesis showed agreement in some systems (Revsbech et al., 1981). The limitations suggested by our first report on the new method seemed largely to be due to the long, 1-min dark incubations we used, during which the oxygen gradients changed considerably. The rate of decrease in oxygen concentration should preferably be read within a few seconds of darkening while the decrease is still linear with time. The spatial resolution of the method was shown to be 0.1 mm when the rate of decrease in oxygen concentration was measured over only 1 sec (Revsbech and Jørgensen, 1983). Using the initial rate for calculating the photosynthesis recommended by Revsbech and Jørgensen (1983) should probably be avoided, as some electrodes are slightly sensitive to light (Baumgärtl and Lübbers, 1983). The small change in the reading because of this artifact occurs instantaneously during light–dark changes and is usually insignificant, but it might cause problems during quantification of very low rates of photosynthesis. The possible effect of light on the oxygen microelectrodes can be checked in media without photosynthetic activity, e.g., sterile water. In addition to the direct effect of light on the response of the oxygen microelectrodes, there may also be an effect from heating. The oxygen microelectrodes are highly sensitive to changes in temperature (Baumgärtl and Lübbers, 1983; Revsbech and Ward, 1983), and a temperature drop associated with darkening may result in a decrease in the signal, which could be misinterpreted as photosynthetic activity. Controls with photosynthesis measurements in inactivated sediment should be carried out to check this source of error.

The advantages of the oxygen microelectrode method for measuring microbial photosynthesis in stagnant media are numerous: No other method has similar spatial and temporal resolutions. The result of the measurement is known within seconds, and a new measurement can be taken as soon as a stable oxygen concentration is reestablished, usually within 0.5–1 min. The method is nondestructive, and repetitive measurements can be made on the same cluster of microalgae. Also, methodological errors are under better control than in other methods (Revsbech *et al.,* 1981).

Two other methods have been used extensively to quantify the photosynthetic activity of benthic and epiphytic microalgae. The ^{14}C method is based on the determination of the assimilation of radioactively labeled bicarbonate, whereas the oxygen exchange method is based on the measurement of the oxygen exchange between the substratum and the overlying medium during light–dark cycles. Both methods suffer from inherent methodological problems.

The ^{14}C method (Steemann-Nielsen, 1952) has been used extensively for the quantification of planktonic primary productivity, and the sources of error inherent in the technique have been investigated thoroughly (e.g., Lean and Burnison, 1979). Additional complications arise, however, when the ^{14}C methodology is adapted for use in stagnant media such as microbial films and sediments. It is very important to know the specific radioactivity of the total pool of dissolved inorganic carbon in the environment where photosynthesis takes place, but unfortunately there is no method to assess accurately the specific activity in the microlayers that are photosynthetically active in such substrates. Revsbech *et al.* (1981)

reported that the specific activity of bicarbonate in the photic zone of sediments could be severalfold lower than in the overlying water.

The oxygen exchange method (e.g., Hunding and Hargrave, 1973) is based on the assumption that the photosynthetic rate equals the difference between the oxygen outflux from the sediment in the light and the oxygen influx to the sediment in the dark. It is, however, likely that the increased oxygen availability in the light (see Fig. 9) increases the rate of oxygen consumption in the light compared to the rate in the dark. Stirring gives a much smaller increase in the oxygen availability to the sediment compared to light (Fig. 9), and stirring alone is able to increase the oxygen uptake of sediments very considerably (Pamatmat, 1971; Jørgensen, 1977). The oxygen exchange method will therefore tend to underestimate the actual rate of photosynthesis.

The oxygen microelectrode method, however, also has its limitations. To be sufficiently sensitive, the method requires high photosynthetic rates per unit volume, preferably >2 mmole O_2/dm^3 per hr. It was not possible, for example, to quantify the photosynthesis within lichens, as these rates were too low (M. Sonesson and N. P. Revsbech, unpublished data). Other methods should be chosen when working with substrata having such low activities. Also, the environmental conditions should be reasonably stable during the experiment. For example, fluctuating light intensities caused by drifting clouds during outdoor experiments will make it virtually impossible to obtain stable readings. Mechanical instability of the substratum is another problem often encountered. It is not an advantage to be able to measure photosynthesis with 0.1 mm spatial resolution if the substratum moves 1 cm up or down during the measurement! Macrofauna in the substratum often create problems as the animals cause instability of the oxygen profiles in the surrounding substratum.

3.2.2c. Examples of Microprofiles of Photosynthesis. Profiles of photosynthetic rate have been measured in a variety of sediments and algal mats (Revsbech *et al.,* 1981, 1983; Jørgensen *et al.,* 1983; Revsbech and Jørgensen, 1983; Revsbech and Ward, 1983; 1984b). Examples of photosynthesis and oxygen profiles in a microbial mat in which filamentous cyanobacteria were the dominating phototrophs are shown in Fig. 14. The photosynthetically active layer was only 0.4 mm thick, but a very high maximum rate of 190 mmole O_2/dm^3 per hr was measured within this narrow layer. The photosynthesis integrated over all layers was 37 mmole O_2/m^2 per hr. The highest oxygen concentration, 1152 μM, occurred within the photosynthetically active zone, and it corresponded to 97% saturation with pure oxygen. Supersaturation with pure oxygen (PO_2 up to 1.4 atm) has often been observed in such dense cyanobacterial mats. The very compact cyanobacterial layer at the surface of this micro-

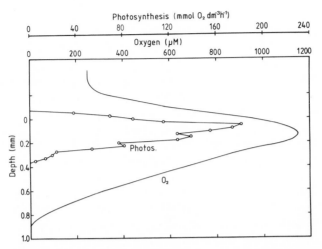

Figure 14. Profiles of oxygen and photosynthesis in an illuminated (250 μEinst/m^2 per sec) cyanobacterial mat from a shallow bay of Limfjorden, Denmark. The cyanobacterial mat was found at the surface of a muddy sediment containing decomposing eelgrass and high sulfide concentrations. Temperature, 21°C.

bial mat caused the light available for oxygenic photosynthesis to be attenuated at 0.4 mm depth. Less dense photosynthetic communities allow a deeper light penetration, for example, diatom communities in coastal sandy sediments where the photic zone may be 2–4 mm thick. An inverse relationship between the rate of photosynthesis per unit volume and the depth of the photosynthetically active layer has been found among microbial mats (Jørgensen *et al.,* 1983; Revsbech and Ward, 1984b).

The high spatial resolution of the photosynthesis determinations by microelectrodes opens possibilities for very detailed analysis of the interactions between environmental factors and the photosynthetic activity in layers or clusters of organisms. The photosynthesis profile can be closely related to the vertical distribution of microorganisms (Jørgensen *et al.,* 1983; Revsbech and Ward, 1984a). In addition, the spatial heterogeneity in the distribution of the photosynthetically active microorganisms can be analyzed in great detail (Revsbech and Jørgensen, 1983).

The effect of incident light intensity on the total photosynthetic rate per unit area, as well as on the vertical distribution of photosynthesis, was analyzed by Revsbech *et al.* (1983) and Revsbech and Ward (1983). These studies indicated that photoinhibition or photosaturation of the investigated microbial mat communities did not occur even at very high light intensities (about 2000 μEinst/m^2 per sec). The surface layers of the microbial mats were photosaturated, but deeper layers continued to increase in activity with increasing light intensity.

3.2.2d. Computer Modeling of Photosynthetic Rates. As mentioned before, the spatial resolution of the photosynthesis measurement is about 0.1 mm when the rate of decrease in oxygen concentration within 1 sec is used to calculate the photosynthetic rate. A resolution of 0.1 mm may seem adequate for most purposes, but compared to the (e.g., 0.4 mm thickness of the photic layer in dense cyanobacterial mats) 0.1 mm spatial resolution is still rather crude.

The limited spatial resolution of the photosynthesis measurements results in a profile of photosynthesis that is smoothed out when compared to the actual profile. It can be calculated (Revsbech *et al.,* 1986) that the measured rate of photosynthesis at depth x, $P_1(x)$, is related to the actual photosynthetic rate, $P_0(x)$, as follows:

$$P_1(x) = \int_{-\infty}^{+\infty} P_0(y)(4\pi D_s)^{-0.5} \exp\left[-\frac{(x-y)^2}{4D_s}\right] dy \qquad (6)$$

where D_s is the apparent diffusion coefficient for oxygen in the substratum. It is not possible to calculate the actual P_0 values from Eq. (6), but the values can be approximated by computer simulation of a P_0 profile. Experimentally derived rates of photosynthesis are shown in Fig. 15 (solid circles). The straight lines represent an estimated profile of photosynthesis, and the thin curve shows the profile of photosynthesis that would have been measured (P_1) if the estimated (P_0) profile had been the actual profile of photosynthesis. The estimated profile seems to be a good estimate of the actual profile, since the thin curve represents a good fit to

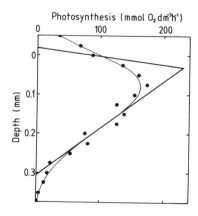

Figure 15. Estimated (P_0) profile of photosynthesis (straight lines) that would have resulted in a measured (P_1) profile of photosynthesis (thin curve) closely simulating the experimentally derived rates of photosynthesis (solid circles, also shown in Fig. 14).

the experimentally derived photosynthetic rates. It is shown in Fig. 15 that the photosynthetic zone as measured by the microelectrodes extends beyond the actual photosynthetic layer. This has also been demonstrated experimentally. It was thus possible to measure photosynthesis in the water above microbial mats (Revsbech and Jørgensen, 1983) if the distance to the photosynthetically active layers was less than 0.1 mm. This reflects why the spatial resolution is only 0.1 mm. Another effect of the limited spatial resolution is that the measured maximum photosynthetic rate is lower than the actual maximum.

3.3. Microbial Respiration

3.3.1. Total Oxygen Uptake

The oxygen uptake of sediments (or biofilms, detritus, etc.), much of which is due to microbial respiration (Dale, 1978), can be calculated from the slope of the linear oxygen concentration profile through the true diffusive boundary layer (Section 3.1.3) above the sediment surface, using the one-dimensional version of Fick's first law of diffusion (Berner, 1980):

$$J = -\phi D_s \, \delta C(x)/\delta x \qquad (7)$$

where J is the flux of molecules through a unit area per unit time, D_s is the apparent (Berner, 1980) diffusion coefficient of oxygen in the substratum, ϕ is the porosity ($\phi = 1$ in water), and $C(x)$ is the concentration of molecules at depth x.

As an example, the oxygen gradient just above the sediment surface shown in Fig. 5 indicates an oxygen import to the sediment of 2.6 μmole/cm^2 per day, assuming a diffusion coefficient in the diffusive boundary layer of 2.1 cm^2/sec (Broecker and Peng, 1974). An oxygen uptake rate of 0.2 μmole/cm^2 per day in a deep sea sediment from 3750 m water depth has also been calculated from oxygen microprofiles (Reimers and Smith, 1986). Instead of the oxygen gradient in the diffusive boundary layer above the sediment, these authors used the oxygen gradient immediately below the sediment surface for the flux calculation and therefore needed to estimate the porosity and the apparent diffusion coefficient in the sediment. Their calculated oxygen uptake and the rate of sediment community oxygen consumption determined *in situ* by use of a belljar approach were not significantly different. Similar comparisons between the two methods applied to shallow water sediments usually resulted in the estimates obtained by the flux calculations being 10–50% lower than the figures obtained by belljar methodology (B. B. Jørgensen, unpublished results). This discrepancy is partly caused by the respiration of the abundant infauna inhabiting shallow sediments and by the oxygen consumption of the sediment surfaces within infaunal burrows, but some of the difference may also be due to the irregular surface topography of the sediment surface. In deep sea sediments, the burrowing fauna is much sparser than in shallow sediments, and the additional oxygen consumption caused by infauna seems to be insignificant (Reimers and Smith, 1986).

3.3.2. Depth Profiles of Oxygen Consumption

Unfortunately, a high-resolution method is not readily available for respiration measurements with microelectrodes similar to the one for photosynthesis. It is, however, possible to calculate the oxygen consumption at various depths in a substratum by computer simulation of measured oxygen profiles in light and dark.

The change in the oxygen profile with time is given by an extended version of Fick's second law of diffusion (Berner, 1980):

$$
\frac{\delta C(x,t)}{\delta t} = D_s(x)\frac{\delta^2 C(x,t)}{\delta x^2} + \left[\frac{\delta D_s(x)}{\delta x} + \frac{D_s(x)}{\phi(x)}\frac{\delta\phi(x)}{\delta x} \right]\frac{\delta C(x,t)}{\delta x}
$$
$$
+ P(x,t) - R(x,t) \tag{8}
$$

where $C(x,t)$ is again the oxygen concentration at depth x and time t, $D_s(x)$ is the apparent diffusion coefficient, $\phi(x)$ is the porosity, $R(x,t)$ is the oxygen consumption, and $P(x,t)$ is the oxygen production. Equation (8) is an extension of Eq. (2) and can be used to model the oxygen profiles when the diffusion coefficient and the porosity vary with depth. Equation (8) is a one-dimensional model for the oxygen in the sediment, and it will consequently only give a satisfactory description if lateral heterogeneity is small. The oxygen consumption $R(x,t)$ can be calculated from Eq. (8) if the oxygen profiles are sufficiently well described and if the other parameters in the equation are known.

It was shown in Section 3.2.1. how the vertical profile of photosynthesis $P(x,t)$ can be determined. It is, however, also necessary to know the vertical distribution of diffusion coefficients D_s and porosity ϕ to make use of Eq. (8). The porosity can be calculated from the weight loss of a known volume of sediment after drying, but it is difficult to get a sufficiently good depth resolution of ϕ this way. The diffusion coefficient can be determined by measuring the down-diffusion of oxygen into a sterilized, air-exposed, and fully oxidized sediment core subjected to a sudden change in the overlying atmosphere from air to 100% oxygen. The diffusion coefficient can then be read from an inverse error function plot as described by Duursma and Hoede (1967). This method (Revsbech *et al.,*

1986) gives only one estimate of the diffusion coefficient and not a vertical profile, but such vertical profiles can be obtained by scraping off thin layers of surface sediment and consecutively measuring the diffusion of oxygen into these newly exposed surfaces. The depth resolution of the diffusion coefficient can be about 1 mm using this procedure. New computerized methods are now being developed to improve the estimates of diffusion coefficients and porosities in the surface layers of sediment (N. P. Revsbech, unpublished data).

In the simple case of a steady-state oxygen profile $[\delta C(x,t)/\delta t = 0]$ and no photosynthetic activity $[P(x,t) = 0]$, and when D_s and ϕ are known, it is quite simple to calculate the rate of oxygen consumption at various depths using Eq. (7) or Eq. (8). When the oxygen profile is not in a steady state, it is necessary, however, to perform computer simulation of the oxygen profiles to obtain reasonably accurate estimates of the rates of oxygen consumption in various layers. Very rapid changes in the oxygen profile may occur after changes in the environmental conditions. The development in the oxygen concentration at various depths in a microbial mat after turning off the light is shown in Fig. 16. The steady-state oxygen profile in the light and the measured profile of photosynthesis from this site are shown in Fig. 14. The diffusion coefficient and porosity

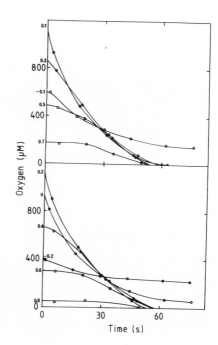

Figure 16. Oxygen concentrations at various depths in the cyanobacterial mat after turning off light. The steady-state starting profile of oxygen during illumination is shown in Fig. 14.

in the uppermost 1 mm layer were found, as described above, to be 1.5
$\times$ 10^{-5} cm²/sec and 0.95, respectively.

The steady-state oxygen profile shown in Fig. 14 was modeled from
Eq. (8) using an iterative computer method [Crank-Nicholson method
(Crank, 1983)]. Satisfactory simulations of both the steady-state oxygen
profile in the light and the transient oxygen profiles developing after darkening were obtained (Fig. 17) by using the P_0 profile of photosynthesis
shown in Fig. 15 and the profile of oxygen consumption shown in Fig.
18. All photosynthesis values shown in Fig. 15, however, had to be multiplied by a factor of 1.248 for the simulation to yield good approxima-
tions to both steady-state and transient profiles. This factor varied from
0.79 to 1.248 when five sets of data like those shown in Figs. 14 and 16
were analyzed (Revsbech *et al.*, 1986). The need to multiply by a factor
different from 1 is an indication of lateral heterogeneity. As a result of the
high spatial resolution of the photosynthesis measurements, only the
microbial mat closer than 0.1 mm to the site being analyzed contributes
significantly to the measured rate of photosynthesis (Section 3.2.2b). The
oxygen concentration profile, however, is influenced by the production
rates of oxygen much further away than 0.1 mm. A three-dimensional
diffusion model could compensate for this, but it would require an
unrealistic amount of experimental data. The multiplication factor mentioned above was introduced to compensate for the inability of a one-
dimensional model to include the effect on the oxygen profile of neigh-

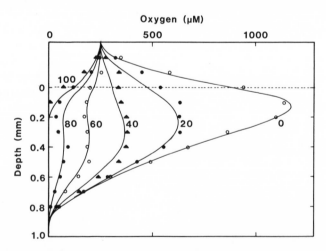

Figure 17. Oxygen profiles after a light–dark shift simulated by use of Eq. (8) and inserting
the values of respiration and photosynthesis shown in Figs. 15 and 18. The solid circles
show the experimentally derived profiles, which were obtained by interpolation between
data points shown in Fig. 16.

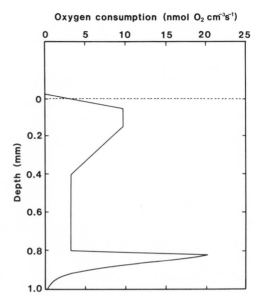

Figure 18. Estimated profile of oxygen consumption that gave a good simulation (Fig. 17) of the oxygen data shown in Figs. 14–16.

boring sites with different net production rates of oxygen. The proximity of the multiplication factors to unity (mean value 1.03) shows that the algal mat was only moderately heterogeneous. This was also suggested by the similarity of the five measured profiles of photosynthesis and oxygen concentration (Revsbech *et al.*, 1986).

Of the five sets of data analyzed, all oxygen consumption profiles showed a maximum of oxygen consumption in the lower part of the photic zone and another maximum at the oxic–anoxic boundary. The upper maximum was probably due to microbial respiration of photosynthates and degradation of senescent microbial cells accumulating in this layer. The lower maximum was associated with the extensive oxidation of reduced inorganic compounds in this layer, especially of sulfide (Revsbech *et al.*, 1983; Jørgensen and Revsbech, 1983).

3.4. Microbial Sulfide Oxidation

Oxygen is a key factor in the environment and exerts great control over all biological processes. In the oxic parts of the world, aerobic respiration is the dominant factor in energy metabolism, and complete oxidative pathways are present within the individual cells. In the anoxic world, bacterial fermentations and anaerobic respiration yield a range of reduced organic or inorganic end products, which are excreted from the cells. The oxic–anoxic interface thereby becomes an important site for the reoxidation of these reduced compounds. With one notable exception, the sulfide electrode, suitable microelectrodes to detect these compounds still need to be developed.

Microorganisms specially adapted to life at the oxic–anoxic interface are well known. Heterotrophic bacteria often reveal the position of the interface by the formation of narrow bands in stagnant liquid cultures at a certain distance from the medium surface. In liquid enrichment cultures with a heavy inoculum of soil or organic debris, such microaerophilic bacteria can sometimes be observed swarming just over the solid substratum. It has been inferred from independent knowledge of the aerobic metabolism of the bacteria and perhaps also of their low oxygen tolerance that their position here was just at the oxic–anoxic interface. Their adaptation to microoxic conditions can now be demonstrated by direct measurements with an oxygen microelectrode.

Aerobic, heterotrophic bacteria, which through the chemotactic response are able to aggregate in the microoxic zone, may also find in this environment the optimal supply of organic substrates, which diffuse out of the anoxic zone. Many chemoautotrophic bacteria are also highly adapted to this environment, where inorganic compounds such as H_2, NH_4^+, H_2S, and Fe^{2+} serve as electron donors for their energy metabolism. The specialized sulfur and iron bacteria, which are adapted to oxidize H_2S and Fe^{2+}, are in the unusual situation of having to compete with a simultaneous chemical oxidation of their energy substrates by oxygen (Jørgensen, 1982). They are therefore closely tied to the oxic–anoxic interface, where they live between the opposed gradients of their electron donor and acceptor. These fascinating gradient bacteria have been difficult to study because of the insufficient spatial resolution of the analytical techniques. Microelectrodes now provide a powerful tool for studying the ecology of the sulfide-oxidizing bacteria.

3.4.1. Life at the O_2–H_2S Interface

Most sulfide oxidation in nature takes place within sediments at depths of several centimeters from the surface, where it is difficult to discriminate the sulfide oxidizers quantitatively from other bacterial populations (Jørgensen, 1983). In sediments with a high organic turnover, however, high rates of oxygen consumption and sulfate reduction may push the O_2–H_2S interface up to the sediment–water interface. Under these conditions, mass development of colorless sulfur bacteria such as the filamentous *Beggiatoa* spp. can sometimes be observed on the sediment surface. An example of such a *Beggiatoa* mat is shown in Fig. 19 together with the oxygen distribution as measured by microelectrodes (Jørgensen and Revsbech, 1983). The two-dimensional map of isopleths

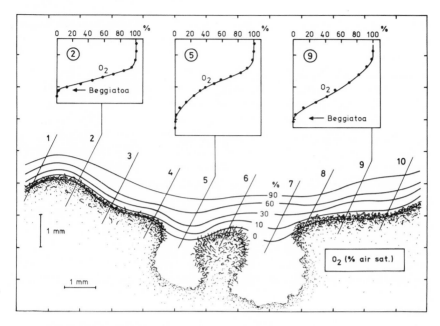

Figure 19. Vertical section through a *Beggiatoa* mat that grew on the sulfide-rich mud in a Danish fjord. Oxic water was flowing over the mud surface. Ten oxygen microprofiles, three of which are shown in insert, were used to construct the O_2 isopleths. Numbers indicate percent air saturation. [From Jørgensen and Revsbech (1983).]

was constructed from an array of vertical O_2 profiles in a similar manner as in Fig. 11.

Although the sediment was covered by circulating, aerated seawater, the bacteria were living under microoxic or anoxic conditions. This is due to the presence of the diffusive boundary layer (Section 3.1.3) and to the high oxygen uptake of the bacteria themselves. The oxygen uptake was limited only by the diffusion flux through the boundary layer, and the potential oxygen uptake of the *Beggiatoa* mat was much higher than the actual uptake. This could be seen either by increasing the water flow and thereby decreasing the thickness of the diffusive boundary layer or by increasing the oxygen concentration in the water. The diffusive flux of oxygen into the bacterial mat simply increased in proportion to the steepness of the oxygen gradient in the boundary layer, while the mat itself remained anoxic.

The combination of an oxygen transport barrier provided by the diffusive boundary layer and the high bacterial metabolism may lead to anoxia in many other environments that one would intuitively expect to be oxic. Thus, the surface of detritus particle and sediments, of biofilms on rocks and plants, etc., may also be microoxic at the very surface (Jørgensen and Revsbech, 1985). This phenomenon can explain why anaerobic bacteria, such as purple sulfur bacteria in a sulfuretum, can grow on surfaces seemingly exposed to high oxygen concentrations. The oxygen isopleths in Fig. 19 show how the diffusive boundary layer covered the bacterial mat like a 0.5- to 1-mm-thick blanket of water that followed the coarser surface topography.

The position of the O_2–H_2S interface within the *Beggiatoa* mat is shown in detail in Fig. 20. The mat was again covered by flowing, aerated seawater with a uniform oxygen distribution. A sharp drop in oxygen concentration was observed within the 0.5-mm-thick diffusive boundary layer of the mat. On the expanded scale (Fig. 20, right) the oxygen is seen to penetrate only 100 μm into the mat. A steep H_2S gradient started from 50 μm below the mat surface and the H_2S level increased to 500–1000 μM a few centimeters deeper in the black mud. The zone in which both oxygen and sulfide coexisted in the mat was thus only 50 μm thick. Within this narrow layer, all sulfide produced in the sediment and diffused to the surface evidently became oxidized. The zone extended only over one-tenth of the total thickness of the 0.5-mm *Beggiatoa* mat, which

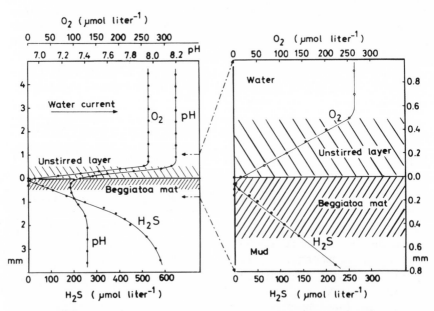

Figure 20. Oxygen, sulfide, and pH microgradients through a *Beggiatoa* mat that grew on a mud surface under flowing, aerated seawater. The expanded depth scale shows the 0.5-mm-thick mat with the narrow O_2–H_2S interface and the overlying diffusive boundary layer. [From Jørgensen and Revsbech (1983).]

meant that only a small fraction of the bacteria had access to both substrates at any given time.

At the time when the measurements described above were first done, chemoautotrophic growth of *Beggiatoa* in pure cultures with H_2S as the sole energy source had not yet been proven. Calculations of the high H_2S turnover rate in the *Beggiatoa* mat in Fig. 20 showed, however, that this process had to be biological rather than chemical. The first unequivocal demonstration of chemoautotrophy in a marine *Beggiatoa* strain was achieved shortly after by Nelson and Jannasch (1983). They also observed this chemoautotrophic metabolism when the organisms were grown in gel-stabilized gradients of O_2 and H_2S. Recent microelectrode studies of such gradient-grown *Beggiatoa* cultures have confirmed the results obtained from the natural mats and have also demonstrated the potential uses of microelectrode techniques for the study of gradient bacteria in general (D. C. Nelson, N. P. Revsbech, and B. B. Jørgensen, in preparation).

One important piece of information that can be derived from such microelectrode data is the overall stoichiometry of the bacterial metabolism. This can be done from simple flux calculations as described in Section 3.3.1. As the calculation of a flux involves the product of the gradient and the molecular diffusion coeffcient of the chemical species, the diffusion coefficient needs to be determined experimentally in the growth medium. This can be done either with microelectrodes or with chemical or radiotracer techniques by tracing the propagation through the medium of an increase or decrease, in concentration (see Section 3.3.2). Other information that can be obtained concerns the interactions between the opposed chemical gradients and the gradient bacteria, the overlap and turnover of the chemical species as a function of the bacterial growth, and even the growth yield of the bacteria.

As an example, results from the gradient culture system described by Nelson and Jannasch (1983) are shown in Fig. 21. The O_2, H_2S, and pH profiles were measured through two test tubes with seawater partly solidifed with agar. The lower 2 cm of the agar initially contained 4 mM H_2S, while the upper agar layer was free of H_2S and only contained mineral salts. One of the tubes was inoculated with a chemoautotrophic strain of marine *Beggiatoa*. The situation after 3–4 days when the H_2S had diffused up to reach the lower boundary of the oxic surface layer is shown in Fig. 21. In the uninoculated tube, O_2 and H_2S overlapped by 5–6 mm and there was only a small pH effect derived from the relatively slow, chemical sulfide oxidation. In the inoculated tube, *Beggiatoa* grew in a 0.4-mm-thick horizontal band just at the sharp O_2–H_2S interface. Coexistence of the two substrates could only be measured in a 50-μm-thick zone, which demonstrated a very efficient uptake and oxidation of H_2S

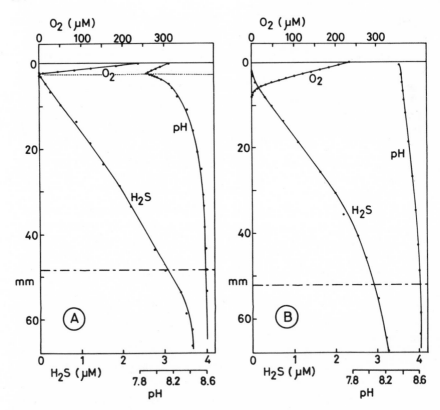

Figure 21. Gradient culture of a marine *Beggiatoa* (strain MS-81-6). (A) The oxygen, sulfide, and pH gradients 3 days after inoculation, showing the sharp O_2–H_2S interface within the *Beggiatoa* mat (dotted line). (B) Uninoculated control after 4 days, showing a broad O_2–H_2S overlap. [Data from D. C. Nelson, N. P. Revsbech, and B. B. Jørgensen (in preparation).]

by the bacteria. The mean residence time of O_2 and H_2S in the overlap zone was only a few seconds. The oxidation of H_2S to sulfuric acid caused a sharp minimum in pH. An oxidation to sulfuric acid was in agreement with the calculated ratio between H_2S and O_2 fluxes.

Throughout the rapid growth phase, which was linear due to constant diffusion limitation, the ratio of H_2S/O_2 consumption in the *Beggiatoa* mat was about 0.6. If all H_2S was oxidized to sulfuric acid, the ratio would be 0.5. The slightly lower oxygen consumption was found not to be due to a lower oxidation state of the sulfur product, i.e., due to the formation of elemental sulfur, which was only a transient oxidation product, but to the reduction of CO_2. Thus, about 15% of the electrons from H_2S were transferred to CO_2 and the rest to O_2. In accordance with this finding, the increase in bacterial biomass corresponded to a growth yield of 8 μg/μmole H_2S. This is slightly higher than for aerobic thiobacilli (Kelly, 1982) and shows that *Beggiatoa* cells are well adapted to a chemoautotrophic mode of life at the O_2–H_2S interface.

3.4.2. H_2S and Photosynthesis

The distribution of oxygen in nature, and thus the position of the O_2–H_2S interface, is highly dependent on light and photosynthetic activity. In shallow waters, benthic microalgae on sediments, rocks, plants, and other solid surfaces produce local oxygen maxima, which push the oxic zone deeper into the substratum as discussed earlier. The diurnal light–dark variations thus create a cyclic pulse of oxygen to which both the photosynthetic organisms and the gradient bacteria need to adapt. In those organic-rich sediments where light penetrates down into the sulfide zone, a similar daily pulse in sulfide concentration may be found due to the activity of photosynthetic sulfur bacteria. As an example of a benthic microalgal community with both high photosynthetic activity and high sulfide concentration, we have studied cyanobacterial mats from the hypersaline Solar Lake in Sinai (Jørgensen *et al.*, 1983; Revsbech *et al.*, 1983). These mats consist mostly of filamentous cyanobacteria, e.g., *Microcoleus* and *Phormidium*, but also contain a variety of coccoid cyanobacteria, heterotrophic bacteria, and colorless and photosynthetic sulfur bacteria. We analyzed the dynamic distributions of O_2, H_2S, and pH in these mats during abrupt light–dark shifts as well as during the natural light variations over a 24-hr cycle. In order to obtain precise depth and time correlations between the three parameters during the light–dark shifts, the three microelectrodes were glued together into one unit with the sensing tips less than 1 mm apart.

Two examples of such simultaneous O_2, H_2S, and pH profiles, one in the dark and one in the light are shown in Fig. 22. The O_2–H_2S interface was situated just 0.5 mm below the mat surface in the dark. The pH decreased steeply from the water value of 8.2 to a mat value of 7.3. Due to the intensive photosynthesis in the light, a sharp maximum of oxygen built up and pushed the O_2–H_2S interface down to a depth of 2 mm in the mat. Concurrently, pH built up due to the photosynthetic CO_2 assimilation to a maximum of 9.3 just below the surface and then dropped by two units over the following 2 mm.

The microgradients in the mat show how dynamic and variable a microbial environment can be. Organisms living at 0.5 mm depth in the mat were exposed to variations from anoxia to over 1 atm partial pressure of oxygen, depending on the light conditions. The corresponding variations at 1.5 mm depth were from 250 μM H_2S in the dark to 500 μM

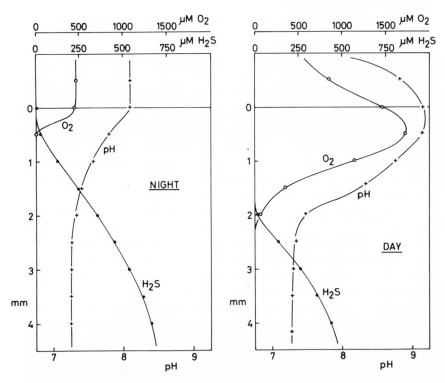

Figure 22. Distribution of O_2, H_2S, and pH in a *Microcoleus* mat from Solar Lake, Sinai, during the night and during the day. [Data from Revsbech *et al.* (1983).]

O_2 in the light. Gradient bacteria associated with the O_2–H_2S interface experienced a vertical movement of their proper chemical environment over 1.5 mm as a result of the light–dark changes. It is probably because of these moving chemical gradients and changing light intensities that most of the dominant mat organisms are motile and can adjust to the changing conditions by migrating up and down. Others, however, remain fixed in the mat structure and may only experience optimal conditions during a limited part of the daily light–dark cycle.

The strong variations in the chemical environment of the mat organisms may seem extreme, but they are not unusual for benthic, photosynthetic communities. Only after the introduction of microelectrodes have such seemingly harsh microenvironments been demonstrated, although they could in fact have been predicted from crude estimates of metabolic rates and molecular diffusion parameters. If the appropriate microelectrodes were available, similar fluctuations could probably also be demonstrated for many other chemical species, such as

NO_3^-, NH_4^+, HPO_4^{2-}, Fe^{2+}, Mn^{2+}, etc., as well as for organic compounds.

A technique to measure anoxygenic photosynthesis by purple or green sulfur bacteria or by cyanobacteria could theoretically be developed based on the initial disappearance rate of H_2S in the light after steady state in the dark. Our experiments with such a technique in microbenthic communities have not been very promising. One serious problem of the technique is the tendency of H_2S to react with elemental sulfur in such systems. This leads to the formation of polysulfides, which rapidly equilibrate with a changing H_2S pool. Relative rates may, however, still give important information on the type and distribution of H_2S-dependent photosynthesis.

One example of such anoxygenic photosynthesis is shown in Fig. 23. The rate of sulfide disappearance after a shift from steady state in the dark to light was followed at 1 mm depth within the same microbial mat (B. B. Jørgensen, Y. Cohen, and N. P. Revsbech, unpublished data). For the present experiment, however, the mat was covered by H_2S-enriched water. The phototrophic mat community depleted the sulfide pool within 5 min and already after 3 min an O_2 pool started to build up. In order to demonstrate whether the sulfide disappearance was due to an anoxygenic photosynthesis or to a secondary oxidation by O_2 of an initially nonde-

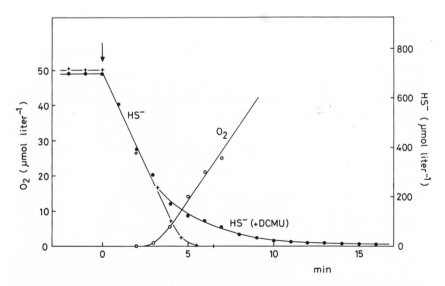

Figure 23. Photosynthetic sulfide oxidation and oxygen production at 1 mm depth in a *Microcoleus* mat after a dark–light shift (arrow). Experiments done without and with DCMU added. [Data from Jørgensen *et al.* (1986).]

tectable concentration, the oxygenic photosynthesis was inhibited by the addition of DCMU [3-(3,4-dichlorophenyl)-1,1-dimethyl-urea]. The initial sulfide disappearence was unaffected by DCMU, which indicated that the phototrophic organisms carried out anoxygenic photosynthesis with H_2S (cf. Cohen et al., 1975a,b). When the sulfide level had dropped below 200 μM, the anoxygenic photosynthesis became inefficient. This was the level at which free O_2 had started to develop in the uninhibited mat. The results of this and other experiments led to the conclusion that oxygenic and anoxygenic photosynthesis may operate in concert in these microbial mats at levels that mostly depend on the sulfide concentration (Cohen, 1984).

By experiments such as the one shown in Fig. 23, the microelectrode techniques allow physiological studies of microbial communities *in situ* without causing physical disturbance to the system.

3.5. Symbiosis

There are many fascinating examples of symbiotic associations between microorganisms and animals. The mutual benefits of the association for the host and symbionts vary, but one classical and widespread example is the symbiosis between microalgae and aquatic invertebrates. This type of association has reached its highest development in tropical, marine environments, where, for example, the growth of coral reefs is dependent on the photosynthetic activity of algal cells imbedded in the tissue of each polyp. Many other examples are known of similar symbiotic associations between microalgae, or just their chloroplasts, the protozoa, sponges, nudibranchs, bivalves, tunicates, etc. (Stanier et al., 1977).

Through the close symbiotic association, an efficient nutrient cycling between algae and the animal host may be achieved, which is especially important in nutrient-poor tropical waters. The algal photosynthesis provides organic carbon, and thus energy, for the host, whereas the host efficiently retains inorganic nutrients, which are recycled to the algae. In addition, the photosynthetic activity may be of significance for the calcification process in hosts that build a calcareous skeleton (Drew, 1973).

Due to the intimate association between autotrophic and heterotrophic processes in algal–animal symbiosis, it is experimentally difficult to analyze the metabolic activity of the two components separately. These problems are in many respects similar to those encountered when other, nonsymbiotic communities are studied. Oxygen and ^{14}C techniques have been used to distinguish respiration and photosynthesis in corals, sponges, foraminifera, etc. Both approaches have serious limitations, however, in the quantitative and spatial separation of the two processes.

Over the last 2 years, we have experimented with the use of micro-electrodes to study symbiotic photosynthesis in subtropical waters. With the oxygen microelectrodes it is possible to analyze the oxygen dynamics in the immediate surroundings of the host animal or even to penetrate into the tissue of the host. By rapid light–dark shifts, we could thereby detect and map the gross rate of photosynthesis of the symbionts, inde-pendent of the surrounding respiratory activity of the host, by the tech-nique described in Section 3.2.

One example of the oxygen dynamics in a polyp of a common reef-building coral, *Acropora* sp., from the Red Sea is shown in Fig. 24. The freshly collected coral branch was kept in an aquarium with running, aer-ated seawater at 24°C and was exposed to alternating darkness and light corresponding to full daylight intensity *in situ*. An oxygen microelectrode was positioned in the polyp tissue, about 1 mm below the tentacles, within the narrow ring of symbiotic zooxanthellae. This could be done with little damage to the tissue and without causing contraction of the polyp. The light–dark variation in oxygen tension was rapid and dra-matic, especially considering that this was within the body of a living ani-mal. In normal daylight, the oxygen tension rose to 80% of the ambient air saturation, 205 μM. In the dark, the animal tissue became nearly anoxic. Oxygen was also measured in other polyps on the same coral branch. Oxygen concentrations of up to 450 μM were measured in the light, and a few measurements showed anoxic conditions in the dark.

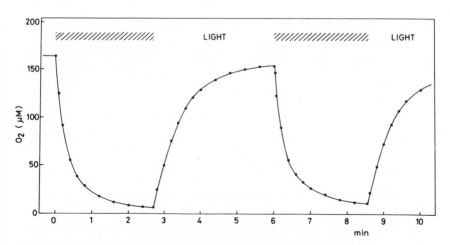

Figure 24. Oxygen concentration in the symbiont-containing polyp tissue of the coral *Acropora* sp. Shifts between darkness and *in situ* daylight (900 μEinst/m^2 per sec) caused rapid variations of PO$_2$ within the animal. [From N. P. Revsbech, B. B. Jørgensen, and Y. Cohen (in preparation).]

From the initial decrease in oxygen after a light–dark shift (Fig. 24), a photosynthetic rate of 410 μM/min could be calculated. Measurements in other polyps showed photosynthesis rates of up to 3300 μM/min. Due to the complex anatomy of corals and the annular positioning of their symbiont population, it is difficult to calculate the quantitative importance of this symbiotic photosynthesis for the energy metabolism of the whole coral.

Planktonic foraminifera, however, with their small size and spherical symmetry, seem to be ideal systems for such a quantitative analysis. Spinose, symbiont-bearing foraminifera are common in the plankton of tropical oceans (Be and Tolderlund, 1971). These 0.1- to 1-mm protozoa build multichambered shells, which house the protoplast with its multitude of intracellular zooxanthellae (Anderson and Be, 1976). The symbionts are actively spread out by the host around the shell during the daytime. They are imbedded in pseudopodia, which glide out along the radial spines. By carefully approaching the animal between the spines with an oxygen or pH microelectrode it was possible to study the chemical microenvironment of the organisms as well as their metabolic activity. An impression of this microenvironment is given in Fig. 25.

The dimensions of the foraminifer *Globigerinoides sacculifer* with its surrounding halo of intracellular symbionts are shown in Fig. 25A. As the oxygen microelectrode radially approached the shell in the light, the oxygen concentration was found to increase steeply to 2.5 times the ambient air saturation of seawater (Fig. 25B). Thus, the total symbiotic system had a high net photosynthesis. The corresponding decrease of oxygen in the dark reflected the combined respiratory activity of host and symbionts. Rapid light–dark shifts showed that the oxygen pool in the microenvironment was even more dynamic than shown for the coral (Fig. 25C). With a photosynthesis rate of 2700 μM/min at the shell surface, as calculated from the initial oxygen decrease in the dark, the mean residence time of oxygen in the light was only 12 sec. This shows that the oxygen environment of the animal changes sufficiently rapidly to follow closely fluctuations in light intensity due, for example, to clouds passing the sun. In addition, the pH of the seawater around the foraminifer changed according to the intensive CO_2 uptake or release in the light or in the dark, respectively (Fig. 25D).

With the oxygen microelectrode it was possible to map the radial distribution of photosynthesis around the shell as well as to penetrate carefully into the shell (Fig. 26). The measurements were taken at 50-μm increments, each measurement being representative of a 50-μm-thick spherical shell concentrically surrounding the host. Connected values of photosynthesis and spherical shell volumes were then multiplied and summed to yield the total gross photosynthesis of the symbiotic system,

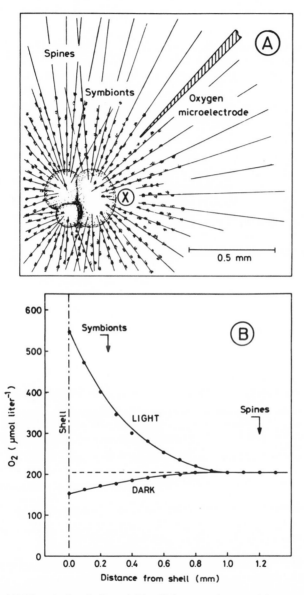

Figure 25. (A) The planktonic foraminifer *Globigerinoides sacculifer,* with its symbiotic zooxanthellae spread out between the spines. The animal was kept in a small glass vessel with water of the Red Sea, where it was collected. (B) Radial gradients of oxygen from the ambient seawater to the shell surface in the dark and in the light. (*continued*)

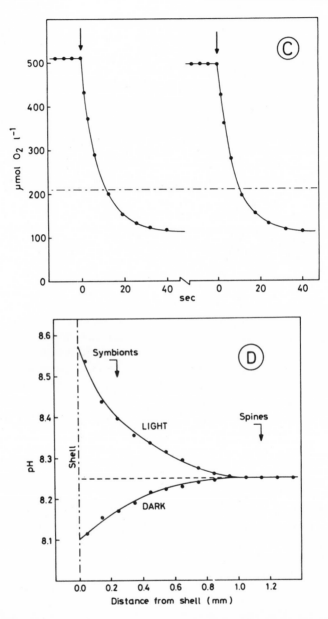

Figure 25 (*continued*). (C) Two experiments showing the rapid decrease in oxygen concentration after a light–dark shift. The electrode tip was positioned at point X shown in A. (D) Radial gradients of pH from the ambient seawater to the shell surface in the dark and in the light. Light intensity was 400 μEinst/m^2 per sec in all experiments; temperature was 24°C. [From Jørgensen *et al.* (1985) and B. B. Jørgensen, J. Erez, and N. P. Revsbech (in preparation).]

18.1 nmole O_2/hr. The net photosynthesis could be calculated from the radial diffusion flux of oxygen over a theoretical sphere of 0.6-mm radius enclosing both host and symbionts. With the O_2 gradient given by dC/dr = 410 μM/mm = 0.41 nmole/mm, the area of the enclosing sphere for r = 0.6 mm of A = $4\pi r^2$ = 4.5 mm^2, and a diffusion coefficient D = 2.35 $\times$ 10^{-5} cm^2/sec = 8.46 mm^2/hr (Broecker and Peng, 1974), one obtains for the net outward diffusion flux of oxygen from the foraminiferan J = $D(dC/dr)A$ = 15.6 nmole O_2/hr, which is close to the gross photosynthesis of 18.1 nmole/hr. This shows that the symbiotic association is highly autotrophic in the light, and that the symbiont productivity could potentially contribute to the nutrition of the host. Flux calculations of O_2 uptake in the dark based on radial gradients also show a comparatively small dark respiration of 2.7 nmole O_2/hr.

This example shows one great advantage of microelectrode techniques. Due to the small dimensions relevant for microbial environments, the important transport process for gases, ions and other dissolved molecules is predominantly molecular diffusion (Vogel, 1981; Boudreau and Guinasso, 1982; Jørgensen and Revsbech, 1985). A simple mapping of the concentration field around microbial communities may therefore

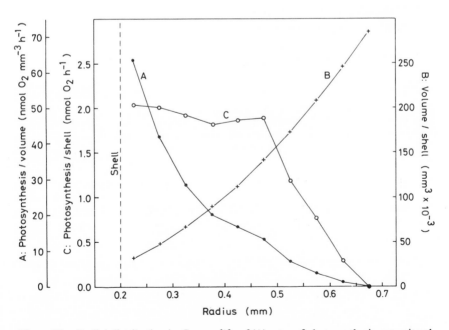

Figure 26. Radial distribution in *G. sacculifer* of (A) rates of photosynthesis per unit volume at 400 μ Einst/m^2 per sec, (B) volume of 50-μm-thick spherical shells, and (C) rates of photosynthesis per spherical shell. [From Jørgensen *et al.* (1985).]

allow the quantitative calculation of their metabolic rates. The planktonic foraminifera provide ideal systems for such studies because of their simple radial symmetry. The laterally uniform, one-dimensional gradients in the sediment systems discussed previously are another example.

Due to the high spatial and temporal resolution of the microelectrode techniques they are also ideally suited for *in situ* studies of the physiology of undisturbed microbial communities. Such an experimental application to the ecological physiology of microorganisms has only just started. In the example of the symbiont-bearing foraminifera, a step in this direction has been the study of photosynthetic adaptation to the ambient light intensity and quality (Jørgensen *et al.*, 1985). These studies were done by keeping the oxygen microelectrode at a fixed position near the shell surface and analyzing the photosynthetic response in relation to the quantum flux and the wavelength of incident light.

The gradual saturation of light-harvesting pigments in the zooxanthellae is shown in Fig. 27. At up to 100 μEinst/m^2 per sec of full-spectrum light, the light reaction in the symbionts limited the photosynthetic rates. Theoretical light saturation I_k was reached at 160 μEinst/m^2 per sec (cf. Harris, 1978). The maximum *in situ* light intensity was reduced to that level only at the 27-m depth in the Red Sea where the animals were collected (Jørgensen *et al.*, 1985). Measurements of radial oxygen micro-

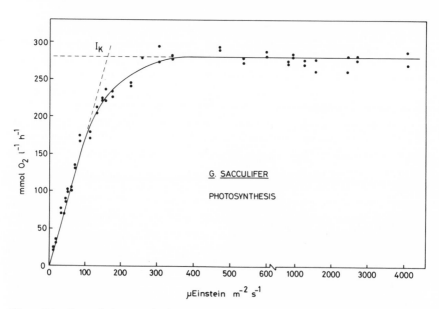

Figure 27. Rate of photosynthesis at the shell surface of a *G. sacculifer* at different light intensities. [From Jørgensen *et al.* (1985).]

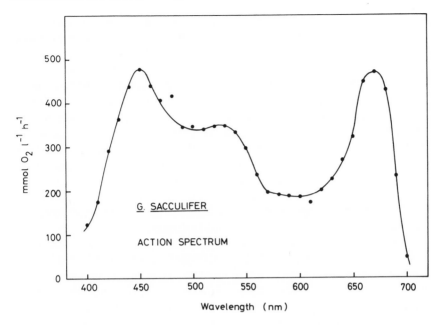

Figure 28. Action spectrum of endosymbiotic zooxanthellae measured at the shell surface of *G. sacculifer*. Photosynthesis maxima show the activity of chlorophyll a and of peridinin. [From Jørgensen *et al.* (1985).]

gradients around the foraminifer at a range of light intensities showed that the compensation light intensity, i.e., the intensity where the gradient was zero due to respiration–photosynthesis balance, was only 30 μEinst/m^2 per sec. This light intensity was reached at 60 m water depth in the Red Sea during midday.

Measurements of photosynthetic rates in monochromatic light at different wavelengths allowed a detailed study of the spectral response of the zooxanthellae. An action spectrum measured at intervals of 10 nm from 400 to 700 nm at a constant quantum flux of 100 μEinst/m^2 per sec. is shown in Fig. 28. Maxima of photosynthetic activity were found at 450 and 670 nm, corresponding to the *in vivo* absorption maxima of chlorophyll a. A broad maximum was also found between 500 and 550 nm. This was due to a protein-bound carotenoid, peridinin, which is an important light-harvesting pigment in dinoflagellates, to which the symbionts belong (Prezelin, 1976; Prezelin *et al.,* 1976; Anderson and Be, 1976). The light absorption by peridinin must be important in the deeper photic zone of the open ocean, where yellow-green light of 500–550 nm is predominant (Jerlov, 1976).

The speed and thus the high analytical capacity of the microelectrode technqiue are important for measurements such as those in Figs. 27 and 28. The collection of detailed data for the determination of light saturation or action spectra, as presented here, would be much more time-consuming by conventional oxygen or ^{14}C techniques. The data sets shown in Figs. 27 and 28 were each obtained over a few hours. Within such a short period, changes in the photocenters and in pigment composition of the microalgae will be minimal. For this reason, microelectrodes may have a wider and more general application in photosynthesis research in the future.

3.6. Miscellaneous Microenvironments

The microenvironments discussed so far were mostly sediments or symbiotic associations of animals and microalgae. Microelectrodes can also give valuable information about the chemistry and metabolic rates in many other microbial systems. Only a few other microenvironments have been analyzed, but the applications will no doubt increase rapidly over the coming few years. Some examples of potential applications are given below to illustrate the types of information that can be gained from work with microelectrodes.

3.6.1. Soils

The oxygen conditions of agricultural soils are important for the turnover of inorganic nutrients and for the yield of crops. Several attempts have been made to determine the oxygen distribution in soils, using simple oxygen macroelectrodes (e.g., Lemon and Erickson, 1952), membrane-covered macroelectrodes (e.g., Fluhler *et al.*, 1976), minielectrodes (Greenwood and Goodman, 1967), and microelectrodes (Sexstone *et al.*, 1985). The measurements performed by the use of macroelectrodes suffered from poor spatial resolution, and information about the conditions within soil aggregates, which is of utmost importance for the understanding of microbial processes in soil, could not be obtained. The measurements of Greenwood and Goodman (1967), who used an electrode with a tip diameter of 0.5 mm (Naylor and Evans, 1960), were able to show an anaerobic center in artificial, 10-mm soil aggregates. They concluded, however, that their electrode was still too large and might have caused significant disturbance of the conditions within the soil crumb.

Sexstone *et al.*, (1985) used microelectrodes with a tip diameter of 50 μm (Revsbech and Ward, 1983) to measure oxygen gradients in natural, water-saturated soil aggregates. They showed that in this specific soil (Muscatine silty clay loam from Iowa), anaerobiosis often occurred within soil aggregates larger than 6 mm diameter. They also found that no denitrification occurred within soil crumbs that were oxic in all parts, whereas it did occur in most partly anoxic crumbs. Microelectrode measurements were also used to determine diffusion coefficients for oxygen within the soil crumbs as described in Section 3.3.2.

The measurements of Sexstone *et al.* (1985) were performed in soil crumbs moistened to field capacity. Considerable difficulties would be associated with measurements in soils held at lower water potentials, as the insertion of the electrode in such semidry soil crumbs might cause introduction of air along the electrode shaft. The electrode would also have to be very sturdy to survive an introduction into the semidry crumbs of most soils.

3.6.2. The Rhizosphere

The rhizosphere around roots in soils and sediments is a microenvironment that highly stimulates microbial activity (Alexander, 1977). There is an uptake by the root of inorganic nutrients from the subtratum, and the root may excrete organic substances and oxygen (e.g., Sand-Jensen *et al.,* 1982). The excretion of oxygen by roots can create an oxic microenvironment with an abundance of aerobic bacteria within an otherwise anoxic soil (Joshi and Hollis, 1977). It is important to understand the factors that regulate the distribution of oxygen around roots and how this affects the redox processes in the surrounding soils or sediments. The *in situ* excretion of oxygen from roots has been quantified by "oxygen availability" (Lemon and Erickson, 1952) measurements using polarized rings of platinum surrounding individual roots (Armstrong, 1967). This method, however, is not very satisfactory in several respects. An accurate calibration of the reading to actual oxygen excretion rates is difficult. The method gives a figure only for the excretion and no information about the extension of the oxic zone. Finally, it it is difficult to place the platinum ring without disturbing the root and its environmental conditions to a significant extent.

The oxygen regime around roots can in theory also be measured *in situ* by use of microelectrodes. The problem is to identify the position of the electrode relative to the roots and to advance the microelectrode to a sufficiently great depth without damaging it. Instead of soil, a preliminary study of oxygen gradients around roots was conducted in agar in which the freshwater macrophyte *Littorella uniflora* was growing (P. B. Christensen, N. P. Revsbech, and J. Sørensen, unpublished data). An oxygen microelectrode, similar to the one shown in Figs. 1B and 2, but with a 10-cm-long tip region of less than 2 mm diameter, was used for the experiments. The data obtained showed that it is possible to measure oxygen gradients accurately deep below the water–substratum interface. The measured oxygen gradients were used to calculate the excretion of oxygen as a function of environmental conditions, such as incident light intensity on the leaves. It was found that the oxygen excretion increased linearly with increasing light intensity up to 70 μEinst/m^2 per sec, where light saturation of the excretion rate (photosynthetic rate?) was reached. The oxygen excretion was calculated from the radial gradients at various distances from the root tip along individual roots. There seemed to be only minor differences in the excretion rate between newer and older root segments.

3.6.3. Rhizobium Nodules

The nitrogen-fixing enzyme of bacteria, nitrogenase, is denatured by oxygen, and it is therefore only active in microenvironments protected against high oxygen concentrations. The oxygen concentration in the most efficient nitrogen-fixing microenvironment known, the *Rhizobium* nodule, was recently measured with a microelectrode (J. F. Witty, L. Skøt, and N. P. Revsbech, unpublished results). The measurements showed that the oxygen partial pressure within the bacteroid-containing tissue of a pea plant nodule was less than 0.0002 atm. The oxygen gradients through the whole nodule were measured as a function of the external oxygen concentration. The oxygen gradients around the symbiont-containing tissue were extremely steep, with oxygen concentrations dropping from 80% air saturation to 0% within 100 μm. The tissue of the nodule consumed all the incoming oxygen and kept the symbiont-containing region at extremely low oxygen concentrations even when the surrounding atmosphere contained significantly higher oxygen levels than found in nature. Only at an oxygen partial pressure as high as 0.8 atm did the diffusive influx exceed the rate of consumption and cause a breakdown of the oxygen barrier. The oxygen concentration then rose simultaneously in all parts of the symbiont-containing region, suggesting a very efficient transport of oxidizing power by the leghemoglobin in the nodule.

3.6.4. Microbial Films from Sewage Treatment Plants

Microbial films or slimes in sewage treatment plants and similar organic-rich environments may have high microbial activities and consequently also contain steep chemical gradients. The gradients of oxygen associated with such films have been studied over the last decade by H. R. Bungay and coworkers.

It was found that the current velocity, organic load, and pH of the overlying water influenced the oxygen profile in microbial slimes (Bungay *et al.,* 1969; Chen and Bungay, 1981). When the organic load was high, oxygen penetrated to a very shallow depth ($\sim$0.15 mm) due to a high respiratory activity, but the oxygen penetration was considerably deeper when the load was low. A similar effect was observed by J. G. Kuenen (unpublished results). He measured an oxygen penetration of 0.4 mm when the slime was covered with pure water, but only 0.1 mm when the organic load was high. Low pH ($<$5) inhibited respiratory activity in the film and therefore caused deep oxygen penetrations (Chen and Bungay, 1981). Photosynthetic oxygen production in biofilms containing microalgae caused high oxygen concentrations and deep oxygen penetrations (Bungay and Chen, 1981; Chen and Bungay, 1981). A photosynthetic rate of 56 mmole O_2/m^2 per hr was measured in a photosynthetic biofilm from a trickling filter (J. G. Kuenen, B. B. Jørgensen, and N. P. Revsbech, unpublished data), but $\sim$70% of this oxygen was concurrently consumed within the biofilm. The oxygen consumption of the slime during illumination was several times as high as the rate measured in the dark. During dark incubation, the oxygen consumption of the biofilm was limited by the rate of oxygen diffusion (Chen and Bungay, 1981), and the photosynthetically produced oxygen relieved this limitation.

3.6.5. Epiphytic Communities

Epiphytic communities occur in both freshwater and seawater environments. Oxygen and pH gradients as well as photosynthesis rates have been measured in such communities (Sand-Jensen *et al.*, 1985, and unpublished results). There was no problem in measuring oxygen and pH, but the photosynthetic activity of the epiphytes could not be distinguished from the activity of the macrophyte in layers closer than 0.1–0.2 mm from the macrophyte surface.

Epiphytes and macrophyte often contributed equally to the combined primary production. As an example, a 1.0- to 1.5-mm-thick epiphyte layer on a leaf of the freshwater angiosperm *Potamogeton crispus* was responsible for 70% of the combined primary production of the association (9 mmole O_2/m^2 per hr) when illuminated at 500 μEinst/m^2 per sec. The investigated epiphytic communities were all photosaturated at relatively low light intensities (100–300 μEinst/m^2 per sec). This is in contrast to microalgal photosynthesis in sediments, which is usually not saturated at even very high light intensities (Section 3.2.2c). The light compensation points of the associations were also low, 20–40 μEinst/m^2 per sec. The epiphytes increased the pH values and the oxygen concentrations at the leaf surface as compared to macrophytes without epiphytes, and this may be as significant as the shading effect often mentioned as the major negative effect of epiphytes on the host macrophyte.

3.6.6. The Air–Water Interface

All exchange of gases between atmosphere and water occurs through the air–water interface. The rate of transfer seems to be governed by molecular diffusion through a diffusive boundary layer (Section 3.1.3) immediately below the surface of the water (e.g., Broecker and Peng, 1974). The oxygen gradients within the diffusive boundary layer can be measured using an oxygen microelectrode where the tip is bent into a U-shape, so that the oxygen gradient can be measured upward. The miniscus formed around the electrode shaft due to the surface tension is then formed around the electrode shaft at such a great distance from the point being analyzed that it does not affect the measured gradients. An example of an oxygen profile down through the boundary layer of nonstirred, partly deoxygenated water is shown in Fig. 29. The oxygen gradient was linear down through the diffusive boundary layer, which was about 500 μm in thickness. At greater than 600 μm depth, the concentration gradient was zero, indicating some turbulence even in the absence of stirring.

4. Future Developments

The technical development of gas- and ion-sensing microelectrodes and of the associated equipment is still in an early phase. Over the coming few years one can expect a number of novel microelectrode types to be developed, some of which exist currently as macroelectrodes. The specificity and sensitivity of these electrodes are rapidly improving and the measuring circuits and data handling systems are becoming more advanced, less expensive, and simpler to use. As the electrodes become more widely applied, more microelectrode types should also become

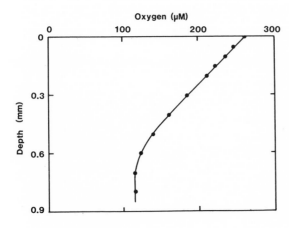

Figure 29. Oxygen profile through the diffusive boundary layer at the air–water interface of a nonstirred column of water. The water was contained in a 100-ml beaker and was partly deoxygenated by N_2-bubbling 15 min before the oxygen profile was recorded. [Data from B. B. Jørgensen and N. P. Revsbech (in preparation).]

commercially available to the occasional user, for whom learning the requisite construction techniques may remain a barrier.

Currently used microelectrodes that should be of interest in microbial ecology besides those sensitive to PO_2, pS^{2-}, and pH, are the CO_2- and Ca^{2+}-sensitive types. Both have been developed to a stage where they can be applied to many microbial systems (Caflish and Carter, 1974; Pucacco and Carter, 1978; Thomas, 1978; Tsien, 1980). The CO_2 microelectrode can be used, as can the O_2 electrode, to analyze respiration and photosynthesis of microorganisms. A disadvantage of the CO_2 microelectrode for photosynthesis studies, however, is, that this process is generally accompanied by high pH, at which little CO_2 is present, and the applicability therefore may be limited due to insufficient sensitivity. Carbonate-sensitive liquid membrane electrodes have also been constructed (Herman and Rechnitz, 1975), but they still have problems of interference from other common ions, which make them impractical for most ecological applications. Calcium-sensitive microelectrodes have important potential applications in the study of calcification and carbonate dissolution.

Many new liquid ion exchangers are developed each year and some are sufficiently specific to be used as membranes in ion-sensitive microelectrodes (Tsien, 1980). Such ion exchangers have already been used in microelectrodes for pH, Na^+, K^+, and Ca^{2+}. In addition, an ammonia-sensitive, potentiometric microprobe has been developed (Pui et al., 1978). A quite different principle is used in the voltammetric microelectrodes that have recently been fabricated with sensing tips of less than 10 μm diameter (Wightman, 1981; Howell and Wightman, 1984). Although these electrodes may not be ideal for many applications, due to problems with their specificity, they can be used to detect a wide range of chemical species in solution.

Novel types of sensors presently being developed as macroelectrodes are based on specific enzyme reactions. Immobilized enzymes at the electrode tip react with specific organic molecules in the surrounding medium and create an electrochemical reaction, which can be recorded. An electrode that responds to, for example, glucose has been constructed according to this principle (Tsuchida and Yoda, 1981). Such electrodes may have a high specificity toward the reacting organic molecules as a result of the high specificity of the enzymatic reactions. In the near future it may be possible to scale down such electrodes to real microscale and make them important new tools for the study of metabolic processes in natural microbial communities. Enzyme-reactive sensors are also being developed based on light detection rather than electrochemical detection (Arnold, 1985). By the use of single optical fibers as light guides, it should be possible to apply similar principles on the microscale.

In addition to the appearance of new types of sensors, the performance of available microelectrodes and measuring equipment is being refined. The new microprocessor technology has made data collection and handling much faster, better, and, at least in some respects, also simpler. For, example, microprocessors can be used to compensate for slow response times of the microelectrodes and for drift in the signal. Electronic compensation for slow response can be especially useful in combination with continuously moving microelectrodes operated by a motor-driven micromanipulator. Semiconductor technology has also provided inexpensive operational amplifiers small enough to attach directly to the microelectrodes, thereby improving the signal-to-noise ratio. Such amplifiers provide more stability and flexibility, especially when operating the microelectrodes in the field, on board a ship, or by remote control.

The introduction and application of microelectrodes in ecological research has invariably led to the wish for additional microsensors. These should measure environmental parameters such as light, temperature, and salinity at a spatial resolution sufficiently high to be directly relevant to the individual microbial populations. Such sensors are also slowly being developed. As one example, a simple light sensor has recently been constructed with a tip diameter of 20–30 μm (B. B. Jørgensen and D. J. Des Marais, in preparation). The probe is based on a single optical fiber of 80 μm diameter with a conical sensing tip. The applied semiconductor detector is a hybrid photodiode/amplifier and the complete unit is directly attached to a micromanipulator, just like a microelectrode.

The range of microbial environments so far studied and the ecological research applications of microelectrodes are still quite limited. For many basic problems and applied aspects of microbial ecology, the use of microelectrodes seems an ideal but yet unexplored possibility. Bacteria often grow on surfaces or at interfaces within a chemical microenvironment quite different from the surrounding medium. These chemical conditions are difficult to analyze with other currently available techniques. Thus, microelectrodes have important future applications in the study of, e.g, the corrosion of off-shore steel constructions, oil degradation in the sea, the develpoment of bacterial plaque on teeth, and the metabolism of immobilized bacterial films in biotechnology.

References

Alexander, M., 1977, *Introduction to Soil Microbiology,* Wiley, New York.
Aller, R. C., 1977, The Influence of Macrobenthos on Chemical Diagenesis of Marine Sediments, Ph.D. thesis, Yale University.
Ammann, D., Lanter, F., Steiner, R. A., Schultess, P., Shijo, Y., and Simon, W., 1981, Neu-

tral carrier based hydrogen ion selective microelectrode for extra- and intracellular studies, *Anal. Chem.* **53**:2267–2269.

Anderson, O. R., and Be, A. W. H., 1976, The ultrastructure of a planktonic foraminifer, *Globigerinoides sacculifer* (Brady), and its symbiotic dinoflagellates, *J. Foram. Res.* **6**:1–21.

Armstrong, W., 1967, The use of polarography in the assay of oxygen diffusing from roots in anaerobic media, *Physiol. Plant.* **20**:540–553.

Arnold, M. A., 1985, Enzyme-based fiber optic sensor, *Anal. Chem.* **57**:565–566.

Baumgärtl, H., and D. W. Lübbers, 1983, Platinum needle electrodes for polarographic measurement of local O_2 pressure in cellular range of living tissue. Its construction and properties, in: *Polarographic Oxygen Sensors: Aquatic and Physiological Applications* (E. Gnaiger and H. Forstner, eds.), pp. 37–65, Springer-Verlage, Heidelberg.

Be, A. W. H., and Tolderlund, D. S., 1971, Distribution and ecology of living planktonic foraminifera in surface waters of the Atlantic and Indian Oceans, in: *Micropaleontology of Oceans* (B. M. Funnell and W. R. Riedel, eds.), pp. 105–149, Cambridge University Press, Cambridge.

Berner, R. A., 1962, Electrode studies of hydrogen sulfide in marine sediments, *Geochim. Cosmochim. Acta* **27**:563–575.

Berner, R. A., 1980, *Early Diagenesis, a Theoretical Approach,* Princeton University Press, Princeton, New Jersey.

Board, P. A., 1976, Anaerobic regulation of atmospheric oxygen, *Atmos. Environ.* **10**:339–342.

Boudreau, B. P., and Guinasso, N. L., 1982, The influence of a diffusive boundary sublayer on accretion, dissolution, and diagenesis at the sea floor, in: *The Dynamic Environment of the Sea Floor* (K. A. Fanning and F. T. Manheim, eds.), pp. 115–145, Lexington Books, Lexington, Massachusetts.

Broecker, W. S., and Peng, T.-H, 1974, Gas exchange rate between sea and air, *Tellus* **26**:21–35.

Bungay, H. R., and Chen, Y. S., 1981, Dissolved oxygen profiles in photosynthetic microbial slimes, *Biotechnol. Bioeng.* **23**:1893–1895.

Bungay, H. R., 3rd, Whalen, W. J., and Sanders, W. M., 1969, Microprobe techniques for determining diffusivities and respiration in microbial slime systems, *Biotechnol. Bioeng.* **11**:765–772.

Caflish, C. R., and Carter, N. W., 1974, A micro PCO_2 electrode, *Anal. Biochem.* **60**:252–257.

Chen, Y. S., and Bungay, H. R., 1981, Microelectrode studies of oxygen transfer in trickling filter slimes, *Biotechnol. Bioeng.* **23**:781–792.

Clark, L. C., Wolf, R., Granger, D., and Taylor, A., 1953, Continuous recording of blood oxygen tension by polarography, *J. Appl. Physiol.* **6**:189–193.

Cohen, Y., 1983, The Solar Lake cyanobacterial mats: Strategies of photosynthetic life under sulfide, in: *Microbial Mats: Stromatolites* (Y. Cohen, R. W. Castenholz, and H. O. Halvorson, eds.), pp. 133–148, Alan R. Liss, New York.

Cohen, Y., Padan, E., and Shilo, M., 1975a, Facultative anoxygenic photosynthesis in the cyanobacterium *Oscillatoria limnetica, J. Bacteriol.* **123**:855–861.

Cohen, Y., Jørgensen, B. B., Padan, E., and Shilo, M., 1975b, Sulfide dependent anoxygenic photosynthesis in the cyanobacterium *Oscillatoria limnetica, Nature* **257**:489–492.

Crank, J., 1983, *The Mathematics of Diffusion,* Oxford University Press, London.

Dale, T., 1978, Total, chemical, and biological oxygen consumption of the sediments in Lindåspollene, Western Norway, *Mar. Biol.* **49**:333–341.

Davis, R. B., 1974, Tubificids alter profiles of redox potential and pH in profundal lake sediment, *Limnol. Oceanogr.* **19**:342–346.

Drew, E. A., 1973, The biology and physiology of alga–invertebrate symbioses. III *In situ* measurements of photosynthesis and calcification in some hermatypic corals, *J. Exp. Mar. Biol. Ecol.* **13**:165–179.

Duursma, E. K., and Hoede, C., 1967, Theoretical, experimental and field studies concerning molecular diffusion of radioisotopes in sediments and suspended solid particles of the sea. Part A: Theories and mathematical calculations, *Neth. J. Sea Res.* **3**:423–457.

Edwards, R. W., 1958, The effect of larvae of *Chironemus riparius* Meigen on the redox potentials of settled activated sludge, *Ann. Appl. Biol..* **46**:457–464.

Fenchel, T., 1969, The ecology of marine microbenthos. 4. Structure and function of the benthic ecosystem, its chemical and physical factors and the meiofauna communities with special reference to the ciliated protozoa, *Ophelia* **6**:1–182.

Fluhler, H., Ardakan, M. S., Szusckiewicz, and Stolzy, L. H., 1976, Field measured nitrous oxide concentrations, redox potentials, oxygen diffusion rates, and oxygen partial pressures in relation to denitrification, *Soil Sci.* **122**:107–114.

Greenwood, D. J., and Goodman, D., 1967, Direct measurement of the distribution of oxygen in soil aggregates and in columns of fine soil crumbs, *J. Soil. Sci.* **18**:182–196.

Guterman, H., and Ben-Yaakov, S., 1983, Determination of total dissolved sulfide in the pH range 7.5 to 11.5 by ion selective electrodes, *Anal. Chem.* **55**:1731–1734.

Harris, G. P., 1978, Photosynthesis, productivity and growth: The physiological ecology of phytoplankton, *Ergeb. Limnol.* **10**:1–171.

Herman, H. B., and Rechnitz, G. A., 1975, Preparation and properties of a carbonate ion-selective membrane electrode, *Anal. Chim. Acta* **76**:155–164.

Howell, J. O., and Wightman, R. M., 1984, Ultrafast voltammetry and voltammetry in highly resistive solutions with microvoltammetric electrodes, *Anal. Chem.* **56**:524–529.

Hunding, C. and Hargrave, B. T., 1973, A comparison of benthic microalgal production measured by C^{14} and oxygen methods, *J. Fish. Res. Board Can.* **30**:309–312.

Jerlov, N., 1976, *Optical Oceanography,* Elsevier, Amsterdam.

Jones, J. G., Gardener, S., and Simon, B. M., 1983, Bacterial reduction of ferric iron in stratified lake, *J. Gen. Microbiol.* **129**:131–139.

Jørgensen, B. B., 1977, The sulfur cycle of a coastal marine sediment (Limfjorden, Denmark), *Limnol. Oceanogr.* **22**:814–832.

Jørgensen, B. B., 1982, Ecology of the bacteria of the sulfur cycle with special reference to anoxic–oxic interface environments, *Phil. Trans. R. Soc. Lond. B* **298**:543–561.

Jørgensen, B. B., 1983, The microbial sulfur cycle, in: *Microbial Geochemistry* (W. E. Krumbein, ed.), pp. 91–124, Blackwell, Oxford.

Jørgensen, B. B., and Revsbech, N. P., 1983, Colorless sulfur bacteria, *Beggiatoa* spp. and *Thiovolum* spp. in O_2 and H_2S micrograients, *Appl. Environ. Microbiol.* **45**:1261–1270.

Jørgensen, B. B., and Revsbech, N. P., 1985, Diffusive boundary layers and the oxygen uptake of sediments and detritus, *Limnol. Oceanogr.* **30**:11–21.

Jørgensen, B. B., Revsbech, N. P., Blackburn, T. H., and Cohen, Y., 1979, Diurnal cycle of oxygen and sulfide microgradients and microbial photosynthesis in a cyanobacterial mat sediment, *Appl. Environ. Microbiol.* **38**:46–58.

Jørgensen, B. B., Revsbech, N. P., and Cohen, Y., 1983, Photosynthesis and structure of benthic microbial mats: Microelectrode and SEM studies of four cyanobacterial communities, *Limnol. Oceanogr.* **28**:1075–1093.

Jørgensen, B. B., Erez, J., Revsbech, N. P., and Cohen, Y., 1985, Symbiotic photosynthesis in planktonic foraminifera, *Globigerinoides sacculifer* (Brady), studied with microelectrodes, *Limnol. Oceanogr.* **30**:1253–1267.

Jørgensen, B. B., Cohen, Y., and Revsbech, N. P., 1986, Transition from anoxygenic to oxygenic photosynthesis in a microcoleus chtonoplastes Cyanobacterial mat, *Appl. Environ. Microbiol.* **51** (2) (in press).

Joshi, M. M., and Hollis, J. P., 1977, Interaction of *Beggiatoa* and rice plant: Detoxification of hydrogen sulfide in the rice rhizosphere, *Science* 195:179–180.

Kelly, D. P., 1982, Biogeochemistry of the chemolithotrophic oxidation of inorganic sulfur, *Phil. Trans. R. Soc. Lond. B* 298:499–528.

Lean, D. R. S., and Burnison, B. K., 1979, An evaluation of errors in the ^{14}C method of primary production measurement, *Limnol. Oceanogr.* 24:917–928.

Lemon, E. R., and Erickson, A. E., 1952, The measurement of oxygen diffusion in soil with a platinum microelectrode, *Soil Sci. Soc. Am. Proc.* 16:160–163.

Lindeboom, H. J., and Sandee, A. J. J., 1984, The effect of coastal engineering projects on microgradients and mineralization reactions in sediments, *Water Sci. Technol.* 16:87–94.

Murray, J. W., and Grundmanis, V., 1980, Oxygen consumption in pelagic marine sediments, *Science* 209:1527–1530.

Naylor, P. F. D., and Evans, N. T. S., 1960, An electrode for measuring absolute oxygen tension in tissues, *J. Polarogr. Soc.* 2:22–24.

Nelson, D. C., and Jannasch, H. W., 1983, Chemoautotrophic growth of a marine *Beggiatoa* in sulfide-gradient cultures, *Arch. Microbiol.* 136:262–269.

Pamatmat, M. M., 1971, Oxygen consumption by the seabed. IV Shipboard and laboratory experiments, *Limnol. Oceanogr.* 16:536–550.

Prezelin, B. B., 1976, The role of peridinin-chlorophyll a-protein in the photosynthetic light adaptation of the marine dinoflagellate, *Glenodinium* sp., *Planta* 130:225–233.

Prezelin, B. B., Ley, A. C., and Haxo, F. T., 1976, Effects of growth irradiance on the photosynthetic action spectra of the marine dinoflagellate, *Glenodinium* sp., *Planta* 130:251–256.

Pucacco, L. R., and Carter, N. W., 1978, An improved PCO_2 microelectrode, *Anal. Biochem.* 90:427–434.

Pui, C. P., Rechnitz, G. A., and Miller, R. F., 1978, Micro-size potentiometric probes for gas and substrate sensing, *Anal. Chem.* 50:330–333.

Purcell, E. M., 1977, Life at low Reynolds number, *Am. J. Phys.* 45(1):3–11.

Reimers, C. E., and Smith, K. L., 1986, Reconciling measured and predicted fluxes of oxygen across the deep sea sediment–water interface, *Limnol. Oceanogr.*

Reimers, C. E., Kalhorn, S., Emerson, S. R., and Nealson, K. H., 1984, Oxygen consumption rates in pelagic sediments from the Central Pacific: First estimates from microelectrode profiles, *Geochim. Cosmochim. Acta* 48:903–911.

Revsbech, N. P., 1983, *In situ* measurement of oxygen profiles of sediments by use of oxygen microelectrodes, in: *Polarographic Oxygen Sensors: Aquatic and Physiological Applications* (E. Gnaiger and H. Forstner, eds.), pp. 265–273, Springer, Heidelberg.

Revsbech, N. P., and Jørgensen, B. B., 1983, Photosynthesis of benthic microflora measured with high spatial resolution by the oxygen microprofile method: Capabilities and limitations of the method, *Limnol. Oceanogr.* 28:749–756.

Revsbech, N. P., and Ward, D. M., 1983, Oxygen microelectrode that is insensitive to medium chemical composition: Use in an acid microbial mat dominated by *Cyanidium caldarium*, *Appl. Environ. Microbiol.* 45:755–759.

Revsbech, N. P., and Ward, D. M., 1984a, Microprofiles of dissolved substances and photosynthesis in microbial mats measured with microelectrodes, in: *Microbial Mats: Stromatolites* (Y. Cohen, R. W. Castenholz, and H. O. Halvorson, eds.), pp. 171–188, Alan R. Liss, New York.

Revsbech, N. P., and Ward, D. M., 1984b, Microelectrode studies of interstitial water chemistry and photosynthetic activity in a hot spring microbial mat, *Appl. Environ. Microbiol.* 48:270–275.

Revsbech, N. P., Jørgensen, B. B., and Blackburn, T. H., 1980a, Oxygen in the seabottom measured with a microelectrode, *Science* 207:1355–1356.

Revsbech, N. P., Sørensen, J., Blackburn, T. H., and Lomholt, J. P., 1980b, Distribution of oxygen in marine sediments measured with microelectrodes, *Limnol. Oceanogr.* 25:403–411.

Revsbech, N. P., Jørgensen, B. B., and Brix, O., 1981, Primary production of microalgae in sediments measured by oxygen microprofile, $H^{14}CO_3^-$ fixation and oxygen exchange methods, *Limnol. Oceanogr.* 26:717–730.

Revsbech, N. P., Jørgensen, B. B., Blackburn, T. H., and Cohen, Y., 1983, Microelectrode studies of photosynthesis and O_2, H_2S, and pH profiles of a microbial mat, *Limnol. Oceanogr.* 28:1062–1074.

Revsbech, N. P., Madsen, B., and Jørgensen, B. B., 1986, Oxygen production and consumption in sediments determined at high spatial resolution by computer simulation of oxygen microelectrode data, *Limnol. Oceanogr.* (in press).

Sand-Jensen, K., Prahl, C., and Stockholm, H., 1982, Oxygen release from roots of submerged aquatic macrophytes, *Oikos* 38:349–354.

Sand-Jensen, K., Revsbech, N. P., and Jørgensen, B. B., 1985, Microprofiles of oxygen in epiphyte communities on submerged macrophytes, *Mar. Biol.* 89:55–62.

Santschi, P. H., Bower, P., Nyffeler, U. P., Azvedo, A., and Broecker, W. S., 1983, Estimates of the resistance of chemical transport posed by the deep-sea boundary layer, *Limnol. Oceanogr.* 28:899–912.

Sexstone, A. J., Revsbech, N. P., Parkin, T. B., and Tiedje, J. M., 1985, Direct measurement of oxygen profiles and denitrification rates in soil aggregates, *Soil Sci. Soc. Am. J.* 49:645–651.

Shiver, D. F., 1969, *The Manipulation of Air-Sensitive Compounds,* McGraw-Hill, New York.

Smith, K. L., Jr., and Baldwin, R. J., 1984, Seasonal fluctuation in deep-sea sediment community respiration: Central and eastern North Pacific, *Nature* 307:624–626.

Smith, K. L., Jr., and Hinga, K. R., 1983, Sediment community respiration in the deep sea, in: *The Sea* (G. T. Rowe, ed.), Vol. 8, pp. 331–370, Wiley, New York.

Sørensen, J., 1984, A headspace technique for oxygen measurement in deep-sea sediment cores, *Limnol. Oceanogr.* 29:650–652.

Sørensen, J., Jørgensen, B. B., and Revsbech, N. P., 1979, A comparison of oxygen, nitrate, and sulfate respiration in coastal marine sediments, *Microb. Ecol.* 5:105–115.

Stanier, R. Y., Adelberg, E. A., and Ingraham, J. L., 1977, *General Microbiology,* 4th ed., Macmillan, London.

Steemann-Nielsen, E., 1952, Use of radioactive carbon (C^{14}) for measuring organic production in the sea, *J. Cons. Cons. Int. Explor. Mer* 18:117–140.

Thomas, R. C., 1978, *Ion-Sensitive Intracellular Microelectrodes, How to Make and Use Them,* Academic Press, London.

Tsien, R. Y., 1980, Liquid sensors for ion-selective microelectrodes. *Trends Neurosci.* 3:219–221.

Tsuchida, T., and Yoda, K., 1981, Immobilization of D-glucose oxidase onto a hydrogen peroxide permselective membrane and application for an enzyme electrode, *Enzyme Microb. Technol.* 3:326–330.

Vogel, S., 1981, *Life in Moving Fluids,* Willard Grant, Boston.

Ward, D. M., Beck, E., Revsbech, N. P., Sandbeck, K. A., and Winfrey, M. R., 1984, Decomposition of hot spring microbial mats, in: *Microbial Mats: Stromatolites* (Y. Cohen, R. W. Castenholz, and H. O. Halvorson, eds.), pp. 191–214, Alan R. Liss, New York.

Whalen, W. J., Riley, J., and Nair, P., 1967, A microelectrode for measuring intracellular PO_2, *J. Appl. Physiol.* 23:798–801.

Wightman, R. M., 1981, Microvoltammetric electrodes, *Anal. Chem.* 53:1125A–1130A.

Wimbush, M., 1976, The physics of the benthic boundary layer, in: *The Benthic Boundary Layer* (I. N. McCave, ed.), pp. 3–10, Plenum Press, New York.

8

Hydrophobic Interactions: Role in Bacterial Adhesion

MEL ROSENBERG and STAFFAN KJELLEBERG

1. Introduction

The last decade has seen a dramatic rise in scientific interest in the field of bacterial adhesion and related subjects. Much of this interest has been directed in search of "specific interactions" between bacterium and substratum, such as those that are inhibited by specific sugar moieties. A second avenue of research has been the study of the role of less specific, hydrophobic interactions. Since it appears that this second avenue of research is being increasingly followed, reviews dealing with bacterial hydrophobicity and adhesion should be of use. The aims of the present chapter are to (1) present some relevant aspects of hydrophobic interactions; (2) describe the methodology available for measurements related to bacterial cell-surface hydrophobicity and the parameters they may measure; (3) discuss investigations dealing with surface components that promote or reduce bacterial hydrophobicity; and (4) survey studies related to the role of hydrophobic interactions in mediating bacterial adhesion to interfaces, traversing areas of environmental and medical interest.

The importance of bacterial hydrophobicity in adhesion to liquid–liquid interfaces and the possible relationship between cell-surface hydro-

MEL ROSENBERG • School of Dental Medicine and Department of Human Microbiology, Sackler Faculty of Medicine, Tel-Aviv University, Ramat-Aviv 69978, Israel.
STAFFAN KJELLEBERG • Department of Marine Microbiology, University of Göteborg, S-413 19 Göteborg, Sweden.

phobicity and phagocytosis were reported over 60 years ago by Mudd and Mudd (1924a,b). These authors observed that certain microorganisms (e.g., *Erythrobacillus prodigiosus,* currently known as *Serratia marcescens*) tend to partition at the water–oil interface, and provided a theoretical definition of this phenomenon in terms of bacteria–oil, bacteria–water, and oil–water interfacial energies. Reed and Rice (1931) made an unsuccessful attempt to quantify bacterial partitioning at the oil–water interface, presumably because of their limited choice of bacterial strains. However, these authors were able to show that certain bacteria (e.g., mycobacteria) are able to pass from the bulk aqueous phase to the bulk oil phase upon vigorous mixing.

In the subsequent four decades, relatively little scientific interest was directed toward the study of bacterial hydrophobicity, with notable exceptions in the work of Dyar (1948) and Hill *et al.* (1963) and studies on bacterial interactions at the air–water interface (e.g., Boyles and Lincoln, 1958). Interest in bacterial hydrophobicity and adhesion was renewed with the work of Marshall and coworkers (Marshall and Cruickshank, 1973; Marshall *et al.,* 1975) and van Oss and colleagues [for a review see van Oss (1978)] in the early 1970s.

During the past decade, a variety of techniques for measuring bacterial hydrophobicity have been proposed, accompanied by a surge in research, as can be discerned from Fig. 1.

Hydrophobic interactions have been implicated in a wide array of adhesion phenomena, including partitioning at the oil–water (M. Rosenberg, 1984a) and air–water (Kjelleberg, 1985) interfaces and adhesion to plastics (Fletcher, 1976; Fletcher and Marshall, 1982b), epithelial cells (M. Rosenberg *et al.,* 1981; E. Rosenberg, *et al.,* 1983a), phagocytes (van Oss, 1978), teeth (M. Rosenberg *et al.,* 1983a,c; Nesbitt *et al.,* 1982), sub-

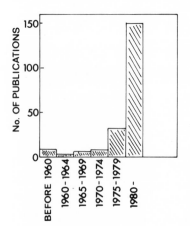

Figure 1. Hydrophobicity- and adhesion-related papers as a function of publication period. As an indication of interest in hydrophobicity and adhesion, the distribution of the publications referred to in this review into publication periods is illustrated. A burst in interest over the past decade is evident.

merged aquatic surfaces (Fattom and Shilo, 1984), sulfur granules (Bryant *et al.,* 1984), mineral surfaces (Stenstrom and Kjelleberg, 1985), and to one another (Stanley and Rose, 1967; Moore and Marshall, 1981; Faris *et al.,* 1983). In several cases, surface components contributing to bacterial hydrophobicity have been identified. Bacterial hydrophobicity may also be an important factor in other adhesion-related phenomena, such as gliding motility (Dworkin *et al.,* 1983; Humphrey *et al.,* 1979; R. Wolkin and J. L. Pate, in preparation) and colony spreading (Gibbons *et al.,* 1983).

Despite the apparent importance of hydrophobic interactions in participating in such a wide range of phenomena, the term "bacterial hydrophobicity" has no unequivocal definition, nor a definitive scale of values. This is not surprising in light of current disagreements among scientists on the precise nature of hydrophobic interactions (e.g., Hildebrand, 1979; Israelachvili and Pashley, 1984) and the lack of consensus on a definitive method for measuring bacterial hydrophobicity (see Section 3). Thus, the literature is replete with confusing terminologies ("pronounced" versus "slight" cell-surface hydrophobicity) and results from various laboratories that cannot be compared with one another.

Are we then to conclude that the plurarity of methods for measuring bacterial hydrophobicity is detrimental? Actually, the opposite is probably the case. The nonuniformity of the bacterial cell surface confounds attempts at a universal criterion: a bacterium that is hydrophilic all over, save for a hydrophobic tip (Marshall and Cruickshank, 1973), may be hydrophilic in one assay and hydrophobic in another. Moreover, the various techniques have given researchers flexibility in approaching various adhesion problems.

Cell-surface hydrophobicity is by no means a subject of interest restricted to bacteriology, and has been implicated in a wide variety of eucaryotic cell interactions and viral adhesion as well. Although such phenomena are beyond the scope of this review, we have listed some examples in Table I.

2. The Hydrophobic Effect

The general concept of hydrophobic interactions in adhesion and other biological systems has been extensively reviewed in recent years (Magnusson, 1980, 1982; Rutter, 1980; Edebo *et al.,* 1980; Ochoa, 1978). Several alternate approaches to explaining hydrophobic interactions have been taken by various investigators. The present section is thus an attempt to illustrate several viewpoints that have gained some consensus.

One definition for hydrophobic interactions involves the free energy

Table I. Hydrophobicity and Adhesion in Nonbacterial Systems

Phenomenon	Reference
Adhesion of poliovirus to membrane filters	Shields and Farrah (1983)
Foam separation of virus	Morrow (1969)
Adhesion of fungal spores *(Colletotrichum lindemuthianum)* to polystyrene and plant surfaces	Young and Kauss (1984)
Adhesion and growth of yeast on hydrocarbons	Nakahara *et al.* (1981), Miura *et al.* (1977), McLee and Davies (1972), Kappeli and Fiechter (1976)
Sexual interactions of *Paramecium caudatum*	Kitamura (1984)
Invasion of *Plasmodium falciparum* into erythrocytes	Breuer *et al.* (1984)
Adhesion to connective tissue	Wang (1974)
Adhesion properties of tissue culture cells	Malmqvist *et al.* (1984), Reuveny *et al.* (1983)
Adhesion of tumor cells	Hills (1984b)
Adhesion and surface properties of erythrocytes	Todd and Gingell (1980), Neumann *et al.* (1983)
Partitioning of sake yeast at the air–water interface	Ouchi and Akiyama (1971)

change ΔG associated with the process of bringing two entities from infinite separation to distances of the order of magnitude of molecular dimensions (Ben-Naim, 1977). In our case, this free energy change relates to the layering of water molecules at cell and substratum surfaces. According to Ben-Naim (1977), a nonpolar moiety introduced into an aqueous phase will be surrounded by water molecules whose structure is greater than that of molecules in the bulk aqueous phase. Such structured water layer(s) are associated with a decrease in entropy ($\Delta S < 0$) related to the increased order of the system. Thus, in order to solubilize nonpolar solutes, a "hole" must be created in the water structure of the bulk phase; the ensuing breaking of bonds between water molecules demands an input of energy. The formation of new hydrogen bonds in such a process leads to a negative but very small enthalpy change ΔH. Thus, the energy input required to carry out this process may be attributed to the negative entropy change.

What occurs when two hydrophobic surfaces covered by structured water molecules approach one another? In such a case, layers of structured water molecules are released into the bulk aqueous phase from the contacting surfaces. This situation is illustrated schematically in Fig. 2. The freeing of such molecules brings about the reverse situation to that described above: a decrease in free energy due to reduction of the overall nonpolar–water interface by the transfer of water molecules to the less ordered bulk water state (i.e., increase in ΔS). This premise, based on the behavior of water molecules surrounding apolar "cavities" (Arakawa *et al.*, 1979), rests on experimental evidence suggesting that contributions from formation of surface–surface and breaking of surface–water interactions can largely be neglected. The free energy for hydrophobic interactions thus has its largest contribution from the cavity effect, i.e., the entropy function.

Several properties of hydrophobic interactions can serve as criteria for their participation in adhesion of two surfaces. Hydrophobic interactions generally increase with ionic strength due to the suppression of electrostatic interactions (Ochoa, 1978). To a certain point, hydrophobic interactions increase with increasing temperature (Ben-Naim, 1977). Temperature may, however, also provoke increased solvation and decreased surface tension, which counteract the hydrophobic interactions. Hydrophobic interactions are also expected to increase with increasing apolar nature of one or both surfaces under consideration. Hydrophobic interactions are also enhanced by salting-out ions, according to the Hofmeister series (Lindahl *et al.*, 1981). At high concentrations of salting-out ions, the solubility of the solute is adversely affected by decreasing availability of water molecules in the bulk and increasing surface tension of water, resulting in an increase in hydrophobic interactions (Tanford, 1973). The increase in surface tension of water in the presence of salting-out agents may be indicative of increased water structure at low-energy (hydrophobic) interfaces (Tanford, 1973); freeing of increasingly structured water upon the approach of two hydrophobic surfaces, as discussed above, would, in turn, be increasingly favored. The effect of structure-breaking (chaotropic) ions may be viewed as the opposite case; the new ordering of water molecules that they induce prevents them from undergoing further positive entropy changes (Ochoa, 1978).

The above descriptions of hydrophobic interactions are derived from considerations of the close approach of extremely apolar moieties, such as two molecules of alkane. The role of hydrophobic interactions in bacterial adhesion is obviously far more complicated. The term "hydropho-

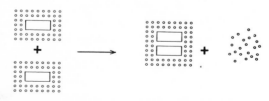

Figure 2. Aggregation of two nonpolar solutes due to hydrophobic interaction. When two separated polar solutes (represented by rectangles, left) covered by ordered layers of water molecules (represented by circles) interact, water molecules are released into the bulk aqueous phase (right). This interaction is accompanied by an increase in ΔS and a smaller increase in ΔH.

bic bacteria" is probably a misnomer: the great majority of bacteria referred to in this chapter are wetted by water and are readily suspended in aqueous medium due to polar, hydrophilic moieties, which abound on the bacterial cell surface. In fact, the ability of certain bacteria to adhere to oil droplets and prevent their coalescence (Rosenberg *et al.*, 1980a) is akin to oil emulsification by amphipathic molecules, which interact with both the water and oil phases. Similarly, the surfaces to which bacteria adhere often contain both polar and nonpolar constituents. The degree of each will influence the extent and structure of the proximal water molecule layers and thus determine the energetic favorability of adhesion, as previously discussed. Thus, it is the interplay between hydrophilic and hydrophobic surface components that determines the overall contribution of hydrophobic interactions to adhesion processes.

The outcome of an adhesion process can be predicted by surface thermodynamics (i.e., the the interfacial tensions between the adhering surfaces), as compared with that of each individual surface with the aqueous medium. Such interfacial tensions are themselves a reflection of the free energy interrelationships between the surfaces in question and the water molecules surrounding them. Development of such a model with respect to bacterial adhesion can be found in Absolom *et al.* (1983). These authors obtained two sets of experimental results; one set consisted of the interfacial tensions of the cells themselves; the second set consisted of adhesion measurements of the cells to surfaces of varying interfacial tensions in the presence of varying concentrations of surfactant (dimethylsulfoxide, DMSO). According to the concept put forward by these investigators, adhesion is independent of substratum surface energy when the surface energies of bacterium and suspending medium are equal. The results obtained for the surface energies of the bacteria tested based on their adhesion correlated well with the surface energies calculated from contact angle measurements. As predicted by their model, adhesion to hydrophilic surfaces was more pronounced only when the bacterial surface tension was larger than that of the surrounding aqueous medium; conversely, adhesion to increasingly hydrophobic surfaces occurred when the surface tension of the aqueous phase was higher than that of the bacterial cells. In many ecosystems, it is the second situation that prevails. In all experiments, moreover, adhesion was highest for the most hydrophobic bacteria for every aqueous surface tension studied.

3. Methods for Determining Cell-Surface Hydrophobicity

The methodology for studying the cell-surface hydrophobicity of bacterial cells can be divided into several groups, according to the type of parameters measured: (1) methods based on adhesion of cells to solid surfaces, at interfaces, or to one another; (2) methods based on the partitioning of hydrophobic molecules between the cells and surrounding medium; (3) methods based on partitioning of cells between two immiscible liquid phases; (4) techniques based upon properties of cell layers; and (5) indirect methods, such as those based on the inhibition of adhesion by agents that interfere with hydrophobic interactions.

Some general factors pertain to most methods currently employed: (1) all materials should be scrupulously clean and free of surfactant contaminants; (2) growth conditions and preparation of bacterial cells should be carefully monitored; and (3) duration of the assay should be minimal, to prevent release of surface components that may have an inhibitory effect, e.g., by "conditioning" of the attachment surface (Fletcher and Marshall, 1982b), or alternatively, synthesis of cell-surface components that alter surface properties.

3.1. Cell-Surface Hydrophobicity Methods Based on Adhesion

In studies related to the role of hydrophobic interactions in bacterial adhesion, methods that are themselves based on adhesion are sometimes the most pertinent. It should be reemphasized that in cases in which adhesion is mediated by a small portion of the bacterial cell, e.g., in partitioning of *Hyphomicrobium, Flexibacter,* and *Rhizobium* cells at the oil–water interface (Marshall and Cruickshank, 1973, Marshall *et al.,* 1975), the surface properties of the large majority of the cell surface make only a small contribution.

3.1.1. Hydrophobic Interaction Chromatography (HIC)

This method, originally developed for protein separation (Hjertén *et al.,* 1974), has found widespread use in many studies of bacterial hydrophobicity. In this technique, aqueous suspensions of Sepharose beads with covalently bound hydrophobic moieties (e.g., phenyl or octyl groups) are usually packed into small columns and the bacterial suspension is subsequently applied. Retention may be determined by various means, such as turbidimetric readings or measurement of colony-forming units (Smyth *et al.,* 1978) or radioactivity (Hermansson *et al.,* 1982). In many cases, salting-out agents are added to promote adhesion to the gel (Smyth *et al.,* 1978). In some cases, adherent cells can be desorbed by lowering the ionic strength of the eluent or by adding detergent (Magnusson *et al.,* 1979a). Since it has been demonstrated that bacteria can become nonspecifically trapped within the gel matrix or between the gel and column (Olsson and Westergren, 1982), the assay is preferably performed by mixing of the beads with the bacterial cells, followed by separation of free from bound cells. Moreover, controls should be performed with the corresponding untreated Sepharose beads themselves, as in some cases bacteria may bind to them as well.

3.1.2. Bacterial Adhesion to Hydrocarbons (BATH)

As already mentioned, bacterial partitioning at the oil–water interface has been studied by various investigators since 1924 (Mudd and Mudd, 1924a,b; Marshall and Cruickshank, 1973; Marshall et al., 1975), including many researchers in the field of petroleum microbiology (e.g., Neufeld et al., 1980; Kennedy et al., 1975; Kirschner-Zilber et al., 1980, Nakahara et al., 1981). In 1980, bacterial adhesion to hydrocarbons (BATH) was suggested as a simple and convenient assay for measuring bacterial hydrophobicity (M. Rosenberg et al., 1980a). In this test, washed bacterial suspensions are vortexed under controlled conditions with an aliphatic (e.g., hexadecane) or aromatic (e. g., xylene) liquid hydrocarbon. In the case of adhesion, bacteria bind to the oil droplets and rise with them following the mixing procedure, bringing about a concomitant drop in the cell density of the lower bulk aqueous phase. This technique and various modifications have been used to study bacterial hydrophobicity in a wide variety of bacterial species and cell mixtures as a function of growth conditions and other parameters (M. Rosenberg, 1984a). The basic technique has also enabled enrichment of hydrophobic mutants in Acinetobacter calcoaceticus (M. Rosenberg and Rosenberg, 1981), Serratia marcescens (M. Rosenberg, 1984c), and Streptococcus sanguis (Gibbons et al., 1983).

One of the main limitations of the method as employed thus far is its semiquantitative nature. A quantitative approach based on the kinetics of adhesion to hexadecane has recently been employed (D. Lichtenstein, M. Rosenberg, N. Sharfman, and I. Ofek, in preparation).

Results obtained using BATH usually correlate with those obtained using other techniques (M. Rosenberg, 1984a). An interesting anomaly in this regard is that enterobacterial strains, which appear to exhibit cell-surface hydrophobicity with other methods, adhere poorly or not at all to hexadecane. However, it was recently shown that adhesion of a rough Escherichia coli strain to hexadecane was promoted from 0 to 76% by adding ammonium sulfate to the bacterial suspension (M. Rosenberg, 1984b). This finding suggests that many enterobacterial strains purported to possess pronounced cell-surface hydrophobicity are actually much less hydrophobic than strains that adhere to hydrocarbons in the absence of salting-out agents, e.g., Streptococcus pyogenes, Acinetobacter calcoaceticus, and Serratia marcescens.

Several points concerning the BATH assay require further clarification: (1) Might the vortexing procedure in the presence of hydrocarbons extract hydrophobic cell-surface components into the hydrocarbon phase, altering the properties measured? (2) Morisaki (1983, 1984) has reported that hydrophobic substrata can affect bacterial metabolism. Do such effects similarly influence adhesion to hydrocarbons? (3) Does cell–cell cooperativity play a role in BATH?

Adhesion to hydrocarbons is discussed further in Section 5.2 in the context of bacterial growth on oil.

3.1.3. Salting-out Aggregation Test (SAT)

This technique, described by Lindahl et al. (1981), is based on the premise that the same laws governing precipitation of protein molecules from aqueous solution hold true for aggregation of bacterial cells, i.e., that the more hydrophobic the cell, the greater its tendency to precipitate out of the solution (i.e., through cell–cell interactions) at a lower concentration of salting-out agents. In this method, bacterial cells are suspended in dilute phosphate buffer, and ammonium sulfate is added until aggregation occurs. This technique appears to correlate with other methods in most, but not all, instances (Ferreiros and Criado, 1984). Configurational changes of cell-surface structures due to the high salt concentrations may introduce errors in measurement.

3.1.4. Adhesion to Solid Surfaces

Adhesion to solid surfaces of relatively low surface free energy has been employed to test for bacterial hydrophobicity by a variety of investigators. The ability of cells from bacterial colonies to adhere directly to polystyrene has been employed as a qualititative technique for screening hydrophobic colonies (M. Rosenberg, 1981), and, more recently, as a rapid agglutination test (Lachica and Zink, 1984b). Various investigators have used adhesion to polystyrene or similar surfaces as a quantitative measure of bacterial hydrophobicity, measuring the number of adherent cells by a variety of techniques (e.g., direct counting, measurement of absorbed stain). Adhesion to solid surfaces of varying surface energies has also been studied (Gerson and Akit, 1980; Fletcher and Loeb, 1979; Dexter et al., 1975; Dexter, 1979; Pringle and Fletcher, 1983). The latter authors observed that 16 freshly collected bacterial isolates adhering to plastic surfaces in freshwater exhibited the highest attachment to more hydrophobic surfaces. Absolom et al. (1983) have used an elegant technique (originally developed for measuring hydrophobic properties of mammalian cell surfaces) for the measurement of bacterial hydrophobicity (see Section 2). Bacteria suspended in different concentrations of

DMSO are brought into contact with smooth, solid surfaces of varying surface free energies. Using this method, these workers have been able to ascribe surface free energies to several bacteria tested. Unfortunately, the bacterial suspensions all fell within a narrow range of surface energy values; thus, it is difficult to predict whether this technique would be similarly successful with strains generally considered to be substantially more hydrophobic.

3.2. Molecular Probes

Dyar (1948), and later Hill *et al.* (1963), employed charged surfactant molecules in order to determine cell-surface hydrophobicity in staphylococcal and streptococcal cells. In such studies, small concentrations of ionic surfactant (usually negatively charged, to eliminate electrostatic interactions between positively charged probes and the net negative charge of most bacterial surfaces) were added to bacterial suspensions. Changes in the electrophoretic mobility of the cells were attributed to binding of surfactant molecules via hydrophobic interactions to the cell surface.

This principle was employed by Kjelleberg *et al.* (1980), who measured binding of radiolabeled dodecanoic acid to bacterial cells. Analysis of the results from several strains indicated that bacteria may differ in the number of binding sites available for attachment of hydrophobic molecules as well as in the affinity of such bonds. This method can yield important information concerning the nature and number of hydrophobic binding sites on the cell surface. A similar approach using palmitic acid has been proposed by Malmqvist (1983). The main concern in the use of such hydrophobic probes is their small size, with the possibility that they can intercalate within the cell envelope, thus binding to sites that are not normally exposed at the outermost cell surface.

3.3. Contact Angle Measurements (CAM)

Contact angle measurements are routinely employed by surface scientists to measure surface properties of clean, smooth, flat, dry, solid surfaces. Small drops of liquid are applied to the surface under controlled conditions and the angle of contact recorded directly. Recently, van Oss and coworkers have proposed that such measurements, conducted on bacterial layers, can provide quantitative information on the hydrophobicity of the individual cells (van Oss and Gillman, 1972; Cunningham *et al.*, 1975; van Oss, 1978), including a value for the surface energy of the bacteria measured (Absolom *et al.*, 1983). Potential drawbacks of the methodology include: (1) the cells must remain moist, but not wet; thus, results must be obtained within a given time during the drying out process of the layer; (2) the surface, being composed of individual cells, is not flat; and (3) it is not known whether the surface presented by the outermost bacterial layer (facing upward) provides an overall picture or is the result of an alignment of cells, perhaps each projecting a patch of increased or decreased hydrophobicity. Despite these reservations, the results obtained using this method have been corroborated by other, independent techniques (Absolom *et al.*, 1983). Fletcher and Marshall (1982a) have introduced a modification (albeit, for studying substrata) in which the contact angle of air bubbles surrounded by water on the inverted cell surface is measured. This method was employed by Fattom and Shilo (1984) to study the hydrophobicity of Cyanobacteria. Similarly, Neufeld *et al.* (1980) have suggested replacing the air surrounding the water droplet by oil. Whereas most studies have employed drops of water or saline, Busscher *et al.* (1984) calculated the surface energies of oral bacteria based on contact angles of water, water–in-propanol mixtures, and α-bromonapthalene.

3.4. Two-Phase Partitioning (TPP)

Several decades ago, Albertsson (1958) proposed a method for particle separation based on partitioning between two polymeric aqueous phases that are mutually immiscible, such as dextran and polyethylene glycol. Later, this technique was used to separate bacterial cells; e.g., smooth and rough *Salmonella typhimurium* cells [reviewed in Edebo *et al.* (1980)]. Hydrophobicity of the polyethylene glycol phase can be enhanced by the covalent binding of acyl groups (Magnusson and Johansson, 1977), enhancing the affinity of hydrophobic bacteria for this phase as compared to the dextran layer. One possible problem associated with this method is that a portion of the bacteria may become trapped between the two phases (Sacks and Alderton, 1961), complicating recovery and interpretation. Nonetheless, this technique may be viewed as being relatively nondestructive, and avoids the use of high concentrations of salts or organic solvents. Gerson and Akit (1980) reported linear correlations between the TPP and CAM methods, enabling them to grade the relative hydrophobicity of several bacterial species.

3.5. Indirect Methods

Hydrophobic interactions are inhibited by a wide range of molecules, including those that exhibit surface-active and/or chaotropic activities. In some cases, a role for hydrophobic interactions in the mediation of adhesion phenomena may be invoked based on the ability of such compounds to inhibit or reverse adhesion (Erne *et al.*, 1984; Nesbitt *et al.*, 1982). For example, adhesion of *Thiobacillus* to sulfur granules is inhibited by sodium dodecyl sulfate (Bryant *et al.*, 1984). Adhesion of *Pseudomonas aeruginosa* to erythrocytes is inhibited by chaotropic agents, but not by various sugars, suggesting the involvement of hydrophobic interactions (R. J. Doyle, personal communication). Inhibition of bacterial autoaggregation and coaggregation by detergents has underlined the importance of hydrophobicity in these phenomena (McIntire *et al.*, 1982; Stanley and Rose, 1967). Surface-active alcohols are often potent adhesion inhibitors (M. Rosenberg *et al.*, 1980a; Fletcher, 1983). E. Rosenberg *et al.* (1983a) have proposed that the majority of attached normal microbiota on the surface of human buccal epithelial cells adhere via hydrophobic interactions, since they are desorbed by low concentrations of emulsan, an amphipathic polymer.

4. Surface Components that Influence Bacterial Hydrophobicity

In preparing this section, we encountered the problem of how to denote surface components that promote bacterial hydrophobicity. To call them "adhesins" would presuppose a role in adhesion. In at least some cases, this is apparently not the case (Davison and Sanford, 1982; Kozel, 1983; Sobel and Obedeanu, 1984). We have thus elected to use the term *hydrophobins* to denote cell-surface components that contribute to cell-surface hydrophobicity; similarly, surface components that reduce cell-surface hydrophobicity have been termed *hydrophilins*.

Investigations into the hydrophobic surface properties of *Acinetobacter calcoaceticus* RAG-1 illustrate the possible complexity involved in determining the participating cell-surface components. Several lines of evidence establish a role for thin fimbriae as hydrophobins in RAG-1: (1) MR-481, a nonhydrophobic mutant of RAG-1, is deficient in thin fimbriae, (2) partial adhesion revertants possess numerous thin fimbriae, (3) shearing of fimbriae reduces hydrophobicity in RAG-1 cells grown on minimal medium, and (4) strain AB-15, an independent, thin fimbria-less mutant, is nonadherent when grown on minimal medium (M. Rosenberg *et al.*, 1982a). Thin fimbriae, however, are not the whole story, since (1) AB-15 cells are hydrophobic following growth on complex medium (despite their total lack of thin fimbriae); and (2) crossed immunoelectrophoresis of supernatants following sonic oscillation of complex-medium-grown cells reveals that the nonhydrophobic mutant MR-481 lacks at least three surface antigens, in addition to the thin fimbriae, when compared with wild-type RAG-1 cells (E. A. Bayer and M. Rosenberg, unpublished data). These results suggest that thin fimbriae are the main hydrophobin of RAG-1 cells grown on minimal medium, but that additional factor(s) contribute to the hydrophobicity of complex-medium-growth cells. To complicate matters further, the emulsan minicapsule of RAG-1 acts as a hydrophilin (E. Rosenberg et al., 1983b). Complete loss of cell-surface emulsan leads to exposure of hydrophobins, which are apparently masked in wild-type cells (Pines and Gutnick, 1984).

In the light of such investigations, the presumption that cell-surface hydrophobicity is due to a single factor may often be premature. In some instances [e.g., Aeromonas salmonicida (Trust et al., 1983, N. D. Parker and Munn, 1984)], bacterial hydrophobicity may indeed be due to a single hydrophobin, which dominates the outer cell surface. This, however, may be the exception rather than the rule. Bacterial hydrophobicity may often involve a delicate interplay of charged and uncharged, polar and apolar groups, exposed to varying degrees, and with varying densities over the cell surface (Kjelleberg, 1984b). Moreover, it is impossible to determine precisely where the cell surface ends and the surrounding aqueous phase begins. On a molecular level there is no sharp boundary at the outermost cell surface, and water molecules can penetrate the outer bacterial layers to varying extents. In some cases, surface projections, such as various fimbriae, fibrils, holdfasts, and so on, may themselves mediate adhesion to hydrophobic surfaces (Marshall and Cruickshank, 1973), in which case the cell itself may be tethered to the surface.

Components of the bacterial cell surface usually vary greatly as a function of growth conditions, including parameters such as aeration, temperature of growth, growth medium, and age of cells. Thus, whereas some investigators originally proposed that bacterial strains could be assigned cell-surface hydrophobicity values independent of growth conditions (van Oss and Gillman, 1972), more recent studies have shown that hydrophobic surface properties vary greatly as a function of physiological conditions (M. Rosenberg, et al., 1980a, 1981; Kjelleberg et al., 1980; Ofek et al., 1983; Nesbitt et al., 1982; Rogers et al., 1984). The dramatic increase in hydrophobicity of A. calcoaceticus RAG-1 cells over the course of several generations of growth is illustrated in Fig. 3. Starvation effects also have a large influence on cell-surface hydrophobicity (Kjelleberg, 1984a; Kjelleberg et al., 1983; Kjelleberg and Hermansson, 1984). The large differences in hydrophobicity of starved versus fed natural populations are shown in Table II. Finally, it must be emphasized that cell-surface hydrophobicity varies not only from species to species, but also from strain to strain (M. Rosenberg, 1982; Olsson and Westergren, 1982). Thus, within a given species, various combinations of hydrophobins may be responsible for the hydrophobic properties observed. Indeed, this may account for arguments such as whether the hydrophobicity of group A streptococci is mediated by lipoteichoic acid (Miorner et al., 1983, 1984),

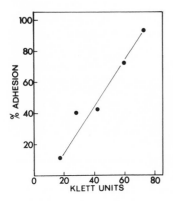

Figure 3. Adhesion to hexadecane as a function of growth. *Acinetobacter calcoaceticus* RAG-1 cells were grown in modified nutrient broth with shaking at 30°C. At various growth culture turbidities samples were removed. Cells were washed and assayed for adhesion to 0.2 ml of hexadecane as described (M. Rosenberg *et al.*, 1980a). Results are expressed as percent adhesion to hexadecane as a function of growth culture turbidity.

M protein (Tylewska *et al.*, 1979), or both (Courtney *et al.*, 1983; Ofek *et al.*, 1983; Wadstrom *et al.*, 1984); all these possibilities, and more, may be found in various strains.

The dependence of hydrophobic properties on physiological conditions can often be correlated to the synthesis, or degree of exposure, of hydrophobins and hydrophilins, as in the case of *Streptococcus pyogenes* (Ofek *et al.*, 1983). The presence of plasmids may similarly affect cell-surface hydrophobicity by allowing expression of fimbriae and other surface components (Lindahl *et al.*, 1981; Lachica and Zink, 1984a; Martinez, 1983; Ferreiros and Criado, 1984). Finally, mechanical, enzymatic, or other destructive treatments of the cell surface can alter hydrophobicity. These include studies in which buried hydrophobins are unmasked (e.g., Mackey, 1983), resulting in increased hydrophobicity, as well as those in which hydrophobins are removed, e.g., by proteolytic enzymes (Ofek *et al.*, 1983) or mechanical means (M. Rosenberg *et al.*, 1982a). The development of cell-surface hydrophobicity in *Escherichia coli* and *Sal-*

Table II. Adhesion to Octyl–Sepharose Beads and Hexadecane of the Total Microbiota of the Colon Content of Fed and Starved Flounder[a]

	Percent adhesion[b]	
Animals	To octyl–sepharose	To hexadecane
Fed	3	16
Starved	74	53

[a]P. L. Conway, J. Maki, R. Mitchell, and S. Kjelleberg (unpublished results).
[b]Percentage adhesion was measured following mixing and equilibration. Results are expressed as the percentage drop in the number of the cells remaining in the aqueous phase following the respective assays.

monella typhimurium following invasion by *Bdellovibrio* cells is an example of bacteria-inflicted damage resulting in changes in surface hydrophobicity (Cover and Rittenberg, 1984).

Several cell-surface components identified as hydrophobins in various bacteria are listed in Table III. The following sections deal with some of these examples in greater detail.

4.1. Fimbriae and Surface Fibrils

Fimbriae (or surface fibrils) have often been cited as hydrophobins. In some cases [e.g., the thick fimbriae of *Acinetobacter calcoaceticus* RAG-1 (M. Rosenberg *et al.,* 1982a)] fimbriae play no role in modulating cell-surface hydrophobicity and may even act as hydrophilins, as in the

Table III. Some Proposed Bacterial Hydrophobins

Component	Bacteria	Reference
Thin fimbriae	*Acinetobacter calcoaceticus* RAG-1	M. Rosenberg *et al.* (1982a)
M protein, lipoteichoic acid, or both	Group A streptococci	Ofek *et al.* (1983), Miorner *et al.* (1983), Wadstrom *et al.,* (1984), Tylewska *et al.* (1979)
A protein	*Staphylococcus aureus*	Jonsson and Wadstrom (1983, 1984)
"A layer" protein	*Aeromonas salmonicida*	Trust *et al.* (1983), N. D. Parker and Munn (1984)
Prodigiosin	Pigmented *Serratia* strains	Blanchard and Syzdek (1982), Hermansson *et al.* (1979), Kjelleberg *et al.* (1980)
Glucosyltransferase	*Streptococcus mutans*	D. Oakley and R. J. Doyle (unpublished results)
Outer membrane proteins	*Vibrio cholerae*	Kabir and Ali (1983)
Surface fibrils	*Streptococcus sanguis* *Yersinia enterocolitica*	Gibbons *et al.* (1983) Lachica and Zink, (1984 a,b), Lachica *et al.* (1984)
Various fimbriae	*Escherichia coli*	Faris *et al.* (1981), Smyth *et al.* (1978), Honda *et al.* (1984), Jann *et al.* (1981), Svanborg-Eden *et al.* (1984), Ohman *et al.* (1982)
Core oligosaccharides or outer membrane lipids	Rough enterobacterial strains	Edebo *et al.* (1980), M. Rosenberg *et al.* (1982b)
Gramicidin S	*Bacillus brevis* spores	E. Rosenberg, *et al.,* 1985.

case of *Serratia* (Hermansson *et al.,* 1982). All, or almost all, of the fimbriae examined to date contain a high proportion of hydrophobic amino acid residues (Jones and Isaacson, 1983). This, however, does not necessarily imply that all fimbriae are hydrophobins, since their ability to undergo hydrophobic interactions would depend on the spatial disposition of these residues (i.e., exposure toward the bulk aqueous phase) rather than their total number. Certainly, among the fimbriae reported to act as hydrophobins, great variations are observed in the cell-surface hydrophobicity that they confer. This, in turn, may also be a function of the number of fimbriae per cell and their topographic distribution.

Of particular interest in this regard are the hydrophobic interactions attributed to mannose-sensitive ("type 1") and various mannose-resistant fimbriae of enterobacterial strains. Although fimbriated *Escherichia coli* strains are usually more hydrophobic than nonfimbriate counterparts, the degree of hydrophobicity that they confer is low. For example, the adhesion of fimbriae-bearing enterobacterial strains in HIC is largely dependent on the presence of high concentrations of salting-out agent (Jann *et al.,* 1981; Ohman *et al.,* 1982; Faris *et al.,* 1981). Moreover, whereas certain fimbriae can promote adhesion to hydrocarbons and polystyrene (M. Rosenberg *et al.,* 1982a), various fimbriated *Escherichia coli* strains do not adhere to either surface (T. Zusman and E. Ron, unpublished results; M. Rosenberg, 1982). Such observations may, however, understate the role of hydrophobic interactions (however weak) on attachment of enterobacteria to host cells, as described in Section 5.7.

The presence of fimbriae is related to bacterial adhesion in natural systems (Pertsovskaya *et al.,* 1973; Romanenko, 1979). Furthermore, it was recently found that hydrophobic fimbriae-bearing *Salmonella typhimurium* cells adhered in higher numbers to several mineral particles than nonfimbriated, less hydrophobic cells (Stenstrom and Kjelleberg, 1985).

4.2. Role of Capsule

In almost all studies [with the exception of some encapsulated staphylococci, according to Hogt *et al.* (1983a)] the presence of bacterial capsule or slime has been shown to decrease surface hydrophobicity. For example, enzymatic removal of the capsule of *Acinetobacter calcoaceticus* BD4 led to an increase in adhesion to hexadecane from 0 to 100% (E. Rosenberg *et al.,* 1983b). Cell-surface capsule reduces cell-surface hydrophobicity in a *Phormidum* sp. (Fattom and Shilo, 1984), and was reported to lower phagocytosis of *Staphylococcus aureus* by rendering the cells more hydrophilic (van Oss and Gillman, 1972). Adhesion of *Streptococcus pyogenes* to hexadecane corresponds to the absence of hyaluronic acid-containing capsule (Ofek *et al.,* 1983). A mutant of a freshwater *Pseudomonas fluorescens* strain producing an exopolysaccharide dominated the aqueous phase, whereas the non-exopolysaccharide-producing wild type was foam-fractionated by the adsorption bubble separation process in a fermentor (Pringle *et al.*, 1983). Recently, Runnels and Moon (1984), employing several strains of encapsulated, fimbriated *Escherichia coli* and their noncapsular mutants, showed that the presence of capsule impeded K99-mediated adhesion.

Rough mutants of smooth enterobacterial strains, such as *Salmonella typhimurium* (Edebo *et al.*, 1980), *E. coli* (M. Rosenberg *et al.*, 1980a; M. Rosenberg, 1984b), and *Proteus mirabilis* (M. Rosenberg *et al.*, 1982b), are more hydrophobic, presumably due to loss of surface oligosaccharides.

Such results suggest the antiadhesive role of capsule, and do not appear to concur with various studies that claim a major role for capsule in mediating adhesion to a wide variety of substrata (Costerton *et al.*, 1981). However, since the latter experiments are usually conducted over the course of several hours or days, the copious amounts of polysaccharide observed at attachment sites may or may not be involved in the initial attachment process. We therefore suggest that whereas capsules may act to bridge and cement layers of adhering bacteria to substrata, they are not necessarily involved in initial adhesion processes. This, however, does not apply for polysaccharide layers undergoing gel formation (Rutter, 1980).

4.3. Prodigiosin and Serratial Hydrophobicity

Prodigiosin is the cell-surface pigment that accounts for the red coloration of *Serratia* strains. Various investigators have found correlations between the appearance of prodigiosin and cell-surface hydrophobicity (M. Rosenberg *et al.*, 1980a; Kjelleberg *et al.*, 1980; Blanchard and Syzdek, 1978; Hermansson *et al.*, 1979; Syzdek, 1985), implicating this molecule as a hydrophobin in pigmented strains. However, several studies have shown that nonpigmented strains and mutants may also be substantially hydrophobic (Hermansson *et al.*, 1979; M. Rosenberg, 1984c; M. Rosenberg and Y. Blumberger, unpublished data). Moreover, a pigmented, nonhydrophobic *S. marcescens* mutant was recently isolated (M. Rosenberg, 1984c). This mutant differs from wild-type cells in the increased translucence of its colonies, suggesting that factor(s) conferring colony opacity are additional (and perhaps crucial) hydrophobins in pigmented cells, and may similarly contribute to cell-surface hydrophobicity in nonpigmented strains. In contrast to reports linking mannose-sensitive adhesins with cell-surface hydrophobicity in *E. coli* (Jann *et al.*, 1981; Ohman *et al.*, 1982), no relationship was found between mannose-sensitive aggregation of yeast cells and hydrophobicity in clinical *Serratia* isolates (M. Rosenberg, and Y. Blumberger, unpublished results).

4.4. Cell-Surface Hydrophobicity in Bacterial Spores

Increased hydrophobicity of *Bacillus* spores as compared to that of vegetative cells has been shown by increased adhesion to hydrocarbon (M. Rosenberg, 1984a; Doyle *et al.*, 1984) and partitioning in an aqueous two-phase system (Sacks and Alderton, 1961) and at the air–water interface (Boyles and Lincoln, 1958; Gaudin *et. al.*, 1960). Cell-surface hydrophobicity in spores may provide a biological advantage in enabling their dispersal via aerosols (see also Section 5.3). However, increased hydrophobicity of spore versus vegetative cells is not universal: Kupfer and Zusman (1984) have recently reported that sporulating *Myxococcus xanthus* cells become less and less hydrophobic as sporulation proceeds.

A possible role for the antibiotic gramicidin S in mediating the hydrophobicity of *B. brevis* spores has been recently demonstrated (E. Rosenberg *et al.*). Whereas wild-type spores, which produce the antibiotic, were hydrophobic, gramicidin S-negative mutant spores were not; however, extraneous addition of gramicidin S increased the hydrophobicity of the mutant spores in a dose-dependent manner. These data suggest that the cationic moiety of gramicidin S binds to anionic receptors on the spore surface, exposing the hydrophobic amino acid residues of the antibiotic molecule on the cell surface.

5. Hydrophobicity and Adhesion in the Environment and Host

5.1. Introduction

In the present section we deal with the role of hydrophobic interactions in bacterial adhesion to a variety of surfaces in the open environment as well as in the host. Whereas most *in vitro* investigations deal with adhesion of washed bacterial suspensions to clean surfaces, the *in situ* situation is usually quite different. Interfaces in aqueous surroundings are rapidly coated with a "conditioning" layer of adsorbed macromolecules and lipids (Fletcher, 1976); Fletcher and Marshall (1982a) proposed a useful method for *in vitro* measurement of the interfacial properties of such films, which may profoundly affect bacterial adhesion (Kjelleberg and Stenstrom, 1980). Morever, bacteria (particularly within animal hosts) adsorb, whether specifically or not (Miorner *et al.*, 1980; Absolom *et al.*, 1982), antibodies and other macromolecules (Wadstrom *et al.*,

1984). This, in turn, either promotes or inhibits the ability of the cell to attach, depending on the cell-surface hydrophobicity of the cell prior to adsorption and the nature of the adsorbing molecules (van Oss, 1978). The *in situ* coating of the bacterial surface and the attachment substratum is generally difficult to estimate, making it hard to extrapolate from laboratory experiments to *in situ* adhesion. Several strategies in this direction would include the following: (1) Comparision of the ability of laboratory strains to adhere to the actual surface under study as a function of their relative hydrophobic surface properties. One such study by Svanberg *et al.* (1984) demonstrated that hydrophobic *Streptococcus mutans* is better able to attach to and colonize oral surfaces of human volunteers as compared to a nonhydrophobic variant; (2) attempting to desorb attached bacteria by means of agents that interfere with the mechanism under study; and (3) direct examination of the adherent properties of bacteria isolated directly from the niche under study. This last point is of great importance in the light of experiments showing that cell-surface hydrophobicity may be dependent on the degree of starvation of cells (Kjelleberg and Hermansson, 1984), strain history (Westergren and Olsson, 1983), and other factors that differ widely between cells in the environment and in the test tube. With few exceptions (e.g., M. Rosenberg *et al.*, 1983c), the cell-surface hydrophobicity of bacteria taken directly from nature and tested without further growth has not been investigated. This is unfortunate, since such experiments are not difficult to perform (some results on hydrophobicity of bacteria taken directly from gastrointestinal systems appear in Table II and in Section 5.6).

5.1.1. "Specific" versus "Nonspecific" Hydrophobicity

Various investigations attest to the nonspecific nature of bacterial adhesion to low-energy surfaces. For example, *A. calcoaceticus* RAG-1 adheres to liquid and solid hydrocarbons, plastic surfaces, human teeth, buccal epithelial cells and air bubbles, exposed metal surfaces, and so on (M. Rosenberg, 1982, and unpublished observations). Thus, it is not surprising that many researchers agree that hydrophobic interactions that mediate bacterial adhesion are not dependent on molecular complementarity between bacterial surface components and receptors on the attachment substratum or on the "molecular texture" of the two surfaces. However, the possibility has been raised that in adhesion to host tissue, specific recognition mechanisms between hydrophobic moieties and stereospecific receptors do occur (Ofek *et al.*, 1983; Simpson *et al.*, 1980), much in the same way that serum albumin binds hydrophobic molecules. Thus, in the case of adhesion of *S. pyogenes* to buccal epithelial cells, specific bacterial hydrophobins (e.g., M protein–lipoteichoic acid complex) would be recognized by epithelial cell receptor(s) (e.g., fibronectin) and adhesion mediated by this specific mechanism (Abraham et al., 1983; Courtney et al., 1983). It is also possible that hydrophobic interactions, in addition to their ability to mediate adhesion per se, may also contribute to stereospecific adhesion mechanisms. In the case of mannose-sensitive adhesion, for example, great importance may be attributed to apolar moieties in the immediate proximity of the mannose-binding site. Firon et al. (1982) have reported that mannosides of an increasingly hydrophobic nature possess much higher affinity for the binding site, similar to results obtained with the lectin concanavalin A (Lis and Sharon, 1977). Similarly, the presence of mannose has been reported to decrease surface hydrophobicity of E. coli bearing type 1 fimbriae, presumably due to some conformational change at the sugar-binding site (Ohman, 1983). Such data suggest that neighboring hydrophobic interactions may promote mannose-sensitive binding in E. coli strains.

5.2. Adhesion and Growth at the Solid–Liquid Interface

The possible role of hydrophobicity and adhesion in the life cycle and survival of bacteria in the environment is discussed in the present section.

Adhesion as a tactic in starvation survival was proposed by Dawson et al. (1981). More recently, the existence of an increased degree of irreversible binding and hydrophobicity upon starvation has been demonstrated for some bacterial species (Kjelleberg and Hermansson, 1984); in other bacteria tested, this trend was not observed (Kjelleberg et al., 1985a). Other researchers (Ellwood et al., 1982), studying a diverse microbial community, showed a correlation between carbon limitation and attachment. In this context, it should be emphasized that each organism has its own specific starvation-survival pattern (R. Y. Morita and R. Jones, personal communication).

The role of interfaces as areas of nutrient supply in nutrient-deprived ecosystems has been highlighted (Marshall, 1979). The question of whether bacterial activity at such surfaces is affected by the surface per se, by nutrients at the interface, or both, is currently under consideration by various investigators (e.g., Morisaki, 1983, 1984; Jeffrey and Paul, 1985; Fletcher, 1985; Gordon and Millero, 1985). It is noteworthy, however, that the scavenging of surface-localized long-chain fatty acids was favored by a hydrophobic S. marcescens strain as compared with its hydrophilic mutant (Kefford et al., 1982). These authors also showed that reversibly bound bacteria, attaching and rapidly desorbing, scavenge surface-localized material. This utilization of surface-immobilized substrates has more recently been shown to be a common feature for a wide range of natural marine isolates (Hermansson and Marshall, 1985); whereas irreversibly bound cells exhibited more efficient utilization, reversibly bound bacteria, which are prevalent in the environment, were also able to utilize surface-immobilized materials.

Based on several investigations (e.g., Dawson *et al.*, 1981; Kjelleberg *et al.*, 1982; Kjelleberg and Hermansson, 1984; Hermansson and Dahlback, 1983; Pedros-Alio and Brock, 1983), Dahlback (1983) suggested a life cycle in which planktonic, starved hydrophobic bacteria adhere readily to available interfaces, efficiently scavenge surface-localized materials, increase in biomass and size, and then leave the interface, usually in the form of a daughter cell following division (Kjelleberg *et al.*, 1982) to face the nutrient-deprived conditions in the planktonic phase. The mechanism for bacterial desorption from the interface may involve changes in cell-surface hydrophobicity or, alternatively, localized bacterial production of amphipathic materials that remove bound cells (E. Rosenberg *et al.*, 1983a).

5.3. Adhesion and Growth at the Oil–Water Interface

Adhesion of bacteria and yeast to hydrocarbons has often been observed by microbiologists interested in the field of oil degradation, but such investigators considered this property to be specific to oil-degrading microorganisms (Kennedy *et al.*, 1975; Miura *et al.*, 1977), ignoring several previous investigations (Mudd and Mudd, 1924a; Marshall and Cruickshank, 1973). M. Rosenberg *et al.* (1980a,b) showed that oil-degrading bacteria may or may not adhere to hydrocarbons, and that nondegraders (e.g. *Serratia marcescens, Staphylococcus aureus*) may adhere in high proportions despite their inability to metabolize hydrocarbon. These results suggested the use of bacterial adhesion to liquid hydrocarbons as a general technique for determining cell-surface hydrophobicity (Section 3.1.2) (M. Rosenberg *et al.*, 1980a,b.) Another interesting result of this study was that adhesion of *Acinetobacter calcoaceticus* RAG-1 is not limited to hydrocarbons that it metabolizes. Later studies (M. Rosenberg and Rosenberg, 1981; M. Rosenberg *et al.*, 1982a) addressed the possible role of adhesion in enabling growth of RAG-1 on hexadecane. RAG-1 cells grew readily on hexadecane under conditions of limited hydrocarbon emulsification (low initial cell density and mild agitation), whereas its nonadherent mutant, MR-481, did not grow unless emulsifier was added. Moreover, prolonged incubation of MR-481 cells in hexadecane growth medium led to the eventual growth of increasingly adherent revertants (M. Rosenberg *et al.*, 1982a; Pines and Gutnick, 1984). In a comparison of the adhesion of RAG-1 with that of an oil-degrading strain of *Pseudomonas aeruginosa*, the latter strain exhibited extremely poor affinity for hydrocarbons, even following growth on hexadecane (M. Rosenberg *et al.,* 1980b; Nakahara *et al.,* 1981). These results suggest that nonadherent strains of bacteria that grow on hydrocarbons in the laboratory (i.e., in closed systems that enable emulsification of the insoluble substrate) will not grow well on oil *in situ.* In the open environment, adhesion to oil is frequently observed (e.g., Wyndham and Costerton, 1982); however, emulsification of oil droplets in the open waters would allow them to diffuse away from the cells. Some investigators (working primarily with *Pseudomonas* strains) have argued that adhesion is not an important factor in growth on oil (Reddy *et al.,* 1982; Chakravarty *et al.,* 1975). This may well be true in the case of growth on aromatic hydrocarbons, which are generally more water-soluble and more toxic than paraffins.

An interesting aspect of growth on hydrocarbons is the relationship between bound and free cells. Kirschner-Zilber *et al.* (1980) studied adhesion and growth of a marine pseudomonad on solid *n*-tetracosane; the investigators proposed that individual cells could alternate between the water–paraffin interface and the bulk aqueous phase, with a continual flux of adhering and desorbing cells. The appearance of free *A. calcoaceticus* RAG-1 cells during growth on hexadecane is not due to their inability to adhere to hexadecane, but probably results from their desorption by emulsan, the amphipathic polysaccharide released by the cells (M. Rosenberg and Rosenberg, 1981; E. Rosenberg *et al.,* 1983a). It has been shown (M. Rosenberg, 1982) that bacteria adhering to hexadecane can be desorbed following freezing and thawing of the hydrocarbon phase (Fig. 4). This phenomenon may reflect a basic difference between adhesion at deformable (liquid–liquid and liquid–gas) and relatively nondeformable (liquid–solid) surfaces (P. Rutter, personal communication). Assuming that bacterial adhesion at the oil–water phase does entail gross deformation of the interface [as proposed by Mudd and Mudd (1924b)], the phase transition from liquid to solid hexadecane may reduce the surface available for interaction and thus decrease the energetic favorability of adhesion, with the resultant desorption of cells. The observation that oils readily remove bound bacteria from polystyrene (M. Rosenberg *et al.,* 1983b) may be related to preferential partitioning of bacteria at liquid–liquid rather than liquid–solid interfaces.

Another interesting observation is that bacteria often tend to adhere to oil droplets in patches, both during growth on oil (E. Rosenberg and M. Rosenberg, unpublished data) and following the adhesion assay. Such patches may arise from cell–cell interactions at the interface. Further evidence for this possibility is that bacteria desorbed from hexadecane following freezing and thawing of the hydrocarbon phase are recovered in the bulk aqueous phase as aggregates, rather than individual cells (M. Rosenberg, 1982).

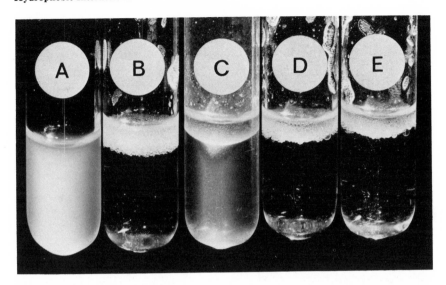

Figure 4. Release of cells bound to hexadecane due to phase transition of hydrocarbon. RAG-1 cells were tested for adhesion to hexadecane (tubes B and C) and octane (tubes D and E) as previously described (M. Rosenberg *et al.*, 1980a). Following the adhesion assay, test tubes C and E were cooled to 12°C for 3 min, then allowed to equilibrate with room temperature (22°C). In the tube containing RAG-1 cells bound to hexadecane (tube C), the hexadecane droplets solidified, and following equilibration at room temperature, melted and coalesced, with concomitant release of clumped RAG-1 cells back into the aqueous phase. No such effect was observed in a similarly treated tube (tube E) containing cells bound to octane (which did not undergo phase transition). Test tube A contains the washed bacterial suspension prior to the adhesion assay. [From M. Rosenberg (1982).]

5.4. Adhesion at Gas–Liquid Interfaces

Mechanisms for adhesion at gas–liquid interfaces have been recently reviewed (Kjelleberg, 1984b, 1985). It appears that cell-surface hydrophobicity is the prime (Hermansson *et al.,* 1982), although not the sole (Kjelleberg and Stenstrom, 1980), criterion in determining attachment at the air–water interface under conditions simulating those present in the environment. A large number of bacteria adhering to the gas–liquid interface of marine waters were tested and found to be significantly more hydrophobic than free-living bulk water bacteria (Dahlback *et al.,* 1980, 1981). Gas–liquid interfaces are not confined in nature to the surface microlayers of oceans and lakes (Norkrans, 1980), but include bubbles rising through the water column and aerosol production by bursting bubbles (Blanchard and Syzdek, 1978, 1982; Bezdek and Carlucci, 1974; B. C. Parker *et al.,* 1983; Wendt *et al.,* 1980; Weber *et al.,* 1983). The significance of such aerosols as microbial carriers has long been recognized (e.g.,

Woodcock, 1948). Not all bacteria are aerosolized with equal efficiency (Hejkal *et al.*, 1980); mediating factors include the sizes of both cell and bubble, and the cell-surface hydrophobicity (Blanchard, 1983). In the laboratory, adhesion at the air–water interface has proven an efficient means of separating bacteria (Boyles and Lincoln, 1958; Gaudin *et al.*, 1960; Rubin *et al.*, 1966), viruses (Morrow, 1969), and yeast (Ouchi and Akiyama, 1971; Iimura *et al.*, 1980) based on relative hydrophobicity. The latter investigators showed correlations between hydrophobicity and surface pellicle formations. Neufeld and Zajic (1984) have recently demonstrated that partitioning of washed suspensions of *A. calcoaceticus* cells at the air–water interface is accompanied by a significant decrease in surface tension. This finding may lead to an additional means for measuring bacterial hydrophobicity.

5.5. Life Style of Cyanobacteria

Fattom and Shilo (1984) have demonstrated a striking correlation between the mode of growth of aquatic cyanobacteria and their ability to adhere to hydrocarbon. Whereas benthic strains, which grow attached to submerged solid surfaces (sediments, stones, plant surfaces), adhered to hexadecane, free-floating (planktonic) types were invariably nonadherent. The hydrophobic surface properties of benthic cyanobacteria were further demonstrated using other techniques for studying cell-surface hydrophobicity: bubble contact angles on cell layers and adhesion to phenyl–Sepharose beads and to polystyrene. Adhesion of benthic cyanobacteria to hexadecane was dependent on the presence of divalent cations in the assay buffer. Fattom and Shilo also isolated mutants of the hydrophilic planktonic strain *Spirulina platensis* that exhibited increased hydrophobicity. Of particular interest are induced changes in hydrophobicity in *Phormidium* that can be correlated ultrastructually to changes in the cell surface (Y. Bar-Or, M. Kessel, and M. Shilo, in preparation).

5.6. Role of Hydrophobic Interactions in the Oral Cavity

The oral cavity has several advantages over other host sites in the study of adhesion phenomena: (1) it is more readily accessible than most other host surfaces; (2) it is usually very heavily colonized with an extremely wide range of adherent bacteria; and (3) since the majority of oral microbiota do not flourish outside the oral cavity, the crucial role of adhesion is patently self-evident (Gibbons, 1980).

Microscopic observation of the adherent normal microbiota of the human mouth reveals that each oral surface is colonized by a different pattern of bacteria. For some time, it was believed that adhesion in the oral cavity was mediated by specific lectin-like recognition mechanisms (Gibbons, 1980) and that such specific adhesion correlated to colonization patterns (Gibbons *et al.*, 1976). However, there has been little success in demonstrating such mechanisms [one exception is the lactose-specific coaggregation of oral bacteria (McIntire *et al.*, 1982]. During the past 4 years, the possibility that nonspecific hydrophobic interactions play a major role in oral adhesion has been subjected to extensive experimentation. Some of the main results may be summarized as follows: (1) Oral bacteria are predominantly hydrophobic, including fresh oral isolates from the tooth surface (Weiss *et al.*, 1982), cariogenic laboratory strains (Nesbitt *et al.*, 1982, Olsson and Westergren, 1982; Gibbons and Etherden, 1983), and plaque bacteria obtained and tested directly from the tooth surface (M. Rosenberg *et al.*, 1983c). Some instances in which pronounced cell-surface hydrophobicity was not observed may be due to excessive strain transfer (Westergren and Olsson, 1983) or choice of growth conditions (Rogers *et al.*, 1984). (2) Adhesion to the model tooth surface made up of saliva-coated hydroxyapatite (s-HA) correlates with cell-surface hydrophobicity in various oral strains tested (Nesbitt *et al.*, 1982; Gibbons and Etherden, 1983; Westergren and Olsson, 1983. (3) The inability of nonhydrophobic mutants to adhere to hydrophobic test surfaces, such as hydrocarbons or polystyrene, correlates with their inability to adhere to human teeth (Weiss *et al.*, 1985) or s-HA (Gibbons *et al.*, 1983) compared to the adherent wild-type cells. (4) A hydrophobic fresh isolate of *S. mutans* is retained in the human oral cavity, whereas a nonhydrophobic variant is not (Svanberg *et al.*, 1984). Moreover, the hydrophobic strain proved more cariogenic in laboratory animals than the nonhydrophobic variant (J. Olsson, personal communication). (5) Emulsan, a bacterial amphiphile which blocks adhesion to hydrocarbons, similarly inhibits adhesion of oral and nonoral strains to human buccal epithelial cells and desorbs most of the adherent oral microbiota as well (E. Rosenberg *et al.*, 1983a). (6) Bacteria obtained directly from dental plaque adhere to human buccal epithelial cells *in vitro;* this adhesion is inhibited by the surfactants sodium dodecyl sulfate and Tween 80 (M. Tal, E. Weiss, and M. Rosenberg, unpublished data).

The critical role that saliva probably plays in influencing oral adhesion via hydrophobic interactions is a complex one. On a thermodynamic level, the surface tension of saliva would be expected to influence adhesion to all oral surfaces bathed by it (Busscher *et al.*, 1984). On a molecular level, salivary glycoproteins, antibodies, and other agglutinins can coat the bacterial surface, the substratum, or both, thus affecting adhesion; varying surface tensions and concentrations of such components from subject to subject complicate matters further (Beighton, 1984;

Abbott and Hayes, 1984). In light of such complexities, simpler model systems may prove preferable to s-HA in understanding saliva-mediated effects.

5.7. Adhesion to Other Host Surfaces

Much investigation into adhesion to host tissue has been directed at the gastrointestinal (GI) tract. As in the case of the oral cavity, there is ample evidence to suggest that hydrophobic interactions play a greater role in adhesion than is generally recognized: (1) Certain surfaces in the GI tract are extremely hydrophobic (Hills *et al.,* 1983); moreover, the mucus layer coating intestinal surfaces possesses hydrophobic properties (Smith and LaMont, 1984). (2) A large proportion of hydrophobic bacteria can be found *in situ,* in human fecal samples (Fig. 5) and in the flounder gut (Table II). (3) Adhesion of rough enterobacterial strains to intestinal mucosal cells in greater than that of less hydrophobic, smooth, wild-type cells (Perers *et al.,* 1977; Edebo *et al.,* 1980). Similarly, adhesion of enteropathogenic *Escherichia coli* strains correlates with increased fimbria-mediated hydrophobicity (Smyth *et al.,* 1978; Lindahl *et al.,* 1981; Faris *et al.,* 1981), although there is much disagreement over which organisms are the more hydrophobic (Jann *et al.,* 1981; Ohman *et al.,* 1982; Lindahl *et al.,* 1981). (4) Many laboratory strains of normal intestinal microbiota as well as pathogens that may be associated with the GI tract exhibit cell-surface hydrophobicity. These include strains of *Yersinia enterocolitica* (Faris *et al.,* 1983; Martinez, 1983; Lachica and Zink, 1984a,b; Lachica *et al.,* 1984), *Vibrio cholerae* (Faris *et al.,* 1982; Kabir and Ali, 1983), *Salmonella* (Jiwa and Mansson, 1981; Xiu *et al.,* 1983), *Bifidobacterium* (H. J. M. Oop den Camp, A. Oosterhof, and J. H. Veerkamp, unpublished data), and *Shigella* (Edebo *et al.,* 1983).

Cell-surface hydrophobicity has been reported in a variety of potentially pathogenic bacteria, including *Legionella pneumophila* (Bohach and Snyder, 1983), *Staphylococcus aureus* (M. Rosenberg *et al.,* 1980a; Jonsson and Wadstrom, 1983, 1984; Malmqvist, 1983), *Streptococcus pyogenes* (M. Rosenberg *et al.,* 1981; Ofek *et al.,* 1983; Miorner *et al.,* 1983, 1984; Tylewska *et al.,* 1979), *Mycobacterium lepraemurium* (Katoh and Matsuo, 1983), *Neisseria gonorrhoeae* (Magnusson *et al.,* 1979a,b), and *Chlamydia psittaci* (Vance and Hatch, 1980).

The adhesion of *Lactobacillus fermentum* to the nonsecreting, keratinized part of the stomach in mice was recently shown to correlate with cell-surface hydrophobicity; the presence of erythrosine in the bacterial growth medium reduced hydrophobicity and abolished adhesion to the mouse stomach tissue (P. O. Conway and R. F. Adams, unpublished results).

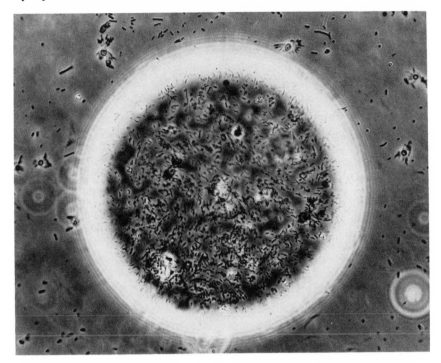

Figure 5. A fresh sample of human feces was washed and suspended in 150 mM phosphate buffer, pH 7.1, to an optical density of 0.9 at 400 nm. Adhesion to 0.2 ml of *n*-hexadecane, *n*-octane, and *p*-xylene was performed as previously described (M. Rosenberg *et al.*, 1983a). Following the mixing, the bottom aqueous phase (1.2 ml) was removed to fresh tubes containing 0.2 ml of the corresponding hydrocarbon, and the phases again mixed. Percent adhesion, as calculated from the drop in turbidity, was: hexadecane, 17%; octane, 45%; xylene, 49%. Shown is an octane droplet with adherent cells, surrounded by aqueous phase, following the mixing procedure. In general, the smaller bacteria tended to adhere at the interface, whereas the larger rods remained in the bulk aqueous phase.

The role of hydrophobic interactions in bacterial adhesion to phagocytic cells has been widely studied in recent years (van Oss, 1978; Absolom *et al.*, 1981, 1982; Edebo *et al.*, 1980); the extent of its participation in relation to more specific mechanisms, however, remains a point of controversy. Hydrophobic interactions play a role in adhesion of *Mycoplasma* to erythrocytes (Banai *et al.*, 1978, 1980; Sobeslavsky *et al.*, 1968), and have probably been overlooked in many cases of bacteria-mediated hemagglutination (R. J. Doyle, personal communication). Similarly, despite the obviously hydrophobic surface properties of human skin and hair, we are not aware of extensive studies on the role of hydrophobic interactions in mediating bacterial adhesion to such surfaces. Do hydrophobic pulmonary surfactants (Cotton and Hills, 1984) mediate or inhibit adhesion to lung tissues? Similarly, what is the role of hydrophobic interactions in mediating adhesion to aging or damaged cells and tissues, e.g., cornea? Is hydrophobicity a factor in adhesion to ear tissue (Hills, 1984a)?

The increasing use of polymeric implants and catheters has led various researchers to study bacterial adhesion to such materials. In many cases, hydrophobic interactions appear to be an important factor (Ashkenazi and Mirelman, 1984; Sugarman, 1982; Hogt et al., 1982, 1983b).

In view of the above, the role of hydrophobic interactions in bacterial pathogenesis is twofold. Cell-surface hydrophobicity may serve to enable initial attachment of bacteria, but may also aid in their removal by phagocytic cells. Ofek et al. (1983) have suggested that pathogens may alter surface hydrophobicity as a function of the stage of infection; S. pyogenes cells, initially hydrophobic, adhere to and then colonize epithelial cells, but following invasion may become hydrophilic and nonadherent through the elaboration of hyaluronic acid capsule. The observation (Y. Blumberger and M. Rosenberg, unpublished data) that clinical Serratia strains become hydrophilic when grown above 37°C is suggestive of a similar example. This phenomenon bears some resemblance to adhesion and desorption of bacteria at marine interfaces (section 5.2).

6. Conclusions

Bacterial adhesion is a multifaceted subject. In some cases (e.g., adhesion of normal human microbiota, cell immobilization, degradation of insoluble pollutants, phagocytosis, nitrogen fixation), adhesion can benefit man. In other situations (colonization of host tissues by pathogens, biofouling) we seek the means to control, inhibit, and desorb adherent microorganisms. Whether trying to promote or prevent bacterial adhesion, an understanding of the underlying mechanisms is of prime importance.

In the present review we have looked at several adhesion phenomena in which hydrophobic interactions appear to play a central role. The possibility exists that the role of hydrophobic interactions in bacterial adhesion is far wider than is generally appreciated. In fact, the participation of hydrophobic interactions in adhesion phenomena is often overlooked, particularly in studies searching for specific (usually sugar-inhibited) adhesion mechanisms. As Kauss and Young (1983) have pointed out, in many investigations a battery of sugars is tested for their ability to inhibit the observed adhesion (if none work, it is often assumed that more exhaustive attempts will eventually uncover the one that does). Only rarely have inhibitors of hydrophobic interactions been compared along-side sugars for their effect on adhesion (e.g., McIntire *et al.,* 1982). Another case in point is the inhibition of adhesion by cell-surface components. It is generally held that the ability of a cell-surface component to inhibit adhesion demonstrates its role as an adhesin in the phenomenon studied. But what if the component is itself an amphipathic molecule, capable of blocking hydrophobic interactions between cell and surface, such as lipopolysaccharide or lipoteichoic acid (Carruthers and Kabat, 1983; Botta, 1981)? In such instances inhibition of adhesion may reflect the participation of hydrophobic interactions rather than a role for the component *per se.* Our appreciation of the extent to which hydrophobic interactions mediate various bacterial adhesion phenomena could be greatly augmented by including agents that interfere with hydrophobic interactions (e.g., surfactants, chaotropic agents) in adhesion assays.

If hydrophobic interactions are, as we submit, chiefly responsible for a great many adhesion phenomena, several corollaries arise. First is the distinct possibility that many bacteria that adhere to surfaces are not too particular about the substrata to which they stick. It can then be argued that the small percentage of cells that do attach to a hospitable surface can quickly proliferate into colonies far outnumbering those that have gone astray. If this is the case, then we must remember when scrutinizing a natural surface that not all the adherent cells we observe may have "meant" to adhere. Similarly, the relative numbers of microorganisms found on a given surface say little about their relative adhesion ability: *A. calcoacaceticus* RAG-1, for example, adheres in much higher numbers to human buccal epithelial cells than does *S. pyogenes,* despite the inability of the former to colonize the oral mucosa (M. Rosenberg *et al.,* 1981; E. Rosenberg *et al.,* 1983a). Similarly, the low relative numbers of filamentous bacteria on buccal epithelial cells as compared with their predominance on the tooth surface (Gibbons, 1980) may be primarily related to their subsequent colonizing ability rather than their initial adhesion. Washed suspensions of filamentous dental plaque bacteria do adhere to buccal epithelial cells *in vitro* in the absence of saliva (M. Tal, E. Weiss, and M. Rosenberg, unpublished results).

Future investigations will be required to deal with the likelihood that the cell-surface hydrophobicity of bacteria *in situ* differs greatly from results obtained using laboratory strains: (1) laboratory strains may lose hydrophobic surface properties upon subculturing; moreover, adherent and aggregating strains, being more difficult to work with in the laboratory, may, over the course of years, have been discarded in favor of less problematic, nonadherent clones; (2) cells in the natural environment may adsorb proteins and other polymers from the surroundings, which in turn affect their surface properties; and (3) bacteria in the environment are often starved; starvation conditions generally result in an increase in hydrophobicity and adhesion.

A problem of increasing interest is the manner in which bacteria modulate cell-surface hydrophobicity in order to detach from, as well as attach to, surfaces. It is clear that bacteria may detach from interfaces as well as adhere to them both in the open environment and in the host. It would be of great interest to learn whether such microorganisms respond to subtle changes in their natural environment which signal them to alter cell-surface hydrophobicity accordingly; in the test tube such changes often appear merely to be related to phase of growth.

How does bacterial hydrophobicity affect other physiological functions? For example, cell-surface hydrophobicity has often been related to the absence of capsule; however, since bacterial capsules may serve to protect the cell from injurious external factors (desiccation, toxic compounds, etc.), the absence of capsule may put the cell at some disadvantage. It has been suggested that cell-surface hydrophobicity is correlated with increased sensitivity to hydrophobic compounds and antibiotics (Miller, 1983; Roantree et al., 1977; Mackey, 1983). However, no correlations between cell-surface hydrophobicity and sensitivity to hydrophobic antibiotics were observed in S. typhimurium (Kjelleberg et al., 1985b) or in A. calcoaceticus RAG-1 (M. Rosenberg, unpublished data), suggesting that the proposed changes are related to permeability and degree of exposure of outer membrane constituents (Roantree et al., 1977; Leive et al., 1984).

Despite the many techniques available for measuring bacterial hydrophobicity, it is probable that we shall see novel and improved methodology in coming years. Hopefully, these will aid researchers in their attempts to answer the many outstanding questions in this field. One of the most exciting prospects is that the novel genetic techniques that have proven so successful in the study of several Escherichia coli adhesion systems will be extended to adhesion phenomena of the type described in this review; some indications of this are already apparent (e.g., Martinez, 1983; Lachica and Zink, 1984a,b; Lachica et al., 1984) in studies on plasmid-mediated hydrophobicity in Yersinia enterocolitica.

In closing, we would like to mention another type of close approach that has been mediated by hydrophobic interactions. Ecologically and medically oriented adhesion researchers, once separated by an abyss of different interests and outlooks, have been drawn together by this common mechanism of bacterial adhesion. Hopefully, the synergism resulting from this merger, already evident in many of the investigations we have discussed here, has only just begun.

ACKNOWLEDGMENTS. The authors are grateful to Patricia L. Conway, Eugene Rosenberg, Ron J. Doyle, Ronit Bar-Ness, and Nachum Kaplan for critical reading of the manuscript. We thank Rita Lazar for expert editorial assistance, and Yardena Mazor and Naomi Regimov for excellent technical help.

References

Abbott, A., and Hayes, M. L., 1984, The conditioning role of saliva in streptococcal attachment to hydroxyapatite surfaces, *J. Gen. Microbiol.* **130**:809–816.

Abraham, S. N., Beachey, E. H., and Simpson, W. A., 1983, Adherence of *Streptococcus pyogenes, Escherichia coli,* and *Pseudomonas aeruginosa* to fibronectin-coated and uncoated epithelial cells, *Infect. Immun.* **41**:1261–1268.

Absolom, D. R., Francis, D. W., Zingg, W., van Oss, C. J., and Neumann, A. W., 1981, Phagocytosis of bacteria by platelets: Surface thermodynamics, *J. Coll. Interface Sci.* **85**:168–177.

Absolom, D. R., van Oss, C. J., Zingg, W., and Neumann, A. W., 1982, Phagocytosis as a surface phenomenon: Opsonization by aspecific adsorption of IgG as a function of bacterial hydrophobicity, *Res. J. Reticuloendothel. Soc.* **31**:59–70.

Absolom, D. R., Lamberti, F. V., Policova, Z., Zingg, W., van Oss, C. J., and Neumann, A. W., 1983, Surface thermodynamics of bacterial adhesion, *Appl. Environ. Microbiol.* **46**:90–97.

Albertsson, P. A., 1958, Particle fractionation in liquid two-phase system. The composition of some phase systems and the behavior of some model particles in them. Application to the isolation of cell walls from microorganisms, *Biochim. Biophys. Acta* **27**:378–395.

Arakawa, K., Tokiwano, K., Ohtomo, N., and Kedaira, H., 1979, A note on the nature of ionic hydrations and hydrophobic interactions in aqueous solutions, *Bull. Chem. Soc. Japan* **52**:2483–2488.

Ashkenazi, S., and Mirelman, D., 1984, Adherence of bacteria to pediatric intravenous catheters and needles and its relation to phlebitis in animals, *Ped. Res.* **18**:1361–1366.

Banai, M., Kahane, I., Razin, S., and Bredt, W., 1978, Adherence of *Mycoplasma gallisepticum* to human erythrocytes, *Infect. Immun.* **21**:365–372.

Banai, M., Razin, S., Bredt, W., and Kahane, I., 1980, Isolation of binding sites to glycophorin from *Mycoplasma pneumoniae* membranes, *Infect. Immun.* **30**:628–634.

Beighton, D., 1984, The influence of saliva on the hydrophobic surface properties of bacteria isolated from oral sites of macaque monkeys, *FEMS Microbiol. Lett.* **21**:239–242.

Ben-Naim, A., 1977, Hydrophobic interaction, in: *l'Eau et les Systemes Biologiques,* Colloques Internationaux de CNRS, no. 246, pp. 215–221, CNRS, Paris.

Bezdek, H. F., and Carlucci, A.F., 1974, Concentration and removal of liquid microlayers from a seawater surface by bursting bubbles, *Limnol. Oceanogr.* **19**:126–132.

Blanchard, D. C., 1983, The production, distribution and bacterial enrichment of the sea-salt aerosol, in: *Air–Sea Exchange of Gases and Particles* (P. S. Liss and W. G. N. Slimm, eds.), pp. 407–454, Reidel, Hingham, Massachusetts.

Blanchard, D. C., and Syzdek, L. D., 1978, Seven problems in bubble and jet drop researches, *Limnol. Oceanogr.* **23**:389–400.

Blanchard, D. C., and Syzdek, L. D., 1982, Water-to-air transfer and enrichment of bacteria in drops from bursting bubbles, *Appl. Environ. Microbiol.* **43**:1001–1005.

Bohach, G. A., and Snyder, I. S., 1983, Characterization of surfaces involved in adherence of *Legionella pneumophila* to *Fischerella* species, *Infect. Immun.* **42**:318–325.

Botta, G. A., 1981, Surface components in adhesion of group A streptococci to pharyngeal epithelial cells, *Curr. Microbiol.* **6**:101–104.

Boyles, W. A., and Lincoln, R. E., 1958, Separation and concentration of bacterial spores and vegetative cells by foam flotation, *Appl. Microbiol.* **6**:327–334.

Breuer, W. V., Ginsburg, H., and Cabantchik, Z. I., 1984, Hydrophobic interactions in *Plasmodium falciparum* invasion into human erythrocytes, *Mol. Biochem. Parasitol.* **12**:125–138.

Bryant, R. D., Costerton, J. W., and Laishley, E. J., 1984, The role of *Thiobacillus albertis* glycocalyx in the adhesion of cells to elemental sulfur, *Can. J. Microbiol.* **30**:81–90.

Busscher, H. J., Weerkamp, A. H., van der Mei, H. C., van Pelt, A. W. J., de Jong, H. P., and Arends, J., 1984, Measurement of the surface free energy of bacterial cell surfaces and its relevance for adhesion, *Appl. Environ. Microbiol.* **48**:980–983.

Carruthers, M. M., and Kabat, W. J., 1983, Mediation of staphylococcal adherence to mucosal cells by lipoteichoic acid, *Infect. Immun.* **40**:444–446.

Chakravarty, M., Singh, H. D., and Baruah, J. N., 1975, A kinetic model for microbial growth on liquid hydrocarbons. *Biotechnol. Bioeng.* **17**:399–412.

Costerton, J. W., Irvin, R. T., and Cheng, K.-J., 1981, The bacterial glycocalyx in nature and disease, *Annu. Rev. Microbiol.* **351**:299–324.

Cotton, D. B., and Hills, B. A., 1984, Pulmonary surfactant: Hydrophobic nature of the mucosal surface of the human amnion, *J. Physiol.* **349**:411–418.

Courtney, H. S., Simpson, W. A., and Beachey, E. H., 1983, Binding of streptococcal lipoteichoic acid to fatty acid-binding sites on human plasma fibronectin, *J. Bacteriol.* **153**:763–770.

Cover, W. H., and Rittenberg, S. C., 1984, Change in the surface hydrophobicity of substrate cells during bdelloplast formation by *Bdellovibrio bacteriovorus* 109J, *J. Bacteriol.* **157**:391–397.

Cunningham, R. K., Soderstrom, T. O., Gillman, C. F., and van Oss C. J., 1975, Phagocytosis as a surface phenomenon. V. Contact angles and phagocytosis of rough and smooth strains of *Salmonella typhimurium,* and the influence of specific antiserum, *Immunol. Comm.* **4**:429–442.

Dahlback, B., 1983, Marine Microorganisms and Interfaces, Ph.D. thesis, University of Goteborg.

Dahlback, B., Hermansson, M., Kjelleberg, S., Norkrans B., and Pedersen, K., 1980, Cell surface hydrophobicity and charge related to the initial microbial adhesion at the air–water interface, in: *Microbial Adhesion to Surfaces* (R. C. W. Berkeley, J. M. Lynch, J. Melling, P. R. Rutter, and B. Vincent, eds.), pp. 540–541, Ellis Horwood, Chichester, England.

Dahlback, B., Hermansson, M., Kjelleberg, S., and Norkrans, B., 1981, The hydrophobicity of bacteria—An important factor in their initial adhesion at the air–water interface, *Arch. Microbiol.* **128**:267–270.

Davison, V. E., and Sanford, B. A., 1982, Factors influencing adherence of *Staphylococcus aureus* to influenza A virus-infected cell cultures, *Infect. Immun.* **37**:946–955.

Dawson, M. P., Humphrey, B. A., and Marshall, K. C., 1981, Adhesion: A tactic in the survival strategy of a marine vibrio during starvation, *Curr. Microbiol.* **6**:195–199.

Dexter, S. C., 1979, Influence of substratum critical surface tension on bacterial adhesion—*In situ* studies, *J. Colloid Interface Sci.* **70**:346–354.

Dexter, S. C., Sullivan, Jr., J. D., Williams III, J., and Watson, S. W., 1975, Influence of substrate wettability on the attachment of marine bacteria to various surfaces, *Appl. Microbiol.* **30**:298–308.

Doyle, R. J., Nedjat-Halem, F., and Singh, J. S., 1984, Hydrophobic characteristics of *Bacillus* spores, *Curr. Microbiol.* **10**:329–332.

Dworkin, M., Keller, K. H., and Weisberg, D., 1983, Experimental observations consistent with a surface tension model of gliding motility of *Myxococcus xanthus, J. Bacteriol.* **155**:1367–1371.

Dyar, M. T., 1948, Electrokinetical studies on bacterial surfaces. II. Studies on surface lipids, amphoteric material and some other surface properties, *J. Bacteriol.* **56**:821–834.

Edebo, L., Kihlstrom, E., Magnusson, K.-E., and Stendahl, O., 1980, The hydrophobic effect and charge effects in the adhesion of enterobacteria to animal cell surfaces and the influences of antibodies of different immunoglobulin classes, in: *Cell Adhesion and Motility* (A. S. G. Curtis and J. D. Pitts, eds.), pp. 65–101, Cambridge University Press, Cambridge.

Edebo, L., Magnusson, K.-E., and Stendahl, O., 1983, Physico-chemical surface properties of *Shigella sonnei, Acta Pathol. Microbiol. Immunol. Scand. B* **91**:101–106.

Ellwood, D. C., Keevil, C. W., Marsh, P. D., Brown, C. M., and Wardell, J. N., 1982, Surface-associated growth, *Phil. Trans. R. Soc. Lond. B* **297**:517–532.

Erne, A. M., Werner, R. G., and Reifenrath, R., 1984, Inhibition of bacterial adhesion by an artificial surfactant, *FEMS Microbiol. Lett.* **23**:205–209.

Faris, A., Wadstrom, T., and Freer, J. H., 1981, Hydrophobic adsorptive and hemagglutinating properties of *Escherichia coli* possessing colonization factor antigens (CFA/I or CFA/II), type 1 pili, or other pili, *Curr. Microbiol.* **5**:67–72.

Faris, A., Lindahl, M., and Wadstrom, T., 1982, High surface hydrophobicity of hemagglutinating *Vibrio cholerae* and other vibrios, *Curr. Microbiol.* **7**:357–362.

Faris, A., Lindahl, M., and Wadstrom, T., 1983, Autoaggregating *Yersinia enterocolitica* express surface fimbriae with high surface hydrophobicity, *J. Appl. Bacteriol.* **55**:97–100.

Fattom, A., and Shilo, M., 1984, Hydrophobicity as an adhesion mechanism of benthic *Cyanobacteria, Appl. Environ. Microbiol.* **47**:135–143.

Ferreiros, C. M., and Criado, M. T., 1984, Expression of surface hydrophobicity encoded by R-plasmids in *Escherichia coli* laboratory strains, *Arch. Microbiol.* **138**:191–194.

Firon, N., Ofek, I., and Sharon, N., 1982, Interaction of mannose-containing oligosaccharides with the fimbrial lectin of *Escherichia coli, Biochem. Biophys. Res. Commun.* **105**:1426–1432.

Fletcher, M., 1976, The effects of proteins of bacterial attachment to polystyrene, *J. Gen. Microbiol.* **94**:400–404.

Fletcher, M., 1983, The effects of methanol, ethanol, propanol and butanol on bacterial attachment to surfaces; *J. Gen. Microbiol.* **129**:633–641.

Fletcher, M., 1985, The utilization of glucose by free-living bacteria and those attached to solid surfaces, in: *Abstracts of the 85th Annual Meeting, American Society for Microbiology*, p. 158.

Fletcher, M., and Loeb, G. I., 1979, Influence of substratum characteristics on the attachment of a marine pseudomonad to solid surfaces, *Appl. Environ. Microbiol.* **37**:67–72.

Fletcher, M., and Marshall, K. C., 1982a, Bubble contact angle method for evaluating substratum interfacial characteristics and its relevance to bacterial attachment, *Appl. Environ. Microbiol.* **44**:184–192.

Fletcher, M., and Marshall, K. C., 1982b, Are solid surfaces of ecological significance to aquatic bacteria?, in: *Advances in Microbial Ecology*, Vol. 6 (K. C. Marshall, ed.), pp. 199–236, Plenum Press, New York.

Gaudin, A. M., Mular, A. L., and O'Connor, R. F., 1960, Separation of microorganisms by flotation. II. Flotation of spores of *Bacillus subtilis* var. *niger, Appl. Microbiol.* **8**:91–97.

Gerson, D. F., and Akit, J., 1980, Cell surface energy, contact angles and phase partition. II. Bacterial cells in biphasic aqueous mixtures, *Biochim. Biophys. Acta* **602**:281–284.

Gibbons, R. J., 1980, Adhesion of bacteria to the surfaces of the mouth, in: *Adsorption of Microorganisms to Surfaces* (R. C. W. Berkeley, J. M. Lynch, J. Melling, P. R. Rutter, and B. Vincent, eds.), pp. 351–388, Ellis Horwood, Chichester, England.

Gibbons, R. J., and Etherden, I., 1983, Comparative hydrophobicities of oral bacteria and their adherence to salivary pellicles, *Infect. Immun.* **49**:1190–1196.

Gibbons, R. J., Spinell, D. M., and Skobe, Z., 1976, Selective adherence as a determinant of the host tropisms of certain indigenous and pathogenic bacteria, *Infect. Immun.* **13**:238–246.

Gibbons, R. J., Etherden, I., and Skobe, Z., 1983, Association of fimbriae with the hydrophobicity of *Streptococcus sanguis* FC-1 and adherence to salivary pellicles, *Infect. Immun.* **41**:414–417.

Gordon, A. S., and Millero, F. J., 1985, Bacterial respiration and assimilation of organic nutrients sorbed to mineral surfaces, in: *Abstracts of the 85th Annual Meeting, American Society for Microbiology*, p. 278.

Hejkal, T. W., LaRock, P. A., and Winchester, J. W., 1980, Water-to-air fractionation of bacteria, *Appl. Environ. Microbiol.* **39**:335–338.

Hermansson, M., and Dahlback, B., 1983, Bacterial activity at the air/water interface, *Microb. Ecol.* **9**:317–328.

Hermansson, M., and Marshall, K. C., 1985, Utilization of surface localized substrate by non-adhesive marine bacteria, *Microb. Ecol.* **11**:91–105.

Hermansson, M., Kjelleberg, S., and Norkrans, B., 1979, Interaction of pigmented wildtype and pigmentless mutant of *Serratia marcescens* with lipid surface film, *FEMS Microbiol. Lett.* **6**:129–132.

Hermansson, M., Kjelleberg, S., Korhonen, T. K., and Stenstrom, T.-A., 1982, Hydrophobic and electrostatic characterization of surface structures of bacteria and its relationship to adhesion at an air–water surface, *Arch. Microbiol.* **131**:308–312.

Hildebrand, J. H., 1979, Is there a "hydrophobic effect"?, *Proc. Natl. Acad. Sci. USA* **76**:194.

Hill, M. J., James, A. M., and Maxted, W. R., 1963, Some physical investigations of the behaviour of bacterial surfaces. X. The occurrence of lipids in the streptococcal cell wall, *Biochim. Biophys. Acta* **75**:414–424.

Hills, B. A., 1984a, Hydrophobic lining of the eustachian tube imparted by surfactant, *Arch. Otolarynogol.* **110**:779–782.

Hills, B. A., 1984b, Surfactant as a release agent opposing the adhesion of tumor cells in determining malignancy, *Med. Hypotheses* **14**:99–110.

Hills, B. A., Butler, B. F., and Lichtenberger, L. M., 1983, Gastric mucosal barrier: Hydrophobic lining to the lumen of the stomach, *Am. Physiol. Soc.* **1983**:G561–G568.

Hjertén, S., Rosengren, J., and Pahlman, S., 1974, Hydrophobic interaction chromatography. The synthesis and the use of some alkyl and aryl derivatives of agarose, *J. Chromatogr.* **101**:281–288.

Hogt, A. H., Dankert, J., Feijen, J., and de Vries, J. A., 1982, Cell surface hydrophobicity of *Staphylococcus* species and adhesion onto biomaterials, *Antoni van Leeuwenhoek J. Microbiol. Soc.* **48**:496–498.

Hogt, A. H., Dankert, J., and Feijen, J., 1983a, Encapsulation, slime production and surface hydrophobicity of coagulase-negative staphylococci, *FEMS Microbiol. Lett.* **18**:211–215.

Hogt, A. H., Dankert, J., de Vries, J. A., and Feijen, J., 1983b, Adhesion of coagulase-negative staphylococci to biomaterials, *J. Gen. Microbiol.* **129**:2959–2968.

Honda, T., Arita, M., and Miwatani, T., 1984, Characterization of new hydrophobic pili of human enterotoxigenic *Escherichia coli:* A possible new factor, *Infect. Immun.* **43**:959–965.

Humphrey, B. A., Dickson, M. R., and Marshall, K. C., 1979, Physicochemical and *in situ* observations on the adhesion of gliding bacteria to surfaces, *Arch. Microbiol.* **120**:231–238.

Iimura, Y., Hara, S., and Otsuka, K., 1980, Cell surface hydrophobicity as a pellicle formation factor in film strain of *Saccharomyces, Agric. Biol. Chem.* **6**:1215–1222.

Israelachvili, J. N., and Pashley, R. M., 1984, Measurement of the hydrophobic interaction between two hydrophobic surfaces in aqueous electrolyte solutions, *J. Colloid Interface Sci.* **98**:500–514.

Jann, K., Schmidt, G., Blumenstock, E., and Vosbeck, K., 1981, *Escherichia coli* adhesion to *Saccharomyces cerevisiae* and mammalian cells: Role of piliation and surface hydrophobicity, *Infect. Immun.* **32**:484–489.

Jeffrey, W. H., and Paul, J. H., 1985, The effect of attachment to polystyrene on the activity of marine microfouling bacteria, in: *Abstracts of the 85th Annual Meeting, American Society for Microbiology,* p. 229.

Jiwa, S. F. H., and Mansson, I., 1981, Hemagglutinating and hydrophobic surface properties of salmonellae producing enterotoxin neutralized by cholera anti-toxin, *Veterinary Microbiol.* **8**:443–458.

Jones, G. W., and Isaacson, R. E., 1983, Proteinaceous bacterial adhesins and their receptors, *CRC Crit. Rev. Microbiol.* **10**:229–260.

Jonsson, P., and Wadstrom, T., 1983, High surface hydrophobicity of *Staphylococcus aureus* as revealed by hydrophobic interaction chromatography, *Curr. Microbiol.* **8**:347–353.

Jonsson, P., and Wadstrom, T., 1984, Cell surface hydrophobicity of *Staphylococcus aureus* measured by the salt aggregation test (SAT), *Curr. Microbiol.* **10**:203–210.

Kabir, S., and Ali, S., 1983, Characterization of surface properties of *Vibrio cholerae, Infect. Immun.* **39**:1048–1058.

Kappeli, O., and Fiechter, A., 1976, The mode of interaction between the substrate and the cell surface of the hydrocarbon-utilizing yeast *Candida tropicalis, Biotechnol. Bioeng.* **18**:967–974.

Katoh, M., and Matsuo, Y., 1983, Adherence of *Mycobacterium lepraemurium* to tissue culture cells, *Hiroshima J. Med. Sci.* **32**:285–290.

Kauss, H., and Young, D. H., 1983, Fungal spores are agglutinated by proteins and adhere to bean hypocotyls due to nonspecific binding, in: *Chemical Taxonomy, Molecular Biology, and Function of Plant Lectins,* pp. 187–196, Alan R. Liss, New York.

Kefford, B., Kjelleberg, S., and Marshall, K. C., 1982, Bacterial scavenging: Utilization of fatty acids localized at a solid–liquid interface, *Arch. Microbiol.* **133**:257–260.

Kennedy, R. S., Finnerty, W. R., Sudarsanan, K., and Young, R. A., 1975, Microbial assimilation of hydrocarbon. I. The fine structure of a hydrocarbon oxidizing *Acinetobacter* sp., *Arch. Microbiol.* **102**:75–83.

Kirschner-Zilber, I., Rosenberg, E., and Gutnick, D., 1980, Incorporation of ^{32}P and growth of pseudomonad UP-2 on *n*-tetracosane, *Appl. Environ. Microbiol.* **40**:1086–1093.

Kitamura, A., 1984, Evidence for an increase in the hydrophobicity of the cell surface during sexual interactions of *Paramecium, Cell Struct. Funct.* **9**:91–95.

Kjelleberg, S., 1984a, Effects of interfaces on survival mechanisms of copiotrophic bacteria in low-nutrient habitats, in: *Current Perspectives in Microbial Ecology* (M. J. Klug and C. A. Reddy, eds.), pp. 151–159, American Society for Microbiology, Washington, D.C.

Kjelleberg, S., 1984b, Adhesion to inanimate surfaces, in: *Microbial Adhesion and Aggregation.* (K. C. Marshall, ed.), pp. 51–70, Springer, Berlin.

Kjelleberg, S., 1985, Mechanisms for bacterial adhesion at gas–liquid interfaces, in: *Bacterial Adhesion: Mechanisms and Physiological Significance* (D. C. Savage and M. M. Fletcher, eds.), pp. 163–194, Plenum Press, New York.

Kjelleberg, S., and Hermansson, M., 1984, Starvation induced effects on bacterial surface characteristics, *Appl. Environ. Microbiol.* **48**:497–503.

Kjelleberg, S., and Stenstrom, T. A., 1980, Lipid surface films: Interaction of bacteria with free fatty acids and phospholipids at the air/water interface, *J. Gen. Microbiol.* **116**:417–423.

Kjelleberg, S., Lagercrantz, C., and Larsson, T., 1980, Quantitative analysis of bacterial hydrophobicity studied by the binding of dodecanoic acid, *FEMS Microbiol. Lett.* **7**:41–44.

Kjelleberg, S., Humphrey, B. A., and Marshall, K. C., 1982, The effect of interfaces on small starved marine bacteria, *Appl. Environ. Microbiol.* **43**:1166–1172.

Kjelleberg, S., Humphrey, B. A., and Marshall, K. C., 1983, Initial phases of starvation and activity of bacteria at surfaces, *Appl. Environ. Microbiol.* **46**:978–984.

Kjelleberg, S., Marshall, K. C., and Hermansson, M., 1985a, Oligotrophic and copiotrophic marine bacteria—Observations related to attachment, *FEMS Microbiol. Ecol.* (in press).

Kjelleberg, S., Conway, P., and Stenstrom, T. A., 1985b, Inhibition of the starvation survival process of laboratory bacteria strains and the colon microflora of mice, *Arch. Microbiol.* (in press).

Kozel, T. R., 1983, Dissociation of a hydrophobic surface from phagocytosis of encapsulated and non-encapsulated *Cryptococcus neoformans, Infect. Immun.* **39**:1214–1219.

Kupfer, D., and Zusman, D. R., 1984, Changes in cell surface hydrophobicity of *Myxococcus xanthus* are correlated with sporulation-related events in the developmental program, *J. Bacteriol.* **159**:776–779.

Lachica, R. V., and Zink, D. L., 1984a, Plasmid-associated cell surface charge and hydrophobicity of *Yersinia enterocolitica, Infect. Immun.* **44**:540–543.

Lachica, R. V., and Zink, D. L., 1984b, Determination of plasmid-associated hydrophobicity of *Yersinia enterocolitica* by a latex particle agglutination test, *J. Clin. Microbiol.* **19**:660–663.

Lachica, R. V., Zink, D. L., and Ferris, W. R., 1984, Association of fibril structure formation with cell surface properties of *Yersinia enterocolitica, Infect. Immunol.* **46**:272–275.

Leive, L., Telesetsky, S., Coleman, Jr., W. G. and Carr, D., 1984, Tetracyclines of various hydrophobicities as a probe for permeability of *Escherichia coli* outer membranes, *Antimicrob. Agents Chemother.* **25**:539–544.

Lindahl, M., Faris, A., Wadstrom, T., and Hjerten, S., 1981, A new test based on 'salting out' to measure relative surface hydrophobicity of bacterial cells, *Biochim. Biophys. Acta* **677**:471–476.

Lis, H., and Sharon, N., 1977, Lectins: Their chemistry and application to immunology, in: *The Antigens* (M. Sela, ed.), Vol. 4, pp. 429–529, Academic Press, New York.

Mackey, B. M., 1983, Changes in antibiotic sensitivity and cell surface hydrophobicity in *Escherichia coli* injured by heating, freezing, drying or gamma radiation, *FEMS Microbiol. Lett.* **20**:395–399.

Magnusson, K.-E., 1980, The hydrophobic effect and how it can be measured with relevance for cell–cell interactions, *Scand. J. Infect. Dis. Suppl.* **24**:131–134.

Magnusson, K.-E., 1982, Hydrophobic interaction—A mechanism of bacterial binding, *Scand. J. Infect. Dis. Suppl.* **33**:32–36.

Magnusson, K.-E., and Johansson, G., 1977, Probing the surface of *Salmonella typhimurium* SR and R bacteria by aqueous biphasic partitioning in systems containing hydrophobic and charged polymers, *FEMS Microbiol. Lett.* **2**:225–228.

Magnusson, K.-E., Kihlstrom, E., Norlander, L., Norqvist, A., Davies, J., and Normark, S., 1979a, Effect of colony type and pH on surface charge and hydrophobicity of *Neisseria gonorrhoeae, Infect. Immun.* **26**:397–401.

Magnusson, K.-E., Kihlstrom, E., Norqvist, A., Davies, J., and Normark, S., 1979b, Effect of iron on surface charge and hydrophobicity *Neisseria gonorrhoeae, Infect. Immun.* **26**:402–407.

Malmqvist, T., 1983, Bacterial hydrophobicity measured as partition of palmitic acid between the two immiscible phases of cell surface and buffer, *Acta Pathol. Microbiol. Immunol. Scand. B* **91**:69–73.

Malmqvist, T., Thelestam, M., and Mollby, R., 1984, Hydrophobicity of cultured mammalian cells and some effects of bacterial phospholipases C, *Acta Pathol. Microbiol. Immunol. Scand. B* **92**:127–133.

Marshall, K. C., 1979, Growth at interfaces, in: *Strategies of Microbial Life in Extreme Environments* (M. Shilo, ed.), pp. 281–290, Verlag Chemie, Weinheim.

Marshall, K. C., and Cruickshank, R. H., 1973, Cell surface hydrophobicity and the orientation of certain bacteria at interfaces, *Arch. Mikrobiol.* **91**:29–40.

Marshall, K. C., Cruickshank, R. H., and Bushby, H. V. A., 1975, The orientation of certain root-nodule bacteria at interfaces, including root-hair surfaces, *J. Gen. Microbiol.* **91**:198–200.

Martinez, R. J., 1983, Plasmid-mediated and temperature-regulated surface properties of *Yersinia enterocolitica, Infect. Immun.* **41**:921–930.

McIntire, F. C., Crosby, L. K., and Vatter, A. E., 1982, Inhibitors of coaggregation between *Actinomyces viscosus* T14V and *Streptococcus sanguis* 34: β-Galactosides, related sugars, and anionic amphipathic compounds, *Infect. Immun.* **36**:371–378.

McLee, A. G., and Davies, S. L., 1972, Linear growth of a *Torulopsis* sp. on *n*-alkanes, *Can. J. Microbiol.* **18**:315–319.

Miller, R. D., 1983, *Legionella pneumophila* cell envelope: Permeability to hydrophobic molecules, *Curr. Microbiol.* **9**:349–354.

Miorner, H., Myhre, E., Bjorck L., and Kronvall, G., 1980, Effect of specific binding of human albumin, fibrinogen, and immunoglobulin G on surface characteristics of bacterial strains as revealed by partition experiments in polymer phase systems, *Infect. Immun.* **29**:879–885.

Miorner, H., Johansson, G., and Kronvall, G., 1983, Lipoteichoic acid is the major cell wall component responsible for surface hydrophobicity of group A streptococci, *Infect. Immun.* **39**:336–343.

Miorner, H., Havlicek, J., and Kronvall, G., 1984, Surface characteristics of group A streptococci with and without M-protein, *Acta Pathol. Microbiol. Immunol. Scand. B* **92**:23–30.

Miura, Y., Okazaki, M., Hamada, S.-I., Murakawa, S.-I., and Yugen, R., 1977, Assimilation of liquid hydrocarbon by microorganisms. I. Mechanism of hydrocarbon uptake, *Biotechnol. Bioeng.* **19**:701–714.

Moore, R. L., and Marshall, K. C., 1981, Attachment and rosette formation by hyphomicrobia, *Appl. Environ. Microbiol.* **42**:751–757.

Morisaki, H., 1983, Effect of solid–liquid interface on metabolic activity of *Escherichia coli, J. Gen. Appl. Microbiol.* **29**:195–204.

Morisaki, H., 1984, Effect of a liquid–liquid interface on the metabolic activity of *Escherichia coli, J. Gen. Appl. Microbiol.* **30**:35–42.

Morrow, A. W., 1969, Concentration of the virus of foot and mouth disease by foam flotation, *Nature* **222**:489–490.

Mudd, S., and Mudd, E. B. H., 1924a, The penetration of bacteria through capillary spaces. IV. A kinetic mechanism in interfaces, *J. Exp. Med.* **40**:633–645.

Mudd, S., and Mudd, E. B. H., 1924b, Certain interfacial tension relations and the behaviour of bacteria in films, *J. Exp. Med.* **40**:647–660.

Nakahara, T., Hisatsuka, K., and Minoda, Y., 1981, Effect of hydrocarbon emulsification on growth and respiration of microorganisms in hydrocarbon media, *J. Ferment. Technol.* **59**:415–418.

Nesbitt, W. E., Doyle, R. J., and Taylor, K. G., 1982, Hydrophobic interactions and the adherence of *Streptococcus sanguis* to hydroxylapatite, *Infect. Immun.* **38**:637–644.

Neufeld, R. J., and Zajic, J. E., 1984, The surface activity of *Acinetobacter calcoaceticus* sp. 2CA2, *Biotechnol. Bioeng.* **26**:1108–1113.

Neufeld, R. J., Zajic, J. E., and Gerson, D. F., 1980, Cell surface measurements in hydrocarbon and carbohydrate fermentations, *Appl. Environ. Microbiol.* **39**:511–517.

Neumann, A. W., Absolom, D. R., Francis, D. W., Omenyi, S. N., Spelt, J. K., Policova, Z., Thomson, C., Zingg, W., and van Oss, C. J., 1983, Measurement of surface tensions of blood cells and proteins, in: *Surface Phenomena in Hemotheology: Their Theoretical,*

Experimental and Clinical Aspects (A. L. Copley and G. V. F. Seaman, eds.) *Ann. N.Y. Acad. Sci.* **416**:276–298.

Norkrans, B., 1980, Surface microlayers in aquatic environments, in: *Advances in Microbial Ecology,* Vol. 4. (M. Alexander, ed.), pp. 51–85, Plenum Press, New York.

Ochoa, J.-L., 1978, Hydrophobic (interaction) chromatography, *Biochimie* **60**:1–15.

Ofek, I., Whitnack, E., and Beachey, E. H., 1983. Hydrophobic interactions of group A streptococci with hexadecane droplets, *J. Bacteriol.* **154**:139–145.

Ohman, L., 1983, Interaction between *Escherichia coli* Bacteria and Human Granulocytes—The Role of Hydrophobic and Sugarspecific Surface Properties, Ph.D. thesis, University of Linkoping, Sweden.

Ohman, L., Magnusson, K.-E., and Stendahl, O., 1982, The mannose-specific lectin activity of *Escherichia coli* type 1 fimbriae assayed by agglutination of glycolipid-containing liposomes, erythrocytes, and yeast cells and hydrophobic interaction chromatography, *FEMS Microbiol. Lett.* **14**:149–153.

Olsson, J., and Westergren, G., 1982, Hydrophobic surface properties of oral streptococci, *FEMS Microbiol. Lett.* **15**:319–323.

Ouchi, K., and Akiyama, H., 1971, Non-foaming mutants of Sake yeasts: Selection by cell agglutination method and by froth flotation method, *Agric. Biol. Chem.* **7**:1024–1032.

Parker, B. C., Ford, M. A., Gruft, H., and Falinham III, J. O., 1983, Epidemiology of infection by nontuberculous mycobacteria. IV. Preferential aerosolization of *Mycobacterium intracellulare* from natural waters, *Am. Rev. Respir. Dis.* **128**:652–656.

Parker, N. D., and Munn, C. B., 1984, Increased cell surface hydrophobicity associated with possession of an additional surface protein by *Aeromonas salmonicida, FEMS Microbiol. Lett.* **21**:233–237.

Pedros-Alio, C., and Brock, T. D., 1983, The importance of attachment to particles for plankonic bacteria, *Arch. Hydrobiol.* **98**:354–379.

Perers, L., Andaker, L., Edebo, L., Stendahl, O., and Tagesson, C., 1977, Association of some enterobacteria with the intestinal mucosa of mouse in relation to their partition in aqueous polymer two-phase systems, *Acta Pathol. Microbiol. Scand. B* **85**:305–316.

Pertsovskaya, A. F., Duda, V. I., and Zvyagintsev, D. G., 1973, Surface ultrastructures of adsorbed microorganisms, *Sov. Soil Sci.* **4**:684–689.

Pines, O., and Gutnick, D., 1984, Alternate hydrophobic sites on the cell surface of *Acinetobacter calcoaceticus* RAG-1, *FEMS Microbiol. Lett.* **22**:307–311.

Pringle, J. H., and Fletcher, M., 1983, Influence of substratum wettability on attachment of freshwater bacteria to solid surfaces, *Appl. Environ. Microbiol.* **45**:811–817.

Pringle, J. H., Fletcher, M., and Ellwood, D. C., 1983, Selection of attachment mutants during the continuous culture of *Pseudomonas fluorescens* and relationship between attachment ability and surface composition, *J. Gen. Microbiol.* **129**:2557–2559.

Reddy, P. G., Singh, H. D., Roy, P. K., and Baruah, J. N., 1982, Predominant role of hydrocarbon solubilization in the microbial uptake of hydrocarbons, *Biotechnol. Bioeng.* **24**:1241–1269.

Reed, G. B., and Rice, C. E., 1931, The behaviour of acid-fast bacteria in oil and water systems, *J. Bacteriol.* **22**:239–247.

Reuveny, S., Mizrahi, A., Kotler, M., and Freeman, A., 1983, Factors effecting cell attachment, spreading and growth on derivatized microcarriers. II. Introduction of hydrophobic elements, *Biotechnol. Bioeng.* **25**:2969–2980.

Roantree, R. J., Kuo, T.-T., and MacPhee, D. G., 1977, The effect of defined lipopolysaccharide core defects upon antibiotic resistances of *Salmonella typhimurium, J. Gen. Microbiol.* **103**:223–234.

Rogers, A. H., Pilowsky, K., and Zilm, P. S., 1984, The effect of growth rate on the adhesion of the oral bacteria *Streptococcus mutans* and *Streptococcus milleri, Arch. Oral Biol.* **29**:147–150.

Romanenko, V. I., 1979, Bacterial growth on slides and electron microscope grids in surface water films and ooze deposits, *Microbiology* **48:**105–109.

Rosenberg, E., Gottlieb, A., and Rosenberg, M., 1983a, Inhibition of bacterial adherence to epithelial cells and hydrocarbons by emulsan, *Infect. Immun.* **39:**1024–1028.

Rosenberg, E., Kaplan, N., Pines, O., Rosenberg, M., and Gutnick, D., 1983b, Capsular polysaccharides interfere with adherence of *Acinetobacter calcoaceticus* to hydrocarbons, *FEMS Microbiol. Lett.* **17:**157–160.

Rosenberg, E., Brown, D. R., and Demain, A. L., 1985, The influence of gramicidin S on hydrophobicity of germinating *Bacillus brevis* spores, *Arch. Microbiol.* **142:**51–54.

Rosenberg, M., 1981, Bacterial adherence to polystyrene: A replica method of screening for bacterial hydrophobicity, *Appl. Environ. Microbiol.* **42:**375–377.

Rosenberg, M., 1982, Adherence of Bacteria to Hydrocarbons and Other Hydrophobic Surfaces, Ph.D. thesis, University of Tel-Aviv, Israel.

Rosenberg, M., 1984a, Bacterial adherence to hydrocarbons: A useful technique for studying cell surface hydrophobicity, *FEMS Microbiol. Lett.* **22:**289–295.

Rosenberg, M., 1984b, Ammonium sulphate enhances adherence of *Escherichia coli* J-5 to hydrocarbon and polystyrene, *FEMS Microbiol. Lett.* **25:**41–45.

Rosenberg, M., 1984c, Isolation of pigmented and nonpigmented mutants of *Serratia marcescens* with reduced cell surface hydrophobicity, *J. Bacteriol.* **160:**480–482.

Rosenberg, M., and Rosenberg, E., 1981, Role of adherence in growth of *Acinetobacter calcoaceticus* on hexadecane, *J. Bacteriol.* **148:**51–57.

Rosenberg, M., Gutnick, D., and Rosenberg, E., 1980a, Adherence of bacteria to hydrocarbons: A simple method for measuring cell-surface hydrophobicity, *FEMS Microbiol. Lett.* **9:**29–33.

Rosenberg, M., Rosenberg, E., and Gutnick, D., 1980b, Bacterial adherence to hydrocarbons, in: *Microbial Adhesion to Surfaces* (R. C. W. Berkeley, J. M. Lynch, J. Melling, P. R. Rutter, and B. Vincent eds.), pp. 541–542, Ellis Horwood, Chichester, England.

Rosenberg, M., Perry, A., Bayer, E. A., Gutnick, D. L., Rosenberg, E., and Ofek, I., 1981, Adherence of *Acinetobacter calcoaceticus* RAG-1 to human epithelial cells and to hexadecane, *Infect. Immun.* **33:**29–33.

Rosenberg, M. Bayer, E. A., Delarea, J., and Rosenberg, E., 1982a, Role of thin fimbriae in adherence and growth of *Acinetobacter calcoaceticus* on hexadecane, *Appl. Environ. Microbiol.* **44:**929–937.

Rosenberg, M., Rottem, S., and Rosenberg, E., 1982b, Cell surface hydrophobicity of smooth and rough *Proteus mirabilis* strains as determined by adherence to hydrocarbons, *FEMS Microbiol. Lett.* **13:**167–169.

Rosenberg, M., Rosenberg, E., Judes, H., and Weiss, E., 1983a, Bacterial adherence to hydrocarbons and to surfaces in the oral cavity, *FEMS Microbiol. Lett.* **20:**1–5.

Rosenberg, M., Judes, H., and Weiss, E., 1983b, Desorption of adherent bacteria from a solid hydrophobic surface by oil, *J. Microbiol. Meth.* **1:**239–244.

Rosenberg, M., Judes, H., and Weiss, E., 1983c, Cell surface hydrophobicity of dental plaque microorganisms *in situ*, *Infect. Immun.* **42:**831–834.

Rubin, A. J., Cassel, E. A., Henderson, O., Johnson, J. D., and Lamb III, J. C., 1966, Microflotation: New low gas-flow rate foam separation technique for bacteria and algae, *Biotechnol. Bioeng.* **8:**135–151.

Runnels, P. L., and Moon, H. W., 1984, Capsule reduces adherence of enterotoxigenic *Escherichia coli* to isolated intestinal epithelial cells of pigs, *Infect. Immun.* **45:**737–740.

Rutter, P. R., 1980, The physical chemistry of adhesion of bacteria and other cells, in: *Cell Adhesion and Motility* (A. S. G. Curtis and J. D. Pitts, eds.), pp. 103–135, Cambridge University Press, Cambridge.

Sacks, L. E., and Alderton, G., 1961, Behavior of bacterial spores in aqueous polymer two-phase systems, *J. Bacteriol.* **82:**331–340.

Shields, P. A., and Farrah, S. R., 1983, Influence of salts on electrostatic interactions between poliovirus and membrane filters, *Appl. Environ. Microbiol.* **45:**526–531.

Simpson, W. A., Ofek, I., and Beachey, E. H., 1980, Fatty acid binding sites of serum albumin as membrane receptor analogs for streptococcal lipoteichoic acid, *Infect. Immun.* **29:**119–122.

Smith, B. F., and LaMont, J. T., 1984, Hydrophobic binding properties of bovine gallbladder mucin, *J. Biol. Chem.* **259:**12170–12177.

Smyth, C. J., Jonsson, P., Olsson, E., Soderlind, O., Rosengren, J., Hjerten, S., and Wadstrom, T., 1978, Differences in hydrophobic surface characteristics of porcine enteropathogenic *Escherichia coli* with or without K88 antigen as revealed by hydrophobic interaction chromatography, *Infect. Immun.* **22:**462–472.

Sobel, J. D., and Obedeanu, N., 1984, Role of hydrophobicity in adherence of gram-negative bacteria to epithelial cells, *Ann. Clin. Lab. Sci.* **14:**216–224.

Sobeslavsky, O., Prescott, B., and Chanock, R. M., 1968, Adsorption of *Mycoplasma pneumoniae* to neuraminic acid receptors of various cells and possible role in virulence, *J. Bacteriol.* **96:**695–705.

Stanley, S. O., and Rose, A. H., 1967, On the clumping of *Corynebacterium xeroxis* as affected by temperature, *J. Gen. Microbiol.* **115:**509–512.

Stenstrom, T.-A., and Kjelleberg, S., 1985, Fimbriae mediated nonspecific adhesion of *Salmonella typhimurium* to mineral particles, *Arch. Microbiol.* (in press).

Sugarman, B., 1982, *In vitro* adherence of bacteria to prosthetic vascular grafts, *Infection* **10:**9–16.

Svanberg, M., Westergren, G., and Olsson, J., 1984, Oral implantation in humans of *Streptococcus mutans* strains with different degrees of hydrophobicity, *Infect. Immun.* **43:**817–821.

Svanborg-Eden, C. S., Bjursten, L.-M., Hull, R., Hull, S. Magnusson, K.-E., Moldovano, Z., and Leffler, H., 1984, Influence of adhesins on the interaction of *Escherichia coli* with human phagocytes, *Infect. Immun.* **44:**672–680.

Syzdek, L. D., 1985, Influence of *Serratia marcescens* pigmentation on cell concentrations in aerosols produced by bursting bubbles, *Appl. Environ. Microbiol.* **49:**173–178.

Tanford, C., 1973, *the Hydrophobic Effect: Formation of Micelles and Biological Membranes,* Wiley-Interscience, New York.

Tanford, C., 1979, Interfacial free energy and the hydrophobic effect, *Proc. Natl. Acad. Sci. USA* **76:**4175–4176.

Todd, I., and Gingell, D., 1980, Red blood cell adhesion. I. Determination of the ionic conditions for adhesion to an oil–water interface, *J. Cell Sci.* **41:**125–133.

Trust, T. J., Kay, W. W., and Ishiguro, E. E., 1983, Cell surface hydrophobicity and macrophage association of *Aeromonas salmonicida, Curr. Microbiol.* **9:**315–318, 1983.

Tylewska, S. K., Hjerten, S., and Wadstrom, T., 1979, Contribution of M protein to the hydrophobic surface properties of *Streptococcus pyogenes, FEMS Microbiol. Lett.* **6:**249–253.

Vance, D. W., and Hatch, T. P., 1980, Surface properties of *Chlamydia psittaci, Infect. Immun.* **29:**175–180.

Van Oss, C. J., 1978, Phagocytosis as a surface phenomenon, *Annu. Rev. Microbiol.* **32:**19–39.

Van Oss, C. J., and Gillman, C. F., 1972, Phagocytosis as a surface phenomenon. I. Contact angles and phagocytosis of non-opsonized bacteria, *Res. J. Reticuloendothel. Soc.* **12:**283–292.

Wadstrom, T., Schmidt, K.-H., Kuhnemund, O., Havlicek, J., and Kohler, W., 1984, Comparative studies on surface hydrophobicity of streptococcal strains of groups A, B, C, D and G, *J. Gen. Microbiol.* **130:**657–664.

Wang, P. Y., 1974, Evidence of hydrophobic interaction in adhesion to tissue, *Nature* **249**:367–368.

Weber, M. E., Blanchard, D. C., and Syzdek, L. D., 1983, The mechanism of scavenging of waterborne bacteria by a rising bubble, *Limnol. Oceanogr.* **28**:101–105.

Weiss, E., Rosenberg, M., Judes, H., and Rosenberg, E., 1982, Cell surface hydrophobicity of adherent oral bacteria, *Curr. Microbiol.* **7**:125–128.

Weiss, E., Judes, H., and Rosenberg, M., 1985, Adherence of a non-oral hydrophobic bacterium to the human tooth surface, *Dental Med.* **3**:11–13.

Wendt, S. L., George, K. L., Parker, B. C., Gruft, H., and Falkinham III, J. O., 1980, Epidemiology of infection by nontuberculous mycobacteria. III. Isolation of potentially pathogenic mycobacteria from aerosols, *Am. Rev. Respir. Dis.* **122**:159–263.

Westergren, G., and Olsson, J., 1983, Hydrophobicity and adherence of oral streptococci after repeated subculture *in vitro, Infect. Immun.* **40**:432–435.

Woodcock, A. H., 1948, Note concerning human respiratory irritation associated with high concentrations of plankton and mass mortality of marine organisms, *J. Mar. Res.* **7**:56–62.

Wyndham, R. C., and Costerton, J. W., 1982, Bacterioneuston involved in the oxidation of hydrocarbons at the air–water interface, *J. Great Lakes Res.* **8**:316–322.

Xiu, J. H., Magnusson, K.-E., Stendahl, O., and Edebo, L., 1983, Physicochemical surface properties and phagocytosis by polymorphonuclear leucocytes of different serogroups of *Salmonella, J. Gen. Microbiol.* **129**:3075–3084.

Young, D. H., and Kauss, H., 1984, Adhesion of *Colletotrichum lindemuthianum* spores to *Phaseolus vulgaris* hypocotyls and to polystyrene, *Appl. Environ. Microbiol.* **47**:616–619.

Index